Applications of Random Process Excursion Analysis

Applications of Random Process Excursion Analysis

Irina S. Brainina

Povolzhye State University of Telecommunications and Informatics (PSUTI), Russia

Translated by

Dmitri Arch

ELSEVIER

AMSTERDAM • BOSTON • HEIDELBERG • LONDON NEW YORK • OXFORD
PARIS • SAN DIEGO • SAN FRANCISCO • SINGAPORE • SYDNEY • TOKYO

Elsevier
32 Jamestown Road, London NW1 7BY
225 Wyman Street, Waltham, MA 02451, USA

British Library Cataloguing-in-Publication Data
A catalogue record for this book is available from the British Library

Library of Congress Cataloging-in-Publication Data
A catalog record for this book is available from the Library of Congress

ISBN: 978-0-12-409501-4

For information on all Elsevier publications
visit our website at store.elsevier.com

This book has been manufactured using Print On Demand technology. Each copy is produced to order and is limited to black ink. The online version of this book will show color figures where appropriate.

I dedicate this book to the memory of my dear and
beloved teacher Professor Daniel D. Klovsky

Contents

Preface

The author of this monograph took interest in the theory of excursions for random processes and its applications in early 1970s. The research on an adaptive radio communication system with a shortwave channel affected by multipath frequency-selective fading revealed the need to estimate efficiency loss resulting from finite fading rate. The problem of finding out, by way of periodic sounding of the communication channel, which of the two frequency subchannels available provides the best quality of transmission was naturally reduced to the estimation of the distribution law for time intervals preceding the first deterioration in the condition of the previously selected subchannel.

However, the literature dedicated to the theory of excursions for random processes does not seem to provide final solutions to the problem. The distribution of intervals between level crossings suggested in 1945 by S.O. Rice turned out to be a slowly converging series of multiple integrals whose multiplicity grows without limitation. Even in case of Gaussian random processes for which only rough approximations were obtained, those approximations are only true for the initial parts of the distribution.

The problems related to excursions of broadband random processes are extremely complicated; modern computers are not powerful enough to obtain accurate general solutions, especially in case of non-Gaussian processes.

Under such circumstances, it is appropriate to try and solve the problems by using approximate methods suitable for engineering applications. The approximations that have been proposed so far provide low accuracy or require cumbersome calculations. Besides, in most cases, the authors have not estimated approximation errors, nor have they determined the areas for which the errors remain within given limits.

At the same time, determining probabilistic properties of excursions in random processes is a highly topical issue for experts not only in radio engineering and communication but also in other fields, since knowing such properties enables solving a large number of applied problems.

This monograph presents the results of the author's many years of research into the theory of excursions. Among other things, the author managed to apply calculation methods chosen for Gaussian processes to a larger class of non-Gaussian random processes.

The formulae obtained—both precise and approximate—are rather simple and have a wide range of applications for engineering. For example, they can be used for the implementation of adaptive signal processing devices capable of overcoming a considerable uncertainty at the beginning of a self-instruction process. Many problems related to the detection or recognition of random and deterministic signals against the background of noise with *a priori* unknown statistic properties are solved with the help of special devices which adapt to the most informative parameters of the excursions of received signal-plus-noise waveform.

This monograph contains the results of many years of the author's work in the Research Department of Samara Institute of Telecommunications, presently the Povolzhye State University of Telecommunications and Informatics (PSUTI). The research was, for the most part, carried out at the chairs of multichannel telecommunication and theoretical foundations of radio engineering and communication (TFRC).

In 2002, Moscow-based *Radio i sviaz* publishing house published the author's monograph entitled *Adaptive Digital Processing of Telecommunication Signals*, Using the Applications of Random Process Excursion Theory. The first two chapters of this book also describe some of the new results obtained in the field of the excursion theory, which have been expanded and updated by the author in this monograph.

The author is heartily grateful to her teacher D. D. Klovsky, D. Sc. (Engineering), full Professor, Head of Chair (TFRC), for his interest in this work and continuous support.

Big thanks go to Dmitri Arch who translated the book from the Russian and made it accessible to an international audience.

The author is grateful to associate Professor Oleg Petrov of TFRC chair, who is a talented programmer, for his large contribution to the development of a suite of programs for simulating random process excursion.

Deep gratitude is also due to Professor Vladimir Lyandres of Ben-Gurion University of the Negev, Department of Electrical and Computer Engineering, and Dr. Roman Goot of Academic Technological Institute in Holon, Israel, for their interest in this work and their support.

The author also expresses her thanks to the unknown benevolent reviewer for giving useful comments and recommendations regarding this monograph; all of them were taken into consideration in the process of preparing the book for publishing.

Irina S. Brainina

Introduction

In spite of the fact that for decades excursions of stationary random processes have been the focus of attention of a number of researchers, the development of excursion theory is far from its completion. The reason for that is the difficulty of describing the subtle nature of random processes by means of mathematics, which would allow us to estimate a variety of distribution functions, for example, the distribution of widths of above-threshold excursions, the distribution of areas enveloped by the signal excursions, the distribution of time periods required for a signal to reach certain limits, and the distribution of extrema amplitudes.

Of all characteristics of random process excursions, the first approximations of excursion duration distribution density, the mean duration of excursions, and, in some cases, variance in their duration are, perhaps, the ones that have been studied the most thoroughly.

The researchers who made the largest contributions to the development of random process excursion theory include S.O. Rice, H. Cramer, M.R. Leadbetter, J.A. McFadden, D. Slepian, R.L. Stratonovich, V.I. Tikhonov, B.R. Levin, and Yu.K. Belyaev, to mention a few.

The first chapter of this monograph starts with a historical sketch of methods applied for studying parameters of excursions in broadband random processes (the methods are briefly described in Section 1.1).

Section 1.2 of the book provides some examples of using excursion statistics for practical purposes.

The random process excursions theory has already proved very useful and the number of its applications is likely to grow in the future.

In view of the difficulty of finding strict solutions of many problems in the field of excursion theory, we found particularly useful the approximations that are accurate enough for engineering applications.

A number of new results obtained in the excursion theory are presented in the Chapters 2–6 of this monograph.

Some methods of both estimation and accurate calculation of informative parameters for excursions in Gaussian stationary random processes have been developed. Such parameters include, for example, the variance in the number of points where the signal crosses the predetermined level, or the variance in the number of above-threshold excursions within the time interval of length τ (both variances are defined in Chapter 2 of this book—see Section 2.1).

In Sections 2.2 and 2.3 we obtain upper and lower estimates for distribution functions of time intervals between an arbitrarily chosen moment and the first zero crossing of the random process (Section 2.2) or the first moment when zero level is crossed in a given direction (Section 2.3). Similar estimates for excursion duration distribution densities are obtained in Section 3.1 (Chapter 3).

Having assumed that any two successive excursions above the predetermined threshold are statistically independent, we obtained recurrent formulae that allow calculating more accurately the densities of distribution for excursion durations and for durations of intervals between excursions (Section 3.2). The same assumption was used as a basis for the estimation of variances in excursion durations and in the length of intervals between excursions (Section 4.1).

Methods of extending application of the results obtained to certain classes of non-Gaussian random processes are presented in Section 3.3. They suggest using nonlinear noninertial transformation or a non-Gaussian process with a predetermined correlation function into the equivalent Gaussian process (it will have a different correlation function, but the same threshold-crossing parameters). It is shown that the distribution law for momentary values of centered non-Gaussian processes does not have any significant effect on the characteristics of excursions, which are mainly determined by the waveform and root-mean-square frequency of the process power spectrum.

In Section 4.2 we estimate parameters for exponential tails of distribution curves representing the distribution of such excursions that are much longer than the process time correlation interval (in this case, we are dealing with the tails of distribution density curves representing the distribution of durations of excursions above a given standardized level).

Section 5.1 explains how the generation sequence obtained by bilateral clipping on level zero of a centered random process can be used. The points where the pulse sequence changes its sign from negative to positive or vice versa coincide with those of crossing level zero by the process. This pulse sequence, therefore, carries all the information on the distribution law for time intervals between successive crossings of level zero by the process. It is called "the generation sequence" because we can obtain from it a family of pulse sequences by adding up any number $i = 2,3,4,\ldots k$ of successive intervals between zero crossings. Each sequence so obtained carries the information on the duration of a combination of i successive intervals. This additional information allowed us to reduce as much as we wanted the error of calculating the parameters of zero crossings, by increasing the total number k of sequences.

Formulae are provided for estimating the cumulative function and distribution density of zero-crossing intervals. In those formulae, the functions estimated are expressed through correlation functions. Using the readily available software for computing correlation functions saves us a lot of time and efforts that would be necessary if we were to simulate zero-crossing statistics.

In Chapter 5 (Section 5.2), we show how the method based on the use of a generating pulse sequence can be extended to random processes of any kind including the noncentered and even non-Gaussian ones. Our approach allows choosing any threshold for the analysis of above-threshold excursions, as long as it remains within the range of momentary values of the process under analysis.

In order to obtain the distribution laws for durations of above- or below-threshold excursions in a broadband random process for the case of a randomly chosen standardized threshold $(X_0/\sigma) \neq 0$ (here σ is the root-mean-square value of the process), the generating sequence is formed from the flow of positive pulses which represent excursions above the given level (X_0/σ).

After inverting every other pulse and removing spaces between pulses, we obtained a symmetrical bipolar pulse sequence. The time intervals between zero crossings in the generating sequence carry all the information we need on the duration of the process excursions above the predetermined relative value.

Similarly, the generating sequence can be used to obtain a family of i auxiliary sequences and then apply the same method for calculating the parameters of excursions as the one that was applied in the case of level zero.

By inverting pulses representing the random signal excursions above the predetermined level another generating sequence can be obtained which carries information on the time intervals during which the signal level remains below the predetermined level.

For any given two threshold levels one can make six generating pulse sequences, two of which will represent excursions above and below the first threshold, respectively, two others will represent excursions above and below the second threshold, and the last two will represent time periods during which the signal level remains within or outside the corridor limited by the two thresholds, respectively.

The problems such as the one mentioned above often arise when there is a need to control tolerances and check for compliance various technical specifications and norms. Information on the parameters of excursions may also become useful, for instance, in the processing of medical data such as cardiograms, encephalograms, tomographic images, in recognition of images and sounds, the studies of the distribution of intervals between events when analyzing the type of telecommunication traffic, the studies of signal fading caused by multipath propagation of radio waves, and in many other fields.

What is especially important is that the proposed method of calculating distribution laws for the durations of random process excursions above or below any arbitrarily chosen relative value is applicable to any processes, and not just to centered Gaussian ones. For example, it can be applied to Rayleigh distribution, Rice distribution, distribution χ^2, and uniform or rectangular distribution, to name a few.

In Section 5.3, we show how the distribution of the total duration of i successive intervals is calculated. The distribution function is expressed through the correlation functions of a family of pulse curves obtained for $i = 2, 3, 4, \ldots k$.

Chapter 6 of this monograph focuses on two hitherto poorly studied characteristics of random process excursions.

Section 6.1 describes one method allowing to estimate, on the basis of results obtained in Chapter 5, the cumulative function and distribution density of time intervals between an arbitrarily chosen moment and the next moment when the stationary differentiable random process reaches the set level (whether it reaches it from above or from below is irrelevant). So far this problem has only been solved for nondifferentiable Markov processes.

Computer simulation results presented in Appendix 2 prove that the above method of estimating the cumulative function and distribution density of time to the next level crossing is pretty accurate.

Section 6.2 studies areas enveloped by the random process excursion curves. Some sources [4, 6, 53] give approximate probability distribution of areas under the curves that represent short excursions above high standardized levels. Due to their complicated nature, the generic problems of calculating the distribution of areas enveloped by excursion curves still remain to be solved.

The tables in Section 6.2 show the estimates obtained for average areas enveloped by excursion curves (both above- and below-threshold excursions are considered). The values presented in the tables were obtained by multiplying the known average durations of excursions by their average amplitudes. The calculated average values of all excursion parameters are given for two kinds of random processes—a Gaussian one and a Rayleigh one.

The calculated average areas closely match the results obtained by computer simulation (see Appendix 3).

Section 6.2 describes one method (also proved accurate through simulation) of estimating exponential tail parameters for distribution of areas enveloped by relatively long excursions.

Chapter 7 of this monograph presents algorithms for building certain digital adaptive analyzers of stationary random process characteristics and provides estimates of statistical errors and measurement times. Instrumental analysis of real random processes combined with computer simulations allows conducting experimental studies of a number of characteristics of excursions whose theoretical analysis still presents major problems. One example of such studies is the analysis of the distribution of areas under excursions above a given threshold. Due to a complicated nature of the problem, the theoretical solution to it has not yet been found.

All the adaptive digital analyzers that we proposed function on the basis of the same principle. In the process of adaptation of a particular analyzer, the average values of its settings are quickly estimated, which then serve as a basis for the automated process of finding the analyzer's optimal settings. The settings that are subject to optimization include some important parameters such as sampling rate, amplitude quantization increment, signal gain/attenuation factor, the number of channels available in the analyzer, and measurement duration for the given maximum acceptable values of statistical and instrumental errors.

By doing preliminary "training" which does not take very long, we managed to build our analyzers so that their adaptation ("training") time is automatically set in inverse proportion to the effective width of the power spectrum of the process. This is achieved by accumulating the predetermined number of excursions during adaptation or measurement (N and E, respectively). The counter capacity E is chosen on the basis of a compromise between the given maximum statistical error of measurement and the maximum acceptable measurement duration. As in most cases, the adaptation does not have to be very accurate; it does not take a long time in proportion to the total measurement session duration (the adaptation time accounts for no more than a few percentage points of the total measurement time).

Knowing average values of measured parameters obtained in the process of adaptation enables choosing the best sampling rate (this rate is in inverse proportion to the average duration of excursions). By choosing the best sampling rate, we automatically choose the optimal time sampling interval, which is directly proportionate to average values of the measured parameters and minimize the root-square-mean error of measurement.

The need to use various tools to analyze the distribution of excursion duration values in random processes often arises in many fields including radiolocation, telecommunication, meteorology, manufacturing of medical instruments (the processing of cardiograms, encephalograms, tomograms), etc. The use of such adaptive analyzers linked to microprocessor-based personal computers enables automation of random signal processing.

Some results obtained with statistical computer simulations of excursion durations for two broadband random processes (one Gaussian and one Rayleigh) with various power spectra are given in Appendix 1 of this monograph. By using simulations we managed to verify experimentally the quality and reliability of theoretical results discussed in Chapters 2–6 of the book.

Our findings support the hypothesis that there is virtually no correlation between successive excursions above a given threshold, and all the approximations based on this hypothesis prove to be effective.

Simulations also allowed us to estimate the applicability limits of and calculation errors for excursion duration distribution laws. Our estimates are based on the use of a family of pulse sequences obtained from one generating sequence. Experiments confirmed a very close match (with the accuracy determined by the statistical error of measurement) between the calculated figures and simulation results for a wide range of excursion duration distribution densities (including exponential tails).

Theory-based estimates of parameters for exponential tails of excursion duration distributions also closely matched the results obtained by means of computer simulations for distribution density curve tails—not only for all Gaussian processes but also for Rayleigh ones.

This book is addressed to researchers, engineers, and senior university students who are interested in the theory of random process excursions and its practical engineering applications.

1 Probability Characteristics of Random Processes

1.1 A Historical Sketch of Methods Applied for Studying Parameters of Excursions in Broadband Random Processes

Statistics of above-threshold excursions of a random process are useful in solving many problems of practical importance.

Figure 1.1 shows a realization of a stationary random process $X(t)$ on a time interval of length T.

All real physical processes are continuous functions of time. On a finite time interval T such functions have a finite number of maxima and minima whose magnitude H varies. As we can see in Figure 1.1, at the time t_m the process $X(t)$ reaches its absolute maximum H_m, the so-called maximum maximorum.

On the time interval T, the process $X(t)$ up-crosses a given level X_0 several times with the first up-crossing happening at the time τ_0. The time τ_0 is therefore called the first passage time.

If τ is the time interval between an up-crossing and the subsequent down-crossing of the given threshold X_0 by the random process $X(t)$, the process is said to have made a positive excursion of duration τ. Respectively, if θ is the time interval between a down-crossing and the subsequent up-crossing of the threshold X_0 by the same random process (such intervals are somctimes referred to as intervals between excursions), the process is said to have made a negative excursion of duration θ.

In the general case, we can say that on the time interval T, a random process $X(t)$ has n positive/negative excursions above/below the threshold X_0 (in Figure 1.1, the number of excursions is three).

For any realization of a particular random process on a given time interval, the variables H, τ, and θ may take several different values (their exact number depends on X_0 and T). Along with the variables n, τ_0, and H_m, these variables change randomly as we switch from one realization of the random process $X(t)$ to another.

Apart from the random variables mentioned above, there are other important parameters, such as areas S_+ under the positive excursion curves and areas S_- above the negative ones.

By taking into consideration all these random variables, we can describe the process in great detail. In fact, they provide much more information on the process than do such integral characteristics as the power spectrum and correlation function.

Applications of Random Process Excursion Analysis. DOI: http://dx.doi.org/10.1016/B978-0-12-409501-4.00001-2

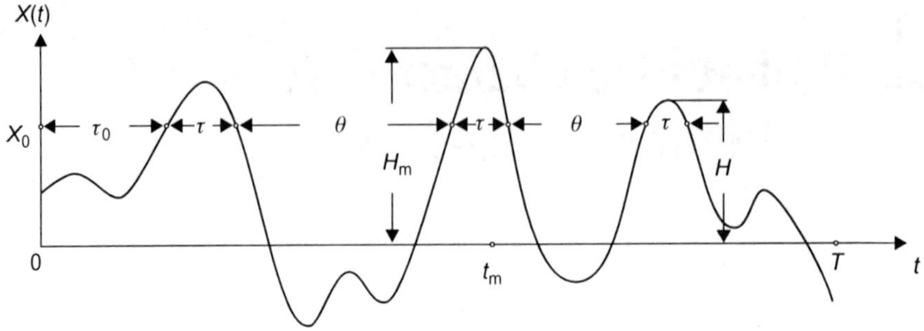

Figure 1.1 A stationary random process $X(t)$ on a time interval of length T.

In the next section of this chapter, we are going to see some examples of how excursion characteristics are used for solving practical problems.

The importance of studying excursion characteristics was recognized quite a long time ago. The first works on the subject were published in 1930s. They presented theoretical and experimental studies on the accidental exposure of nonlinear oscillating systems.

The development of the theory of random process excursions was started by a US scientist Stephen Oswald Rice.[1] He was the first to come up with a formula for the average rate of level crossings of a random process as far back as 1936. In 1939 he described the distribution of maxima of a random curve. Another early paper on the theory of excursions was written by a US mathematician M. Katz, and it was published in 1943.

Various later studies of excursion parameters were based on a remarkable theoretical work by S.O. Rice published in 1945.[2] In that work, exact formulas for the average rate of level crossings and the average rate of maxima of a random curve were obtained for certain types of random processes. The same work presents an approximate formula for the distribution density of excursion durations. The formula was found to be fairly accurate in the cases of short-time excursions whose duration is comparable with the random process time correlation interval.

The following years saw the publication of a number of theoretical and experimental works dealing with random process excursions. The theory of random process excursions was developed by prominent American and Russian scholars.

In the United States, important results were obtained by A.J. Siegert,[91] J.A. McFadden,[40] D. Slepian,[68,92] D. Middleton,[93] C.W. Helstrom,[94] A.J. Rainal,[48,69] and others.

In Russia the first work in the field of excursion theory was published by V.I. Bunimovich and M.A. Leontovich.[3] Russian scholars V.I. Bunimovich, R.L. Stratonovich,[4,5] V.I. Tikhonov,[6,7] and B.R. Levin[8] made a significant contribution to the development of the theory.

One chapter in the book on radio engineering statistics by R.L. Stratonovich[4] is dedicated to the theory of excursions. The author developed a number of unique methods for calculating the average number of excursions and finding their duration probability distribution. The methods are based on the theories of Markov processes, smoothed random processes, and correlated random points. The book also presents a method for estimating the distribution of under-curve areas for high-level excursions of a Gaussian random process for various high-standardized thresholds.

A further impulse to the development of the theory of excursions was given by V.I. Tikhonov and his disciples. Years 1970 and 1987 saw the publication in Russia of two books entirely dedicated to the theory of excursions.[6,7] The books, which were unequaled anywhere in the world, summarized the most important results of theoretical and experimental studies in the field.

An effective method of solving problems related to the theory of random process excursions was proposed by B.R. Levin[8] and his postgraduate student A.Ya. Fomin.[9] The authors called their method, which is based on time sampling of processes, the reference pulse method.

Until recently, researchers focused on finding precise formulas that can be used to solve practical problems, but at present, the main trend in the study of excursion distribution functions is to develop approximate methods and experiment broadly with the use of personal computers.

The rapid development and accessibility of computer technology enables the use of computers for modeling any statistical parameters of excursions in random processes. Due to their high speed of operation and large memory, modern PCs can work through large volumes of data on samples of a random process in a short period of time. This reduces significantly both the errors and time of measuring statistics of excursions.

However, the fact that modeling techniques have high potential does not mean that the search for good approximate methods for the analysis of excursion parameters should stop. Both approaches to obtaining excursion statistics are worth following, as they may complement each other. The effectiveness and validity of approximate methods, as well as their applicability limits, can only be tested by comparing calculated results with experimental measurements.

All the data presented in this monograph that had been obtained by approximate methods have been thoroughly checked. Statistical errors of the data obtained in modeling experiments rarely exceeded a few tenths of a percentage point, reaching a few percentage points in the worst-case scenario where the measured parameter took small values.

1.2 The Use of Excursion Statistics for Practical Purposes

Parameters of excursions in random processes are increasingly used to solve various practical problems. New areas of application for excursion characteristics are found continuously. The most important areas of application include telecommunications, radio positioning, signal-to-noise ratio measurement, calculating parameters of tracking systems, medical instrumentation industry, estimating excessive load in mechanical structures, estimating the impact of ionospheric and tropospheric factors on radio wave propagation, surface-finish monitoring, ocean wave parameter measurement, speech signal recognition, and identifying noise sources and measuring noise in semiconductor devices, to mention just a few.

Let us consider some examples of using specific parameters of excursions in random processes for various applications.

To start with, let us look at the main areas of the author's expertise, i.e., statistical radio engineering and telecommunications theory. The author has also worked on problems related to multipath fading of desired signal transmitted over a decametric or shorter-wave radio channel.

Problems encountered in statistical radio engineering are closely connected with probabilistic problems solved by designers of devices that process signals on the background of random noise. For over a century such problems have been the focus of attention of researchers from all over the world, and intensive research has been done in this area.

As far back as mid-1930s, scientists realized that transmission and reception of radio signals can only be successful if they take into account statistical nature of various processes related to random changes (fading) in the received signal, fluctuation noise, and narrow-band interference.

One of the key problems faced by scientists during World War II was that of maximizing sensitivity of radar receivers, which suffered from both accidental and intentional interference. For radars, higher sensitivity meant higher range, which made them more effective in the Allied military operations against Nazi Germany.[1]

Detection of weak signals received on the background of strong noise is a very complicated process. Solving the numerous problems it presents took the efforts of many prominent scientists and engineers from the United States, Russia, and the United Kingdom. The solution was found due to the use of statistics of excursions in random interference, which is, in many ways, different from the desired signal.

The design of jam-resistant communication systems in radio channels with frequency-selective fading is closely linked with the study of distribution laws for fading time θ of the signal (the time during which the signal drops below a given threshold level X_0). During that time, the signal-to-noise ratio drops sharply and communication actually becomes impossible. Fading is typical not only for short-wave radio channels, but for some other channels as well, including satellite ones.[10–21]

In broadband satellite channels, which are used for mobile communication, we observe multipath propagation of radio waves. It results in the appearance of a number of narrow frequency bands within a common broad frequency band of the radio channel, which fade independently in a selective manner. This feature offers an opportunity to redistribute in the best possible way the frequency and power resources of any given multiuser satellite communication system. This is done by directing the satellite-mounted transmitter power to the best, that is, fading-free frequency sub-bands, the ones that are not suffering from significant interference.[22,23]

The recent years saw the development of communication systems that make a good alternative to creating costly satellite communication channels. The so-called high-altitude platform (HAP) systems use a platform with transmitting and receiving equipment mounted on it, which is raised to the elevation of about 20 km in the stratosphere by means of a maneuverable aerostat or a zeppelin, or an airplane. The equipment is either powered by an onboard fuel engine, or, as in the case of a satellite, by solar batteries, or uses a mixed power supply system. The broadcasting area radius with such systems ranges from 200 to 300 km.

The main advantages of HAP system over a satellite communication system include the ease of setting up (just think of launching a rocket as compared to taking-off of an airplane), the ease of maintenance and repair, and local nature of deployment: in the case of HAP all equipment and the users are located in a relatively small geographic area, therefore, no international agreements are required.

In Refs 24 and 25, it is shown that such systems have a very flexible mechanism of distributing transmitted power between a large number of users.

Adaptive systems of this kind constantly monitor the broadband channel quality by measuring regularly the signal level for every frequency sub-band allocated to a user.

In the generic case, a successful design of a HAP communication system as well the system's efficiency analysis is hardly possible without knowing in advance the distributions of fading time θ (i.e., the duration of negative excursions in the signal envelope) and the area S_- bounded by the threshold line and the excursion part of the envelope curve. With this information at hand, one can determine the best interval for monitoring individual channels and the maximum number of subscribers the system can service, for a given acceptable probability of communication loss caused by a short-term power cut.

The design of a modern adaptive broadband communication system will, therefore, have to be based on the preliminary study of radio wave propagation conditions in newly allocated frequency bands. It is necessary to gather enough statistical information on distribution laws for the fading envelope of the high-frequency sine wave output, as well as the distributions of excursion durations and under-curve or above-curve areas bounded by the envelope and the horizontal line that represents the set standardized threshold.

Wireless devices widely use tracking systems such as an automatic range finder, phase and frequency locked loops, or an automatic gain control circuit. Each tracking system is designed to track a particular parameter of the desired signal and ensure that the tracking error is admissible.[1]

In the presence of fluctuating noise, the tracking error is random and occasionally gets high enough to cause tracking loss. One can judge about the performance of a tracking system on the basis of the average first passage time τ_0, that is, the first moment since the starting point when the error reaches the set boundary (Figure 1.1).

Statistical characteristics of excursions are used in structural mechanics for strength calculations.[26] Considering that external loads on structural elements are often of random nature, it is essential to calculate the maximum admissible excessive load H_m that will not destroy the element or lead to its permanent residual deformation. Before putting any particular element through extreme structural testing, it is necessary to determine the number n and the size H_{mm} of excursions in the curve that represents the random forces causing irreversible deformation. This is done, for example, in the process of strength calculation for the wings of an aircraft whose lift is affected by random atmospheric vortices, or the hull of a marine vessel caught in a severe storm.

Certain parameters of excursions allow a fairly accurate assessment of surface-finish quality for various parts. The residual surface roughness is assessed on the basis of the surface profile curve showing how the height of irregularities (peaks and valleys) depends on the profile abscissa. Let us assume, for example, that the profile curve of a polished surface reminds that of a Gaussian process. In this case, the profile is best described by three parameters, of which the first two represent the average numbers of points where the curve crosses two standardized levels and the third one is the number of maxima on the given section of the profile.[27]

Problems similar to the above are encountered in the assessment of smoothness of riding surfaces (pavement rideability), in holography when the granular structure of the recording material is analyzed, and in some other applications.

By measuring statistical properties of low-frequency noise in semiconductor devices, one can obtain information on physical processes in them. Investigation into the properties of low-frequency noise also helps assess the maximum sensitivity of modern radioelectronic devices.

In Ref. 28, a method for exploring electrical noise in light emitting diodes is described, which is based on the estimated number n of above-threshold excursions in noise voltage curve which occur during the time T, and the estimated distribution of values for their duration τ.

One area for which the use of random process excursion parameters has especially high potential is medical diagnostics. In electrocardiography, for example, heart action currents are recorded. The useful information on the patient's health is confined in such random parameters as period, shape, duration, and magnitude of pulses (excursions) representing heart action currents. In fact, in this case we are dealing with a pulse random process whose realization can be recorded and analyzed by a special computer program. The nature and magnitude of deviations of any of a particular parameter from its standard value allow diagnosing the cardiac pathology.

The same principle can form the basis for the analysis of encephalograms and tomograms of various body parts, with feeding the primary information to a personal computer. Advantages of this approach include high credibility of information, operational efficiency, high potential in terms of the amount of information that can be gathered, and low probability of medical malpractice.

The design of inexpensive small-size medical instruments based on special-purpose microprocessors with storage elements would also be very useful, as it would allow any individual to monitor their health condition and make a fair judgment of it by looking at a number of characteristics.

The detailed account of the principles for building such adaptive devices is given in Chapter 7 of this monograph.

Knowing excursion parameters of a signal mixed with interference enables the recognition of various waveforms of the desired signal against the noise background. Thus, in the simplest case, counting the number n of excursions above the given threshold X_0 over the time T allows estimating the harmonic signal frequency. It should be noted, though, that, in the presence of relatively strong noise, this method becomes inefficient due to its low noise immunity.

One technique that is characterized by higher noise immunity consists in accumulating the data on durations of periods between successive up-crossings the zero level within a given segment of duration T. The duration of each period is obtained as the sum of the sojourn time τ above zero ($X_0 = 0$), and the sojourn time θ below zero of the process (Figure 1.1). It is known that the signal is best protected from noise at the moments when the combination signal-plus-interference changes sign. This explains, among other things, high noise immunity of communication systems using frequency modulation (FM): in such systems important information is carried by the signal frequency (i.e., the rate of zero-crossings). As long as the signal-to-interference power ratio exceeds the threshold value of 1, FM provides higher noise immunity compared to amplitude modulation.

Once we have accumulated enough data on above-zero and below-zero sojourn times of the signal-plus-noise mix within the given time T, we can estimate the period and, therefore, the frequency of a harmonic signal.

This method can also be used for the recognition of the signal type, for example, speech signal or harmonic signals in telephony. In the case of digital transmission, it is sufficient to look at the sign bit of the sample code as it contains all information on zero-crossings.

It is known that the most powerful speech signals are registered in the frequency range from a few tens of Hertz to 3 or 4 kHz. More than half of the power falls on frequencies below 800 Hz. In the case where the speech signal sampling rate is $f_s = 8$ kHz, over one half of intervals between zero-crossings contain five or more periods of frequency f_s. At the same time, the line signaling frequency f_{sign} used in telephony equals to 2.6 kHz, which corresponds to the average of one or two periods of frequency f_s within one interval between successive zero-crossings. Given that the speech signal loses its power rapidly in the high-frequency area, it is very unlikely that we will have a long period of continuous voice signal with the frequency f_{sign} and a large number of short intervals between successive zero-crossings at the same time.

It follows from the above that durations of intervals between successive zero-crossings are distributed very differently in a speech and signaling signals. Thanks to this difference, we can reliably distinguish between speech and signaling in the process segment with duration $T = 100$ ms, and avoid misinterpretation of false hang-up and breakdown signals which would result in premature disconnection.

In fact, in the case presented above, the analysis of excursion durations allows replacing the signal's frequency spectrum with the "spectrum" of its zero-crossing intervals; in other words, it allows switching from the frequency domain to the time domain.[29]

When analog signals are processed, their frequency spectra will naturally be worth looking at, using frequency filtering. As far as digital signals are concerned, it is easier to "accumulate" and process them in the time domain. In the example presented above, the information is derived from the sign bit of the sample code. Keeping only the sign bit is equivalent to signal clipping which is often used for increasing the system's immunity to additive noise.

The need to analyze excursions in random processes arises in solving some of the problems posed by the theory of reliability, especially when we deal with the reliability of sophisticated equipment such as radioelectronic equipment with a large number of structure components.[30]

The number n of excursions above a given threshold X_0 in Figure 1.1 may correspond to the number of failures of the equipment components over the time T. In this case, τ_0 will be the time of the first failure since the start of the equipment operation, τ will be the time required to correct the fault or the time taken to put a backup component into operation, θ will be the time of no-failure operation after fixing the latest failure. The standard practice has been to measure the average values of these random variables.

The key issue, which is to find the original random process $X(t)$ and determine the threshold X_0 that describe in the best manner the reliability of a particular structure component of the equipment in question, is solved by gathering and processing a large amount of experimental data on the duration of fail-free operation, mean time to the first failure, failure rate, and time required to fix failures.

Effects of random oscillatory processes (vibrations) on living organisms and mechanical structures cannot be assessed without estimating their vibration stability. Random vibrations can be very harmful to people's health; they often speed up the wear and tear on mechanisms, cause accidents, or lead to a failure or destruction. At the same time, in some new fields of engineering such as vibration technology, vibratory transport, vibration-based diagnostics, and vibrometry, people actually benefit from the use of man-made vibrations.[31]

Vibration stability analysis boils down to determining a number of excursion characteristics for a certain random process $X(t)$ representing vibrations. Those characteristics include the number n of excursions above a given threshold X_0 that occur in the vibration curve within a given time T, the distribution of maxima H_m, the value H_{mm} of maximum maximorum in the vibration level, and the estimate of the first passage time τ_0 since the beginning of random vibrations (Figure 1.1).

The classical queuing theory (also known as the mass service theory) and, in particular, the teletraffic theory assume that time intervals between entry requirements are distributed according to the Poisson law. This assumption was true in the case where a normal telephone line traffic load was estimated: it was based on the fact that, on average, for any person engaged in a telephone conversation with another person, pauses during which he or she remains silent and listens to the other take about one half of the conversation time, during which time the communication channel remains unoccupied.

This is not, however, the case with modern computer-based telecommunication networks, as the progress of technology has brought significant changes to teletraffic structure. At present, the speech signal coming from each subscriber is coded, and code words are transmitted in batches over the telecommunication network. The traffic is based on packet switching (IP-telephony, corporate networks with data trunks). Due to the limited buffer space, some packets may be lost, which has a negative effect on the quality of customer service.

The study found that modern traffic models are remarkably different from the Poisson model.[32] Attempts to use classical methodology for estimating the mean waiting time, the required buffer space, or the probability of losing data packets bring inaccurate and too optimistic results.

In the classical model, time intervals θ between entry requirements are distributed exponentially, whereas non-Poisson traffic models are characterized by distribution curves with the so-called heavy tails. In the tails area (i.e., in the area of large values for θ), the distribution function decreases according to a power law, i.e., slower, than the exponential function. In this case, we are dealing with long memory self-similar processes.

The desired quality of service can only be ensured by a closer analysis of entry requirements. The purpose of such analysis is to predict such requirements and enable non-Poisson traffic monitoring. A well-organized computer network must ensure low blocking percentage and high channel occupancy. This is only possible if computers can quickly analyze the information on the distribution of interval durations θ. Once the current relative occupancy of a communication channel and the distribution of interval durations θ have become known, it is possible to increase, if necessary, the buffer space, or decrease the probability of losing data packets or the waiting time.

So far, we have discussed only a few examples of using random process excursion characteristics for solving problems of practical importance. However, even those few examples can be useful in the study of excursion characteristics.

2 Study of Informative Parameters of Excursions in Stationary Random Processes

2.1 Relation Between Two Variances: That of the Number of Level Crossings and That of the Number of Excursions in an Interval of Length τ

For many engineering applications, it is important to know parameters of points where a wideband stationary random process curve crosses a given level. Among the parameters which are of interest to researchers are the distribution of the number of excursions above a given level in interval $(t_0, t_0 + \tau)$, the distribution of the total number of level crossings in an interval of length τ, the distribution of excursion durations, or intervals between an arbitrary time moment and the first subsequent moment when the process function crosses a given level in a certain or any direction.[38,84,85,87,89]

So far the accurate formulae for the above-mentioned distributions have not yet been obtained even for Gaussian random processes. There are a number of publications[7–9] whose authors proposed various methods for the estimation of level crossing parameters. Those methods are usually inconvenient, either because they require cumbersome calculations or because of not being sufficiently accurate. In Ref. 33, it is proved that as $\tau \to \infty$, the tails of curves which describe the distribution of intervals between level crossings tend to get closer and closer to being exponential, but the authors do not propose any methods for calculating the distribution parameters.

Experimentally obtained estimates of parameters related to level crossing[7,34,35] are few. This is only natural, given the complexity of computer-simulation-based analysis of random processes. Promising as it may seem, the use of PC simulations for studying distribution functions of broadband random processes[7,9,36] can be very difficult due to a high requirement to computer speed and memory, and due to the necessity to process a large volume of statistical data.

In view of the above, any new findings related to the theory of excursions in random processes are topical. In particular, it would be interesting to obtain, relying on theory alone, the relation between two variances: that of the number of level crossings in a stationary random process and that of the number of excursions

Applications of Random Process Excursion Analysis. DOI: http://dx.doi.org/10.1016/B978-0-12-409501-4.00002-4

above a given level in the same process in an interval of length τ. Knowing the relation between the two variances will make it easier to estimate each.[29,37] Further in this chapter, it will be shown that this relation can also be useful in the estimation of distribution of time interval between an arbitrary moment in a broadband stationary random process and the first subsequent moment when the process function crosses a given level, in a certain or any direction.

Let p_0 be the conventional probability of the absence of any crossings of level x_0 by process $X(t)$ in an interval of length τ, which begins at arbitrary time t_0, and let p_1, p_2, etc. be, respectively, the conventional probabilities of occurrence of one, two, etc. crossings of level x_0 by process $X(t)$ in the same interval for the case where $X(t_0) \geq x_0$.

Let us denote by $\overline{p_0}$, $\overline{p_1}$, $\overline{p_2}$, etc. similar conventional probabilities for the case where $X(t_0) < x_0$. Let us now write a system of equations:

$$P(x_0) * \sum_{i=0}^{\infty} p_i + [1 - P(x_0)] * \sum_{i=0}^{\infty} \overline{p_i} = 1 \tag{2.1}$$

where $P(x_0)$ is the probability that $X(t_0) \geq x_0$.

$$P(x_0) * \sum_{i=1}^{\infty} i * p_i + [1 - P(x_0)] * \sum_{i=1}^{\infty} i * \overline{p_i} = \lambda(x_0) * \tau \tag{2.2}$$

$$P(x_0) * \sum_{i=1}^{\infty} i^2 * p_i + [1 - P(x_0)] * \sum_{i=1}^{\infty} i^2 * \overline{p_i} = m_2(\tau, x_0) \tag{2.3}$$

Here, $\lambda(x_0) * \tau = 2 * \lambda_1(x_0) * \tau$ is the average number of crossings of level x_0, from above or from below, in an interval of length τ; $\lambda_1(x_0)$ is the intensity of level crossings with a given direction (upward or downward); $m_2(\tau, x_0) = \sigma^2(\tau, x_0) + [\lambda(x_0) * \tau]^2$ is the second moment of number of level crossings in an interval of length τ; $\sigma^2(\tau, x_0)$ is the variance in the number of crossings of level x_0 in an interval of length τ.

As we are trying to establish the relation between $\sigma^2(\tau, x_0)$ and variance $\sigma_+^2(\tau, x_0)$ of the number of excursions (i.e., level crossings with a given, e.g., positive sign of the first derivative), it is necessary to write equations similar in form to (2.2) and (2.3), which will contain the first moment $m_{1+}(\tau, x_0)$ and the second moment $m_{2+}(\tau, x_0)$ of the number of excursions above level x_0 in an interval of length τ.

Let p_{1+}, p_{2+}, etc. denote the conventional probabilities of one, two, etc. upcrossings of level x_0 in an interval of length τ for the case where $X(t_0) \geq x_0$. It is evident that

$$p_{1+} = p_2 + p_3,\ p_{2+} = p_4 + p_5,\ p_{3+} = p_6 + p_7,\ p_{k+} = p_{2k} + p_{2k+1}$$

Similarly, let $\overline{p_{1+}}$, $\overline{p_{2+}}$, etc. be the conventional probabilities of one, two, etc. upcrossings of level x_0 in an interval of length τ for the case where $X(t_0) < x_0$:

$$\overline{p_{1+}} = \overline{p_1} + \overline{p_2},\ \overline{p_{2+}} = \overline{p_3} + \overline{p_4},\ \overline{p_{3+}} = \overline{p_5} + \overline{p_6},\ \overline{p_{k+}} = \overline{p_{2k-1}} + \overline{p_{2k}}$$

By analogy with Eqs. (2.2) and (2.3), we can write:

$$P(x_0) * \sum_{i=1}^{\infty} i * (p_{2i} + p_{2i+1}) + [1 - P(x_0)] * \sum_{i=1}^{\infty} i * (\overline{p_{2i-1}} + \overline{p_{2i}}) = \lambda_1(x_0) * \tau \tag{2.4}$$

$$P(x_0) * \sum_{i=1}^{\infty} i^2 * (p_{2i} + p_{2i+1}) + [1 - P(x_0)] * \sum_{i=1}^{\infty} i^2 * (\overline{p_{2i-1}} + \overline{p_{2i}}) = m_{2+}(\tau, x_0) \tag{2.5}$$

Applying the same logic to downcrossings of level x_0, let us write:

$$P(x_0) * \sum_{i=1}^{\infty} i * (p_{2i-1} + p_{2i}) + [1 - P(x_0)] * \sum_{i=1}^{\infty} i * (\overline{p_{2i}} + \overline{p_{2i+1}}) = \lambda_1(x_0) * \tau \tag{2.6}$$

$$P(x_0) * \sum_{i=1}^{\infty} i^2 * (p_{2i-1} + p_{2i}) + [1 - P(x_0)] * \sum_{i=1}^{\infty} i^2 * (\overline{p_{2i}} + \overline{p_{2i+1}}) = m_{2+}(\tau, x_0) \tag{2.7}$$

From Eqs. (2.1)–(2.7) with a little algebra, we obtain:

$$2 * P(x_0) - 2 * B(\tau, x_0) = 4 * m_{2+}(\tau, x_0) - m_2(\tau, x_0) \tag{2.8}$$

where

$$B(\tau, x_0) = P(x_0) * \sum_{i=0}^{\infty} p_{2i} \tag{2.9}$$

Let us clarify what function $B(\tau, x_0)$ stands for. Let us assume that initial stationary random process $X(t)$ is put through an ideal limiter with the following characteristic function:

$$f(x) = 1, \quad x \geq x_0$$

$$f(x) = 0, \quad x < x_0$$

At the output of the limiter, we obtain pulse random process $Y(t)$ with correlation function $B(\tau, x_0)$. Indeed, the product of two values of $Y(t)$ which are separated by interval τ equals 1 when the process crosses level x_0 an even number of times in an interval of length τ and equals zero in the case of an odd number of crossings in the same interval. This is only true if $Y(t_0) = 1$, i.e., if $X(t_0) \geq x_0$. But conventional probabilities p_k were introduced especially for this case; therefore, it would be fair to say that formula (2.9) describes the correlation function of a random process which has passed through an ideal limiter with threshold level x_0.

To be able to find $B(\tau, x_0)$, we need to determine two-dimensional distribution law $W_2(x_1, x_2, \tau)$ of random process $X(t)$. From Ref. 8, we obtain:

$$B(\tau, x_0) = \int_{x_0}^{\infty} \int_{x_0}^{\infty} W_2(x_1, x_2, \tau) \mathrm{d}x_1 \, \mathrm{d}x_2 \tag{2.10}$$

Given that $m_2(\tau, x_0) = \sigma^2(\tau, x_0) + [\lambda(x_0) * \tau]^2$ and $m_{2+}(\tau, x_0) = \sigma_+^2(\tau, x_0) + [\lambda_1(x_0) * \tau]^2$, from (2.8) we obtain:

$$\sigma^2(\tau, x_0) = 4 \times \sigma_+^2(\tau, x_0) - 2 * P(x_0) + 2 * B(\tau, x_0) \tag{2.11}$$

Using formula (2.11), we can derive variance $\sigma^2(\tau, x_0)$ of the total number of level crossings from known variance $\sigma_+^2(\tau, x_0)$ of the number of excursions above level x_0 in an interval of length τ in a stationary random signal supplied to the input of an ideal limiter with threshold level x_0 and from known correlation function $B(\tau, x_0)$ of the ideal limiter output process.

It is evident that as $\tau \to \infty$, the number of level crossings gets twice as high as the number of excursions, both variances tend to infinity, and $\sigma^2(\tau, x_0) \to 4 * \sigma_+^2(\tau, x_0)$.

Using relation (2.11) simplifies the calculation of both variances by expressing them through known functions. According to Ref. 8:

$$\begin{aligned} \sigma_+^2(\tau, x_0) &= \lambda_1(x_0) * \tau - [\lambda_1(x_0) * \tau]^2 + 2\tau * \int_0^{\tau} \left(1 - \frac{\alpha}{\tau}\right) * V_2(x_0, x_0, \alpha) \mathrm{d}\alpha \\ V_2(x_0, x_0, \alpha) &= \int_0^{\infty} \int_0^{\infty} y_1 y_2 W_4(x_0, y_1, x_0, y_2, \alpha) \mathrm{d}y_1 \, \mathrm{d}y_2 \end{aligned} \tag{2.12}$$

Here, $W_4(x_0, y_1, x_0, y_2, \alpha)$ is a joint distribution of the process and its first derivative at two moments separated by time interval α.

Similarly, the variance in the total number of level crossings is

$$\sigma^2(\tau, x_0) = 2 * \lambda_1(x_0) * \tau - 4 * [\lambda_1(x_0) * \tau]^2 + 2\tau * \int_0^{\tau} \left(1 - \frac{\alpha}{\tau}\right) * \overline{V}_2(x_0, x_0, \alpha) \mathrm{d}\alpha \tag{2.13}$$

$$\begin{aligned}\overline{V}_2(x_0,x_0,\alpha) &= \int_{-\infty}^{\infty}\int_{-\infty}^{\infty} y_1 y_2 W_4(x_0,y_1,x_0,y_2,\alpha)\mathrm{d}y_1\,\mathrm{d}y_2 \\ &= 2*V_2(x_0,x_0,\alpha)+\lambda_1(x_0)\times[q_1(\alpha,x_0)+\overline{q}_1(\alpha,x_0)]\end{aligned} \tag{2.14}$$

$$q_1(\alpha,x_0) = -\frac{1}{\lambda_1(x_0)}\int_0^{\infty}\int_{-\infty}^{0} y_1 y_2 W_4(x_0,y_1,x_0,y_2,\alpha)\mathrm{d}y_1\,\mathrm{d}y_2$$

$$\overline{q}_1(\alpha,x_0) = -\frac{1}{\lambda_1(x_0)}\int_{-\infty}^{0}\int_{0}^{\infty} y_1 y_2 W_4(x_0,y_1,x_0,y_2,\alpha)\mathrm{d}y_1\,\mathrm{d}y_2$$

Functions $q_1(\alpha,x_0)$ and $\overline{q}_1(\alpha,x_0)$ are usually called the first approximations of distribution density of duration α of excursions above level $x_0(q_1(\alpha,x_0))$ and of intervals between excursions $(\overline{q}_1(\alpha,x_0))$ in a random process.

In Ref. 8, these functions are found for Gaussian and Rayleigh processes. After the substitution of (2.14) into (2.13), we obtain:

$$\begin{aligned}\sigma^2(\tau,x_0) &= 2\times\lambda_1(x_0)*\tau-4[\lambda_1(x_0)*\tau]^2+4\tau*\int_0^{\tau}\left(1-\frac{\alpha}{\tau}\right)*V_2(x_0,x_0,\alpha)\mathrm{d}\alpha \\ &\quad +2\times\lambda_1(x_0)*\tau*\int_0^{\tau}\left(1-\frac{\alpha}{\tau}\right)[q_1(\alpha,x_0)+\overline{q_1}(\alpha,x_0)]\mathrm{d}\alpha\end{aligned} \tag{2.15}$$

Using Eqs. (2.11), (2.12) and (2.15), we obtain:

$$\begin{aligned}\sigma_+^2(\tau,x_0) &= P(x_0)-B(\tau,x_0)-[\lambda_1(x_0)*\tau]^2 \\ &\quad +\lambda_1(x_0)*\tau*\int_0^{\tau}\left(1-\frac{\alpha}{\tau}\right)[q_1(\alpha,x_0)+\overline{q_1}(\alpha,x_0)]\mathrm{d}\alpha\end{aligned} \tag{2.16}$$

All functions in formula (2.16) have been found for Gaussian and Rayleigh processes.

To verify the validity of (2.16), we carried out numerical integration for a Gaussian process where $R(\tau)=\exp(-\tau^2/\tau_0^2)$ and for several values of normalized level $x_0/\sigma=0,1,2,3$. The comparison of our results with those reported in Ref. 38, which were obtained by applying much more difficult methods of computing, shows that they are absolutely identical.

Let us now determine the formulae for variances $\sigma_+^2(\tau,x_0)$ and $\sigma^2(\tau,x_0)$ for the case where the length of observation interval tends to zero $(\tau/\tau_0\to 0)$ and for the case where it tends to infinity $(\tau/\tau_0\to\infty)$, where τ_0 is the time correlation interval of the process. By substituting $B(\tau,x_0)$, $q_1(\alpha,x_0)$, and $\overline{q_1}(\alpha,x_0)$ in formula (2.16) for the case where $\tau\to 0,\alpha\to 0$, after a few transformations we obtain:

$$\sigma_+^2(\tau,x_0)\approx\lambda_1(x_0)*\tau[1-\lambda_1(x_0)*\tau],\quad \text{for}\left(\frac{\tau}{\tau_0}\to 0\right) \tag{2.17}$$

$$\sigma^2(\tau,x_0) \approx 2*\lambda_1(x_0)*\tau*[1-2\times\lambda_1(x_0)*\tau], \quad \text{for}\left(\frac{\tau}{\tau_0}\to 0\right) \tag{2.18}$$

The validity of Eqs. (2.17) and (2.18) can also be verified by using accurate formulae from Ref. 38 for calculations.

In view of the fact that $\lim q_1(\tau,x_0) = \lim \overline{q_1}(\tau,x_0) = \lambda_1(x_0)$.

For $\tau/\tau_0 \to \infty$, after introducing the following denotations: $\Delta(\tau,x_0) = q_1(\tau,x_0) - \lambda_1(x_0)$, $\overline{\Delta}(\tau,x_0) = \overline{q_1}(\tau,x_0) - \lambda_1(x_0)$, we can rewrite (2.16) as follows:

$$\sigma_+^2(\tau,x_0) \approx P(x_0)*[1-P(x_0)] - \gamma(x_0)*\lambda_1(x_0) + \kappa(x_0)*\lambda_1(x_0)*\tau, \quad \text{for}\left(\frac{\tau}{\tau_0}\to \infty\right) \tag{2.19}$$

Here

$$\begin{aligned}\gamma(x_0) &= \int_0^\infty \alpha*[\Delta(\alpha,x_0)+\overline{\Delta}(\alpha,x_0)]\mathrm{d}\alpha,\\ \kappa(x_0) &= \int_0^\infty [\Delta(\alpha,x_0)+\overline{\Delta}(\alpha,x_0)]\mathrm{d}\alpha\end{aligned} \tag{2.20}$$

Constant coefficients $\gamma(x_0)$ and $\kappa(x_0)$ do not depend on τ and are determined entirely by the form of the first approximations $q_1(\alpha,x_0)$ and $\overline{q}_1(\alpha,x_0)$. Similarly, from (2.11) we obtain:

$$\begin{aligned}\sigma^2(\tau,x_0) \approx\ & 2P(x_0)*[1-P(x_0)] - 4*\gamma(x_0)\times\lambda_1(x_0)\\ &+4*\kappa(x_0)*\lambda_1(x_0)*\tau, \quad \text{for}\left(\frac{\tau}{\tau_0}\to\infty\right)\end{aligned} \tag{2.21}$$

The values of variances obtained by using formulae (2.19) and (2.21) can be a little overestimated, but as can τ/τ_0 increases, the estimates become more accurate.

To determine the applicability range of (2.19) and (2.21), we also used the results of Ref. 38, where the values of variances $\sigma^2(\tau,x_0)$ and $\sigma_+^2(\tau,x_0)$ were obtained using accurate formulae for Gaussian processes with several correlation functions, in particular, for processes with Gaussian and uniform power spectrums:

$$R_1(\tau) = \exp\left(-\frac{\tau^2}{\tau_0^2}\right), \quad R_2(\tau) = \frac{\sin(\sqrt{6}*\tau/\tau_0)}{\sqrt{6}*\tau/\tau_0}, \quad \lambda_1(0) = \frac{\sqrt{-R''(0)}}{2\pi} \approx \frac{0.225}{\tau_0}$$

It was shown that for $\tau/\tau_0 \geq 20$, the variance in the number of zero crossings is expressed as follows:

$$\sigma_1^2(\tau,0) = 1.144*\lambda_1(0)*\tau + 0.05818$$

$$\sigma_2^2(\tau,0) = 0.964 * \lambda_1(0) * \tau - 0.1835$$

At the same time, by doing numerical integration we can obtain from formulae (2.20) the following values $\kappa_1(0) = 0.286$; $\kappa_2(0) = 0.241$; $\gamma_1(0) = 0.425\tau_0$; $\gamma_2(0) = 0.7\tau_0$

After substituting them in (2.21), we obtain:

$$\sigma_1^2(\tau,0) = 1.144 * \lambda_1(0) * \tau + 0.12, \quad \sigma_2^2(\tau,0) = 0.964 * \lambda_1(0) * \tau - 0.13$$

The relative error of (2.21) approaches 0.7% for $(\tau/\tau_0) = 35$.

In Ref. 38, it is established that for $\tau/\tau_0 \geq 40$ the variance in the number of excursions of Gaussian noise is proportional to the mean:

$$\sigma_+^2(\tau,x_0) = \kappa(x_0) * \lambda_1(x_0) * \tau \tag{2.22}$$

where coefficient $\kappa(x_0)$ depends on x_0/σ and on the form of $R(\tau)$.

The values of $\kappa(x_0)$ obtained by numerical integration in (2.20) for $R(\tau) = \exp(-\tau^2/\tau_0^2)$ and $x_0/\sigma = 0, 1, 2, 3$ fully coincide with the results of Ref. 38.

Thus, using formulae (2.19) and (2.21) eliminates the need for labor-consuming calculations of variances of the number of level crossings in the case where $\tau/\tau_0 \geq 20$. We have to admit, though, that coefficients in these formulae have to be calculated once and for all for different relative levels x_0/σ and for different $R(\tau)$ of a random process with a given distribution of momentary values. For Gaussian and Rayleigh processes, this presents no problem, as we know the form of the first approximations $q_1(\tau,x_0)$ and $\overline{q_1}(\tau,x_0)$.

If $\tau/\tau_0 \geq 40$, then, according to (2.22), the calculation of the variance in the number of occasions when relative level x_0/σ is crossed in an interval of length τ is simplified and reduced to finding the average number of crossings and proportionality coefficient $\kappa(x_0)$ (the latter is found using formula (2.20)).

2.2 Distribution of Time to the First Zero Crossing

As it was mentioned earlier in this chapter, applying the theory of excursions in random processes for engineering purposes presents certain problems. One of the problems that came in the focus of researchers' attention is finding the distribution of time intervals between an arbitrary moment and the first subsequent moment when a differentiable stationary random process symmetrical about zero crosses zero level from above or from below. Due to insurmountable calculating difficulties arising from attempts to solve this problem using accurate formulae, the preference is given to the use of approximate calculation methods.[8,9,39] In Refs. 8 and 9, the approximation is based on finding multidimensional integral distribution functions of random process. The more accurate is the approximation, the higher is the order of the distribution function. In the meantime, even for a Gaussian process finding the n-dimensional integral function for $n > 2$ requires very cumbersome calculations.

In Ref. 39, the problem is also solved by using approximate calculations. In this case we can judge about the accuracy of approximation from the width of the area limited by two boundary curves, in which lies the real integral distribution of time between an arbitrary moment and the first subsequent moment when the process crosses zero level from above or from below. It is shown that for the essential part of the range of the distribution function values, this area is quite narrow.

Let $P_k(\tau)$ denote the probability that the random process crosses zero level k times in an interval of duration τ beginning at an arbitrary time t_0. In this case, $P_0(\tau)$ will be the probability that no zero crossings occur during time τ.

Let us write a system of equations containing $P_k(\tau)$, and then obtain, by solving this system, both the upper and lower estimates of $P_0(\tau)$. To save space, let us write P_k instead of $P_k(\tau)$.

$$\sum_{k=0}^{\infty} P_k = 1 \tag{2.23}$$

The average number of zeros in an interval of length τ is

$$m_1(\tau, 0) = \sum_{k=0}^{\infty} k * P_k = \lambda(0) * \tau \tag{2.24}$$

where $\lambda(0)$ is the average number of zero crossings by a random process per unit of time (the zero-crossing rate).

The second initial moment of a random number of zero crossings in an interval of length τ is

$$m_2(\tau, 0) = \sum_{k=0}^{\infty} k^2 * P_k = \sigma^2(\tau, 0) + [\lambda(0) * \tau]^2 \tag{2.25}$$

where $\sigma^2(\tau, 0)$ is the variance in the number of zero crossings (both upward and downward) in an interval of length τ.

By analogy with (2.24) and (2.25), we can write expressions for the initial moments $m_n(\tau, 0)$ of higher orders. However, calculating their values would require integrating the $2n$-dimensional function of joint distribution of the process and its first derivative at n moments of time. Although, in principle, it is possible to calculate $m_n(\tau, 0)$ for $n > 2$, such calculation can be quite difficult.

In addition to Eqs. (2.23)–(2.25), there are other relations between P_k to be determined, apart from $m_3(\tau, 0)$, $m_4(\tau, 0)$, etc. To determine those relations, we are going to use correlation function $B(\tau, 0)$ of the process at the output of an ideal limiter with the following characteristics:

$$f(x) = 1, \quad x \geq 0$$
$$f(x) = 0, \quad x < 0$$

As it follows from (2.9), for a random process symmetrical about zero:

$$B(\tau,0) = \frac{1}{2} * \sum_{k=0}^{\infty} \mathrm{P}_{2k} \tag{2.26}$$

From Eqs. (2.23)–(2.26) after a little algebra, we obtain:

$$P_0(\tau) = \frac{1}{2} - \frac{\lambda(0) * \tau}{2} + B(\tau,0) + \sum_{k=1}^{\infty} k * (P_{2k+1} + P_{2k+2}) \tag{2.27}$$

$$P_0(\tau) = \frac{3}{4} + \frac{B(\tau,0)}{2} - \lambda(0) * \tau + \frac{m_2(\tau,0)}{4} - \sum_{k=2}^{\infty} \left[(k-1)^2 * P_{2k} + k * (k-1) * P_{2k+1}\right] \tag{2.28}$$

Hence

$$\mathrm{P}_0(\tau) \geq \frac{1}{2} - \frac{\lambda(0) * \tau}{2} + B(\tau,0) \tag{2.29}$$

$$\mathrm{P}_0(\tau) \leq \frac{3}{4} + \frac{B(\tau,0)}{2} - \lambda(0) * \tau + \frac{m_2(\tau,0)}{4} \tag{2.30}$$

Inequalities (2.29) and (2.30) set the boundaries within which the sought function $\mathrm{P}_0(\tau)$ is found. The width Δ of area between those boundaries is

$$\begin{aligned} \Delta &= \sum_{k=1}^{\infty} k * (P_{2k+1} + P_{2k+2}) \sum_{k=2}^{\infty} [(k-1)^2 * P_{2k} + k * (k-1) * P_{2k+1}] \\ &= P_3 + 2 * P_4 + 4 * P_5 + 6 * P_6 + \cdots \end{aligned}$$

For values of τ comparable with time correlation interval τ_0 of the process, the sought function $\mathrm{P}_0(\tau)$ will not be much different from the upper estimate in (2.30), given that the error can be expressed as

$$\sum_{k=2}^{\infty} \left[(k-1)^2 * P_{2k} + k * (k-1) * P_{2k+1}\right] = P_4 + 2 * P_5 + 4 * P_6 + \cdots$$

and that the probability of four or more zero crossings in an interval of length τ_0 is very small.

Let us find the limits of applicability range for inequalities (2.29) and (2.30) using as an example such processes for which we know the exact value of function $\mathrm{P}_0(\tau)$. Such processes include a random stationary telegraph signal and "clipped" processes

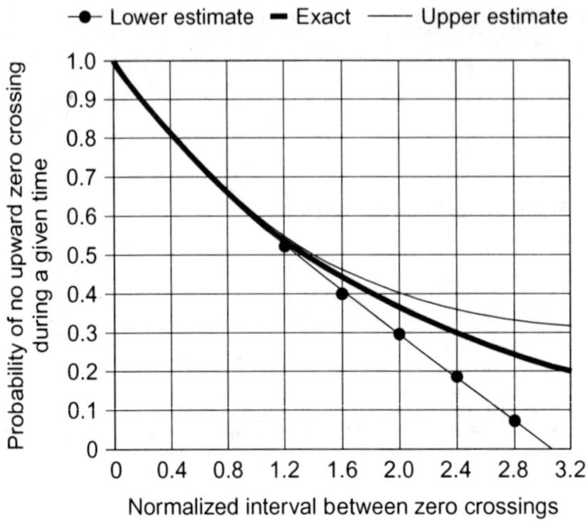

Figure 2.1 Probability of no zero crossings during given time for random stationary telegraph signal.

derived from such signal. For a stationary telegraph signal, the variance in the number of zero crossings equals the mean:

$$\sigma^2(\tau, 0) = \lambda(0) * \tau$$

and the average number of zero crossings per unit of time is $\lambda(0) = (1/2\tau_0)$.

The correlation function of the process at the output of an ideal limiter with zero threshold level is

$$B(\tau, 0) = \frac{1}{4} * \left[1 + \exp\left(-\frac{\tau}{\tau_0}\right)\right]$$

The accurate formula for calculating function $P_0(\tau)$ is known:

$$P_0(\tau) = \exp\left(-\frac{\tau}{2\tau_0}\right) \tag{2.31}$$

The results obtained using formulae (2.29) and (2.30) are shown in Figure 2.1, which also shows as a bold line real function $P_0(\tau)$ calculated using formula (2.31).

If we compare the upper estimate in (2.30) with the accurate value given by formula (2.31), we will see that the relative error of approximation increases with an increase in τ/τ_0 as we would expect. However, for $\tau/\tau_0 \leq 2$ the relative error does not exceed 6.5%. That means that the upper estimate remains sufficiently accurate in a wide range of values of $P_0(\tau)$ ($\exp(-1) \leq P_0(\tau) \leq 1.0$).

The lower estimate (2.29) is applicable for $\tau/\tau_0 \leq 1.2$. In this area, the two approximation curves practically merge and coincide with the curve representing the real values of $P_0(\tau)$ obtained using formula (2.31).

Note that in the example given above, we dealt with the least favorable conditions in terms of the width of applicability range of formulae (2.29) and (2.30). In a stationary telegraph signal, the probability of having three or more zero crossings in an interval of length τ is higher, since zero-crossing points of such a signal are statistically independent. Therefore, in this case, the area limited by the upper and lower estimate curves is wider than it is expected to be in the case of real processes.

To support the above, let us consider another example of a random process; this time let it be a process which is closer to being technically implementable. Let us transform the initial stationary telegraph signal into a pulse process in which the unit length is determined as the length of an interval between the fronts with positive (or negative) values of the first derivative of the initial signal form. A process of this kind is obtained at the output of a trigger flip-flop with a stationary telegraph signal supplied to its input. The zero-crossing rate of this new process will be twice as low as that of the initial process:

$$\lambda_+(0) = \frac{1}{4\tau_0}$$

Earlier we have established the relation between variance $\sigma^2(\tau, x_0)$ of the number of times when stationary random process $x(t)$ crosses level x_0 in an interval of length τ and variance $\sigma_+^2(\tau, x_0)$ of the number of excursions in the same interval.

Knowing $\sigma^2(\tau,0)$ and $B(\tau,0)$ of a stationary telegraph signal, let us now use formula (2.11) to find variance $\sigma_+^2(\tau,0)$ for the new impulse process, bearing in mind that $P(0) = 0.5$:

$$\sigma_+^2(\tau,0) = \frac{\tau}{8\tau_0} + \frac{1-\exp(-\tau/\tau_0)}{8} \tag{2.32}$$

Comparing $\lambda_+(0)$ with $\sigma_+^2(\tau,0)$ shows that, unlike the initial process, this pulse random process does not have a Poisson distribution of the number of zero crossings, and therefore the variance in this number does not equal the mean.

Correlation function $B_+(\tau,0)$ of this clipped process can be obtained from its power spectrum. This spectrum is expressed through distribution $W_+(\tau)$ of zero-crossing interval lengths by formulae from Ref. 8. As we are talking about a succession of intervals in which each interval is a combination of two independent exponentially distributed intervals between zero-crossing points of the initial stationary telegraph signal, we can find distribution $W_+(\tau)$ by using the method of convolution:

$$W_+(\tau) = \frac{\tau}{4\tau_0^2} * \exp\left(-\frac{\tau}{2\tau_0}\right) \tag{2.33}$$

We can now obtain power spectrum $F(\omega)$ of the process, and then, using Wiener–Khinchin formula, the sought expression for correlation function $B_+(\tau,0)$:

$$F(\omega) = \frac{1+2*(\omega\tau_0)^2}{1+4*(\omega\tau_0)^4}$$

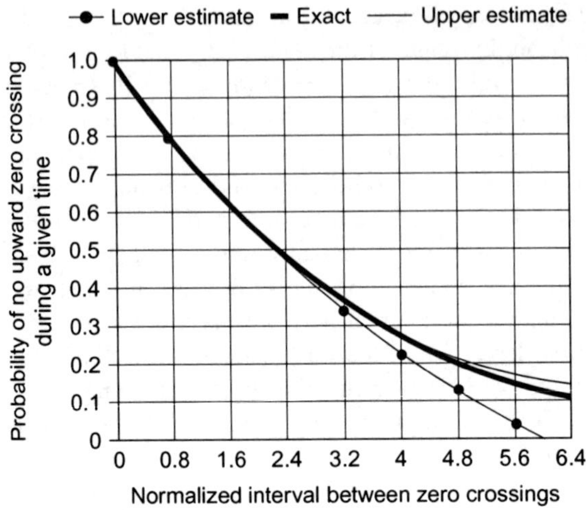

Figure 2.2 Output of a flip-flop, probability of no zero crossings during given time.

$$B_{+}(\tau,0) = \frac{1 + \exp(-\tau/2\tau_0) * \cos(\tau/2\tau_0)}{4} \tag{2.34}$$

To find the accurate value of $P_{0+}(\tau)$, the distribution of time to the first zero crossing, let us use the results of Ref. 8:

$$W(\tau) - \frac{1}{\lambda(0)} * \frac{d^2}{d\tau^2} P_0(\tau) \tag{2.35}$$

Bearing in mind that $P_0(0) = 1$ and $P_0(\infty) = 0$, after substituting (2.33) in (2.35) and integrating the result twice in terms of τ, we obtain:

$$P_{0+}(\tau) = \exp\left(-\frac{\tau}{2\tau_0}\right) * \left(1 + \frac{\tau}{4\tau_0}\right) \tag{2.36}$$

The estimates calculated using formulae (2.29) and (2.30) are shown in Figure 2.2. The curve in bold in the same figure represents the true integral function obtained by using formula (2.36).

Comparing approximations (2.29) and (2.30) with accurate formula (2.36) shows that in the second example their applicability range widened by almost 50%. The upper estimate (2.30) is acceptable for $\tau \le 5\tau_0$; in this case the error remains within 5.5%. The relative error of the rougher lower estimate (2.29) remains $<10\%$ for $\tau \le 3\tau_0$. The two approximation curves practically merge and coincide with the curve representing real function $P_{0+}(\tau)$ for $\tau = 0.6\tau_{av}$, where $\tau_{av} = 1/\lambda_{+}(0) = 4\tau_0$ is the average length of interval between successive zero-crossing points of the process.

The applicability range of the more accurate upper estimate in both examples is almost twice as wide as that of the lower estimate, and covers an essential part in each of the sought distributions (although, it does not cover the distribution tails).

There is every reason to believe that our conclusions about the accuracy of estimates (2.29) and (2.30) are applicable to real random processes. Moreover, for technically implementable random processes we should expect these estimates to have a wider applicability range because of correlation between zero-crossing points in intervals comparable with radius τ_0 of time correlation of the process. In comparison with processes with statistically independent intervals between zero-crossing points, such as Poisson processes and processes formed on their basis, real random processes are characterized by considerably lower probabilities of the occurrence of several short zero-crossing intervals in succession due to correlation between such intervals. As the error of the lower estimate (2.29) is determined as the probability of three and more zero crossings in an interval of length τ, we can expect the accuracy of approximation to be high for all τ that are close enough to average length τ_{av}.

By analogy with the above examples, the applicability range of the upper estimate is expected to extend almost as far as $\tau = 2 * \tau_{av}$.

To verify the above conclusions, let us use formulae (2.29) and (2.30) to calculate two stationary Gaussian random processes, one of which has a Gaussian power spectrum and the other have a uniform one. Both processes are of particular interest to us since they result from the passage of white noise through standard radio engineering sections.

The process with a Gaussian power spectrum has Gaussian autocorrelation function $R(\tau) = \exp(-\tau^2/\tau_0^2)$. For Gaussian processes, the zero-crossing rate is $\lambda(0) = \omega_1/\pi = \sqrt{-R''(0)}/\pi$. In this example, $\lambda(0) = \sqrt{2}/\pi * \tau_0 \approx 0.45/\tau_0$; $\tau_{av} = 1/\lambda(0) \approx 2.221 * \tau_0$.

Correlation function $B(\tau, 0)$ of the process obtained at the output of an ideal limiter with the input signal in the form of a Gaussian process with autocorrelation function $R(\tau)$ is known:[8]

$$B(\tau,0) = \frac{1}{4} * \left[1 + \frac{2}{\pi} * \arcsin R(\tau)\right] \tag{2.37}$$

Variance $\sigma^2(\tau, 0)$ of the number of zero crossings for this process is calculated in Ref. 38, which makes it easy to obtain, using formulae (2.29) and (2.30), the sought estimates of distribution $\mathrm{P}_0(\tau)$ of time to the first zero-crossing point. The results of calculation are shown in Figure 2.3.

We can clearly see that the curves stay close to merging for all τ up to $\tau \approx 1.6 * \tau_0 = 0.72 * \tau_{av}$ and coincide with true curve of function $\mathrm{P}_0(\tau)$. The previous examples show that the applicability range of the upper estimate (2.30) is twice as wide as that of the lower estimate and extends as far as $\tau \approx 1.3 * \tau_{av} \approx 2.9 * \tau_0$.

Figure 2.4 also shows the upper and lower estimate curves for a Gaussian process, but this time it is a process with a uniform power spectrum, whose autocorrelation function has the following form:

$$R(\tau) = \frac{\sin\sqrt{6} * \tau/\tau_0}{\sqrt{6} * \tau/\tau_0}$$

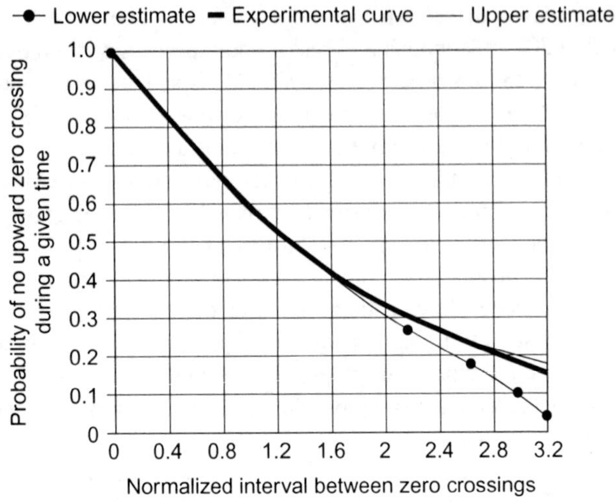

Figure 2.3 Probability of no zero crossings during given time for Gaussian power spectrum.

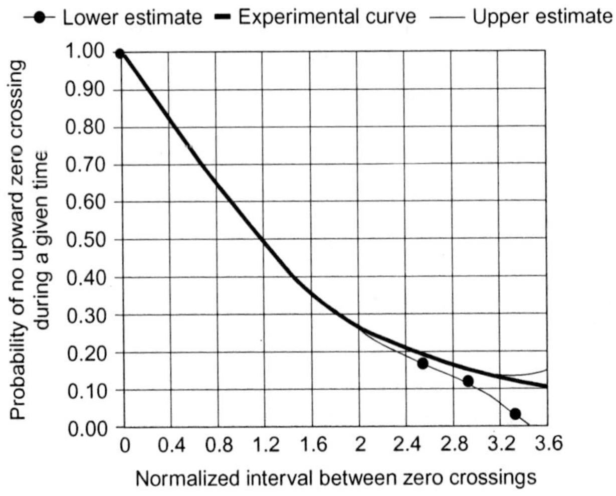

Figure 2.4 Probability of no zero crossings during given time for uniform power spectrum.

Both processes, the one with the Gaussian spectrum and the one with a uniform spectrum, have the same zero-crossing rate $\lambda(0) = \sqrt{2}/\pi * \tau_0 \approx 0.45/\tau_0$. Values of variance $\sigma^2(\tau, 0)$ of the number of zero crossings for the process with a uniform spectrum are also calculated in Ref. 38.

The curves in Figure 2.4 merge for all values of τ which do not exceed $\tau \approx 2 * \tau_0 \approx 0.9 * \tau_{av}$; thus, the applicability range of the upper estimate (2.30) extends as far as $\tau \approx 1.5 * \tau_{av} \approx 3.3 * \tau_0$ and covers an essential part of integral function $P_0(\tau)$. A comparison of Figures 2.3 and 2.4 shows that for $\tau \le \tau_0$ the difference between the two estimates is insignificant. In this area function, $P_0(\tau)$ is mainly determined by zero-crossing rate $\lambda(0)$ of the process which is assumed to be the same for both processes. For $\tau \le \tau_0$, the initial part of function $P_0(\tau)$ approximates well a straight line represented by the following equation:

$$P_0(\tau) \approx 1 - \lambda(0) * \tau = 1 - \frac{0.45}{\tau_0} * \tau$$

If we compare curves in Figures 2.3 and 2.4 for values $\tau > \tau_0$, we will see that by the probability of the occurrence of three or more zero crossings in an interval of length τ is lower in the case of a uniform power spectrum, than in the case of a Gaussian spectrum. Such difference is explained by the presence of high-frequency components in the spectrum of a process with a Gaussian correlation; hence, a larger relative gap between the curves in Figure 2.3 in comparison with Figure 2.4.

High-frequency components of the spectrum weaken as random processes pass through real radio engineering sections with limited bandwidth. Consequently, in the cases that can be interesting to us from a practical viewpoint, the results obtained for the lower and upper estimates for processes with limited spectrum have a sufficiently wide applicability range which covers an essential range of integral distribution function $P_0(\tau)$.

2.3 Distribution of Time to the First Moment When Zero Level Is Crossed in a Given Direction

Consider the problem of finding the distribution of a time interval between an arbitrary moment and the first subsequent moment when a differentiable stationary random process symmetrical about zero crosses the zero level in a given direction (e.g., from down upwards). This kind of a problem arises, for example, when we need to calculate the distribution of frequency excursions of abnormal duration in a frequency-modulated signal received on the background of a Gaussian noise.[34] Abnormal frequency excursions are caused by signal phase jumps of 180° ($\pm\pi$).

Attempts to solve such problems by using accurate calculation methods often fail even in the case of Gaussian random processes; therefore, a number of approximations have been proposed.[9,29,41]

Let us now consider a calculation method which provides the lower and upper estimates of the distribution values (the true values lie somewhere in between those estimates).[29]

By analogy with $P_k(\tau)$ in Section 2.2, let $P_{k+}(\tau)$ denotes the probability of the occurrence of k zero crossings by a random process during an interval of length τ which starts at arbitrary moment t_0, such that all of them occur in a given (e.g., upward) direction. Then $P_{0+}(\tau)$ will be the probability of no upward zero crossings during time τ. Note that $P_{0+}(\tau)$ is the sought integral distribution law of time to the first zero upward crossing (i.e., crossing with a positive derivative).

After writing formulae which establish relations between probabilities P_{k+}, the same way as formulae (2.23)–(2.25) establish relations between P_k, with a little algebra we obtain:

$$P_{0+}(\tau) = 1 - \lambda_1(0) * \tau + \sum_{k=2}^{\infty} (k-1) P_{k+}$$

$$P_{0+}(\tau) = 1 - \frac{3}{2} * \lambda_1(0) * \tau + \frac{m_{2+}(\tau,0)}{2} - \frac{1}{2} * \sum_{k=3}^{\infty} (k-1) * (k-2) P_{k+}$$

Here, $\lambda_1(0)$ and $m_{2+}(\tau,0)$ are, respectively, the zero-crossing rate and the second initial moment of the number of excursions above zero level (i.e., the number of zeros with the positive first derivative) in an interval of length τ.

From the above equations, we obtain a system of inequalities which limit the sought function from below and from above:

$$P_{0+}(\tau) \geq 1 - \lambda_1(0) * \tau \tag{2.38}$$

$$P_{0+}(\tau) \leq 1 - \frac{3}{2} * \lambda_1(0) * \tau + \frac{m_{2+}(\tau,0)}{2} \tag{2.39}$$

Let us now express P_{k+} through P_k, the probabilities of the occurrence of zero crossings of any kind in an interval of length τ. Considering that the processes are symmetrical about zero, we obtain:

$$P_{0+} = P_0 + \frac{P_1}{2}, \quad P_{1+} = P_2 + \frac{P_1 + P_3}{2}, \quad P_{2+} = P_4 + \frac{P_3 + P_5}{2},$$
$$P_{k+} = P_{2k} + \frac{P_{2k-1} + P_{2k+1}}{2}$$

Width Δ_+ of the gap between two estimates, which comprises the sought distribution function, $P_{0+}(\tau)$ is determined as follows:

$$\Delta_+ = P_{2+} + 3 * P_{3+} + 6 * P_{4+} + \cdots = \frac{P_3}{2} + P_4 + 2 * P_5 + 3 * P_6 + \cdots$$

The error of the most accurate of the two estimates, the upper one (2.39), is found as a sum:

$$\sum = P_{3+} + 3 * P_{4+} + 6 * P_{5+} + \cdots = \frac{P_5}{2} + P_6 + 2 * P_7 + 3 * P_8 + \cdots$$

The upper estimate (2.39) of integral function $P_{0+}(\tau)$ will be acceptable as long as we can neglect the probability of the occurrence of five or more zero crossings by the process in an interval of length τ.

The lower estimate represented by a straight line Eq. (2.38) is much less accurate, compared to the upper estimate. The absolute error of lower estimate is determined as the probability of the occurrence of three or more zero crossings in an interval of length τ. At the same time, knowing the lower estimate, we can determine the range of time values $0 \leq \tau \leq \tau_1$, in which the two approximation curves effectively merge and coincide with real function $P_{0+}(\tau)$. Apart from that, knowing the lower estimate (2.38) allows to determine, although indirectly, the applicability range of the upper estimate (2.39). This range is usually twice as wide $0 \leq \tau \leq 2\tau_1$.

To verify the validity of this conclusion, let us find the applicability boundaries of the upper and lower estimates using as an example processes with known distributions $P_{0+}(\tau)$. Let us consider a stationary telegraph signal, for which

$$P_k = \frac{[2 * \lambda_1(0) * \tau]^k}{k!} \times \exp[-2 * \lambda_1(0) * \tau], \quad \lambda_1(0) = \frac{1}{4\tau_0}$$

In this case $P_{0+}(\tau) = P_0 + (P_1/2) = \exp(-\tau/2\tau_0) * (1 + (\tau/4\tau_0))$.

This formula is identical with formula (2.36) obtained earlier. Variance $\sigma_+^2(\tau, 0)$ of the number of zero crossings with a given (upward or downward) direction and the second moment $m_{2+}(\tau, 0) = \sigma_+^2(\tau, 0) + [\lambda_1(0) * \tau]^2$ were determined in the previous section. After substituting them in (2.38) and (2.39), we obtain:

$$P_{0+}(\tau) \geq 1 - \frac{\tau}{4\tau_0}$$

$$P_{0+}(\tau) \leq 1 - \frac{5\tau}{16\tau_0} + \frac{\tau^2}{32\tau_0^2} + \frac{1 - \exp(-\tau/\tau_0)}{16}$$

Figure 2.5 shows the upper and lower estimates of $P_{0+}(\tau)$, as well as the curve of distribution of intervals between positive fronts of a stationary telegraph signal represented by formula (2.36).

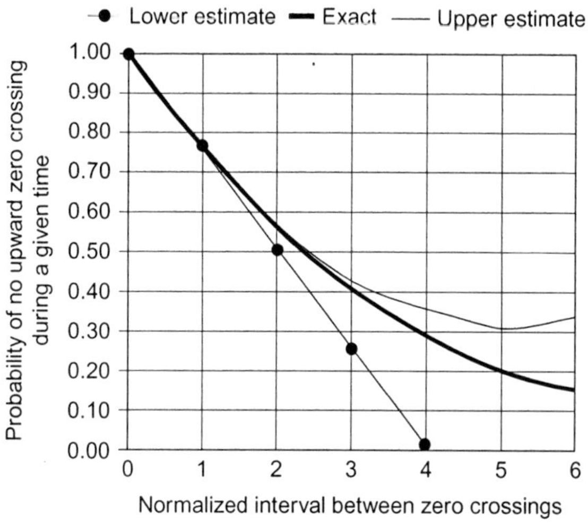

Figure 2.5 The distribution of intervals between positive fronts of a stationary telegraph signal.

As we can see from Figure 1.5, the upper estimate (2.39) is pretty accurate. For $\tau = 3\tau_0 = 0.75 * \tau_{av}$ (where $\tau_{av} = 1/\lambda_1(0) = 4 * \tau_0$ is the average interval between positive fronts of a stationary telegraph signal), the relative error of approximation does not exceed 3.3%.

But with an increase in τ the error increases too, and for $\tau = \tau_{av}$ it amounts to 15%. The applicability range of the simple formula (2.38) is twice as short, for $\tau = 0.5 * \tau_{av}$ the error is approximately 10%.

In the above example, due to statistical independence of successive fronts in a pulse process, both estimates, the upper and the lower, have a narrower applicability range. Therefore, we suggest studying another pulse process, which is obtained from the previous one by halving its frequency with the help of a trigger flip-flop. Let us find the distribution of intervals between successive positive fronts of a signal at the output of the flip-flop. In fact, finding this distribution is equivalent to dividing frequency of the initial stationary telegraph signal by four and studying the distribution of burst lengths or pause lengths in the resulting random process.

For such a process, the average number of zero crossings per unit of time is

$$\lambda(0) = \frac{1}{8\tau_0} = \lambda_{++}(0)$$

Variance $\sigma^2_{++}(\tau, 0)$ of the number of zero crossings can be obtained using formula (2.11) twice in succession. In view of (2.34), for $P(0) = 0.5$ and $x_0 = 0$:

$$\sigma^2_{++}(\tau, 0) = \frac{\tau}{32\tau_0} + \frac{1 - \exp(-\tau/\tau_0)}{32} + \frac{1 - \cos(\tau/2\tau_0) * \exp(-\tau/2\tau_0)}{8}$$

After substituting expressions for $\lambda_{++}(0)$ and $\sigma^2_{++}(\tau, 0)$ in formulae (2.38) and (2.39), we obtain:

$$P_{0+}(\tau) \geq 1 - \frac{\tau}{8\tau_0} \tag{2.40}$$

$$P_{0+}(\tau) \leq 1 - \frac{11\tau}{64\tau_0} + \frac{\tau^2}{128\tau_0^2} + \frac{1 - \exp(-\tau/\tau_0)}{64} + \frac{1 - \cos(\tau/2\tau_0) * \exp(-\tau/2\tau_0)}{16} \tag{2.41}$$

We will obtain function $P_{0+}(\tau)$ after using twice in succession the relation between P_{k+} and P_k, which was introduced earlier in this chapter. After a little algebra, we obtain:

$$P_{0+}(\tau) = \exp\left(-\frac{\tau}{2\tau_0}\right) * \left(1 + \frac{3\tau}{8\tau_0} + \frac{\tau^2}{16\tau_0^2} + \frac{\tau^3}{192\tau_0^3}\right) \tag{2.42}$$

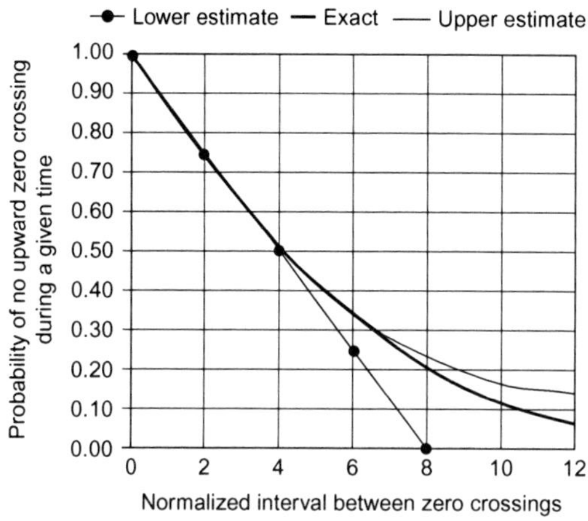

Figure 2.6 The distribution of intervals between successive positive fronts of a signal at the output of the flip-flop.

Figure 2.6 shows the upper (2.41) and the lower (2.40) estimates of $P_{0+}(\tau)$ as well as the real distribution law of intervals between positive fronts of a given pulse random process (the distribution described by (2.42)). As we can see from the graphs, approximation (2.41) is pretty accurate, and for values lower than $\tau = 8\tau_0 = \tau_{av}$ the relative error is $<5\%$. The simple approximation (2.40) is valid for $\tau \leq 4\tau_0$. In this area, the two approximation curves effectively merge and coincide with curve representing real function $P_{0+}(\tau)$ from (2.42).

A comparison of Figures 2.5 and 2.6 shows that the stronger the difference is between the clipped signal (the process after the ideal limitation on the zero level) and the generalized telegraph signal, the wider is the range of values τ/τ_{av}, for which the upper estimate (2.39) remains valid. The reason for that is a relative decrease in the variance in the number of zero crossings in comparison with the mean (we know that in the case of a Poisson distribution the variance equals the mean). An increase in the concentration of values of the random number of zero crossings around the mean is equivalent to a decrease in the probability that some intervals between zero crossings will be essentially shorter than the average. Therefore, we see a decrease in the probabilities of the occurrence of five and more zero crossings during time τ when τ comes close to τ_{av}.

It follows from the above that in real processes the error of estimation (2.39) increases with τ more slowly. For example, in a Poisson process the upper estimate is acceptable for $\tau \leq 0.75\tau_{av}$, whereas in a process obtained by halving the frequency of a pulse signal, the applicability range of approximation widens to the upper limit of $\tau = \tau_{av}$ and covers an important part of integral function (2.42).

Figures 2.7–2.9 show curves which were built using formulae (2.38) and (2.39) for three Gaussian processes. Their respective autocorrelation functions have the following forms:

$$R_7(\tau) = \exp\left(-\frac{\tau}{\tau_0}\right)^2, \; R_8(\tau) = \frac{\sin\sqrt{6} * \tau/\tau_0}{\sqrt{6} * \tau/\tau_0}, \; R_9(\tau) = \exp\left(-\frac{\tau}{\tau_0}\right)^2 * \left(1 - \frac{2 * \tau^2}{\tau_0^2}\right)$$

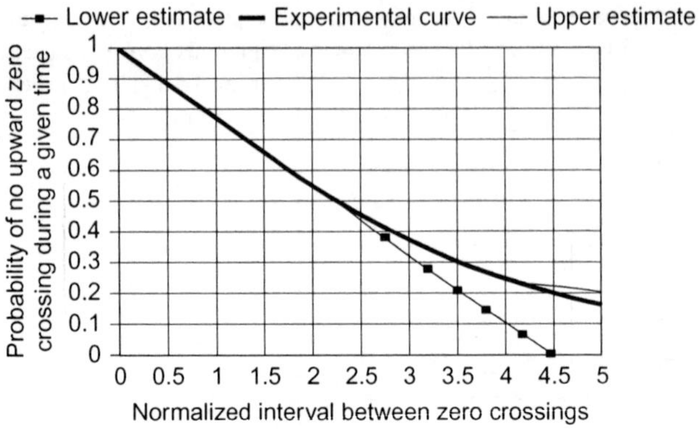

Figure 2.7 Distribution of time to the first moment when zero level is crossed in a given direction for Gaussian processes with a Gaussian power spectrum.

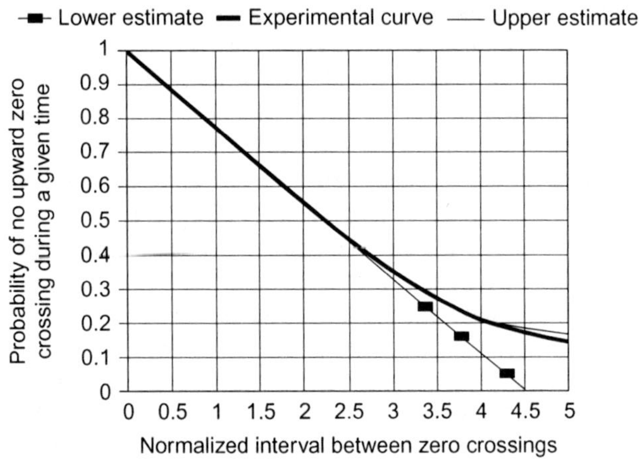

Figure 2.8 Distribution of time to the first moment when zero level is crossed in a given direction for Gaussian processes with a rectangular power spectrum.

The process with correlation function $R_9(\tau)$ is the first derivative of a Gaussian random process with autocorrelation function $R_7(\tau)$.

Processes with correlation functions $R_7(\tau)$ and $R_8(\tau)$ have the same average number of upward zero crossings per unit of time.

The average length of an interval between zero crossings at which the signal changes in the same direction is $\tau_{av} = 1/\lambda_1(0) \approx 4.443\tau_0$.

In a process with correlation function $R_9(\tau)$, zero crossings occur more frequently (their rate is $\sqrt{3}$ times as high):

$$\lambda_1(0) = \frac{\sqrt{3}}{\sqrt{2} * \pi * \tau_0} \approx \frac{0.39}{\tau_0}$$

the average length of interval is $\tau_{av} = 1/\lambda_1(0) \approx 2.565\tau_0$.

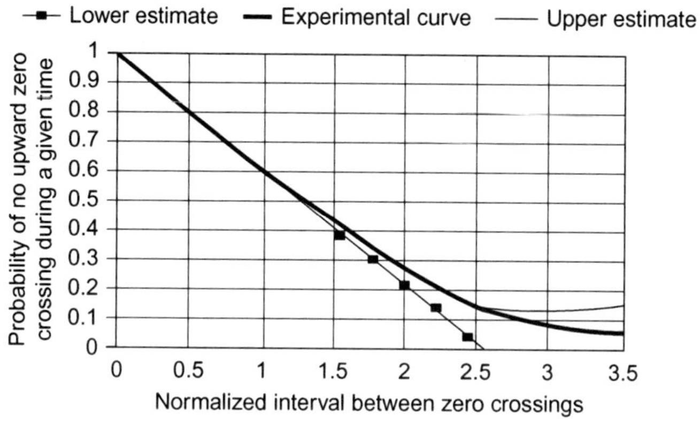

Figure 2.9 Distribution of time to the first moment when zero level is crossed in a given direction for the first derivative of a Gaussian process with a Gaussian power spectrum.

For the first two processes, the second initial moment $m_{2+}(\tau, 0)$ of the number of zeros, or the number of excursions above the zero level in an interval of length τ, is taken from Ref. 38 where the calculations made on a computer are presented.

For the process with correlation function $R_9(\tau)$, the values of $m_{2+}(\tau, 0)$ were obtained using numeric integration according to formula (2.16) for $x_0 = 0$:

$$m_{2+}(\tau,0) = \frac{1}{2} - B(\tau,0) + 2 * \lambda_1(0) * \tau * \int_0^\tau \left(1 - \frac{\alpha}{\tau}\right) q_1(\alpha,0)\mathrm{d}\alpha$$

where $q_1(\alpha, 0)$ is the first approximation of distribution density of excursions of duration α above zero level. All the functions in formula (2.16) can be easily obtained for a Gaussian process with a given correlation function $R(\tau)$.

As we can see from Figure 2.7, the curves begin to diverge when $\tau \approx 0.5\tau_{\text{av}}$. The results obtained earlier lead us to a conclusion that the upper estimate in (2.39) must be valid for any τ lower than $\tau \approx \tau_{\text{av}}$.

Approximations shown in Figure 2.8 have a wider applicability range because the spectrum of a process with correlation function $R_8(\tau)$ is limited. The estimate curves do not start to diverge until $\tau \approx 0.55\tau_{\text{av}}$; the upper estimate is, therefore, acceptable for any τ lower than $\tau \approx 1.1\tau_{\text{av}} \approx 5\tau_0$.

Finally, two approximation curves in Figure 2.9 coincide with real function $P_{0+}(\tau)$ for all values of τ up to $\tau \approx 0.5\tau_{\text{av}}$, which means that for any τ lower than $\tau \approx \tau_{\text{av}} \approx 2.6\tau_0$ the relative error of the upper estimate does not exceed a few percentage points.

From the above cases we can, therefore, draw a conclusion that the proposed method of calculating the integral distribution of interval between an arbitrary time and the first zero crossing at which the signal changes in a given direction is effective. The upper estimate given by formula (2.39) is applicable for all τ lower than

$\tau \approx 1/\lambda_1(0) = \tau_{av}$, that is, in an essential part of the range of function $P_{0+}(\tau)$ (the only parts of the distribution where it is not applicable are the tails, that is the parts where $\tau \gg \tau_0$).

The theory that proves the exponential character of distribution tails is set out in Ref. 33. That work, however, does not describe any effective methods which could be used to calculate the parameters of the exponential functions. Section 4.2 provides a solution to the problem of finding the gradient of exponential functions in limit distributions of durations of excursions above zero level for $\tau \gg \tau_0$.

3 Estimation of Distribution Densities of Excursion Durations for Random Stationary Broadband Signals

3.1 Estimation of Distribution Density of Zero-Crossing Intervals for Random Processes Symmetrical About Zero

Many problems related to detection of signals against a background of noise, to recognition of signals by their form, etc. are reduced to the problems of finding parameters of excursions above a given level (most often, zero level) in random processes. For instance, the detection, against a background of speech, of harmonic single-frequency and multifrequency signals enabling control and interaction in telephony, such as "ring off" or "disconnection" signals, boils down to recognizing their form. In speech, we always find frequency components which coincide with control and interaction tones; hence, the possibility of their erroneous reception and communication failure. If we consider speech as a random process, then designers of noise-resistant receivers of control signals against a background of speech may benefit from information about the distribution of intervals between zero crossings by speech signal. The presence in speech of a sufficiently large number of zero-crossing intervals of "prohibited" length, that is, intervals which are not supposed to occur in periodical harmonic control signals, makes it possible to unmistakably distinguish between the two signals on the basis of their shape. This approach, in fact, suggests carrying out a nonlinear analysis of "spectrum" of zero-crossing intervals instead of a linear spectrum analysis. Such nonlinear analysis turns out to be particularly effective in digital transmission systems, since the sign digit in the signal sample code already carries information about the moments when of the signal crossed the zero level.

The task of finding precise formulae for the sought distributions of zero-crossing intervals[2] is extremely difficult even for Gaussian random processes. Some works[2,9,29,42–44,66–69] describe approximations of such distributions. In Ref. 2, the first approximation of the above-mentioned density $W(\tau, 0)$ is proposed for short time intervals τ. This approximation has the form of differential probability $q_1(\tau, 0)$ that a process which crossed the zero level from below at moment $t = t_0$ will cross the same level from above at moment $t = (t_0 + \tau)$.

Applications of Random Process Excursion Analysis. DOI: http://dx.doi.org/10.1016/B978-0-12-409501-4.00003-6

In actual fact, $q_1(\tau,0)$ also includes as summands differential probabilities of the occurrence of any particular even number of zero level crossings within a predetermined time τ. Therefore, the first approximation $q_1(\tau,0)$ always exceeds true density $W(\tau,0)$ and the error of the first approximation increases with τ. In Ref. 8, it is shown that as $\tau \to \infty$, functions $q_1(\tau,0)$ tend to the average number of upward (or downward) zero crossings per unit of time: $\lambda_1(0) = \lambda(0)/2$.

At the same time, the theory proves that when $\tau \to \infty$, true probability density function $W(\tau,0)$ tends to zero exponentially. Therefore, for $\tau \gg \tau_0$ (where τ_0 is the time correlation interval of the process) the first approximation $q_1(\tau,0)$ does not even come close to true density function $W(\tau,0)$.

In literature on the theory of excursions applicability range of the first approximations is not discussed; the conclusions published so far about the accuracy of approximations do not go beyond stating the fact that approximations obtained are fairly accurate for $\tau < \tau_0$ and rather bad for $\tau > \tau_0$.[7–9]

Using as examples processes with known distributions $W(\tau,0)$, it would be interesting to estimate relative errors of $q_1(\tau,0)$ for different values of τ/τ_0 and to extend the obtained results to processes for which density functions $W(\tau,0)$ are not known.[29,44]

Let us use the results obtained earlier to get the lower and upper estimates of $W(\tau,0)$. From Eqs. (2.29) and (2.30) after double differentiation in terms of τ we obtain the lower estimate:

$$W(\tau,0) = \frac{B''(\tau,0)}{\lambda(0)} + \frac{1}{\lambda(0)} * \frac{d^2}{d\tau^2}\left[\sum_{k=1}^{\infty} k * (P_{2k+1} + P_{2k+2})\right] \tag{3.1}$$

where $B''(\tau,0)$ is the second derivative in terms of τ of correlation function $B(\tau,0)$ of the process obtained at the output of an ideal limiter with zero threshold level.

To obtain the upper estimate, let us write the expression of the second moment $m_2(\tau,0)$ of the number of zero crossings in an interval of length τ. Considering formulae (2.11) and (2.20) we get for $x_0 = 0$ the following expression:

$$m_2(\tau,0) = 1 - 2 * B(\tau,0) + 4 * \lambda(0) * \tau * \int_0^{\tau} \left(1 - \frac{\alpha}{\tau}\right) q_1(\alpha,0) d\alpha \tag{3.2}$$

After substituting Eq. (2.28) in (2.35) we get:

$$W(\tau,0) = q_1(\tau,0) - \frac{1}{\lambda(0)} * \frac{d^2}{d\tau^2}\left(\sum_{k=2}^{\infty}[(k-1)^2 * P_{2k} + k * (k-1) * P_{2k+1}]\right) \tag{3.3}$$

It is not difficult to show that when $\tau \to \infty$, the sum differentiated in Eq. (3.3) tends to $[\lambda(0) * \tau]^2/4$, and $W(\tau,0) \to q_1(\tau,0) - (\lambda(0)/2)$. This bears out the conclusion about the conduct of the first approximation $q_1(\tau,0)$ in the area of big values of τ.

When $\tau \to \infty$, the sum differentiated in Eq. (3.1) tends to $(\lambda(0)/2) * \tau$, and $W(\tau,0) \to (B''(\tau,0)/\lambda(0)) \to 0$.

Equations (3.1) and (3.3) define the boundaries of the area where the sought density $W(\tau,0)$ belongs:

$$W(\tau,0) \geq \frac{B''(\tau,0)}{\lambda(0)} \tag{3.4}$$

$$W(\tau,0) \leq q_1(\tau,0) \tag{3.5}$$

For short intervals ($\tau < \tau_0$) approximation curves (3.4) and (3.5) must merge and coincide with true distribution $W(\tau,0)$ of zero-crossing intervals. Note that, as τ approaches τ_0, distribution $W(\tau,0)$ will get much closer to the upper estimate (3.3) than to the lower one (3.4). Indeed, the error of the first approximation $q_1(\tau,0)$, as we can see from Eq. (3.3), is determined by the rate at which the probabilities of the occurrence of four or more zero crossings in an interval of length τ increase. For $\tau \approx \tau_0$ they increase slowly. The error of the lower estimate (3.4) is bigger than that of the upper one, because, as we can see from expression (3.1), it depends on the rate at which the probabilities of the occurrence of three or more zero crossings in an interval of length τ increase.

Let us find the applicability range of inequalities (3.4) and (3.5) using, as an example, processes for which true distribution density $W(\tau,0)$ is known. Using analogy with the previous sections, let us analyze a stationary telegraph signal and a pulse random process obtained from such a signal with the help of a trigger flip-flop: in our case the stationary telegraph signal is supplied to the input and the pulse random process is obtained at the output of the flip-flop.

After substituting in Eq. (3.1) known expressions $B(\tau,0)$ and $\lambda(0)$ for stationary telegraph signal we obtain:

$$W(\tau,0) \geq \frac{1}{2\tau_0} * \exp\left(-\frac{\tau}{\tau_0}\right)$$

Given that we already know the second moment $m_2(\tau,0)$ of the number of zero crossings, after differentiating twice Eq. (2.28) we obtain:

$$W(\tau,0) \leq \frac{1}{4\tau_0} * \left[1 + \exp\left(-\frac{\tau}{\tau_0}\right)\right]$$

The distribution of zero-crossing intervals in a stationary telegraph signal is known:

$$W(\tau,0) = \frac{1}{2\tau_0} * \exp\left(-\frac{\tau}{2\tau_0}\right) \tag{3.6}$$

Calculations carried out on the basis of formula (3.5) show that for $\tau \le 0.5\tau_0$ the relative error of the first approximation (the upper estimate) does not exceed 3.1%. However, for $\tau = \tau_0 = 0.5\tau_{av}$, the error already amounts to 12.7% and it continues to grow considerably with τ. The applicability range of the lower estimate (3.4) is only half as wide as that of the upper estimate which is much more accurate; thus, the relative error reaches 12% at $\tau = 0.5\tau_0$.

By halving the frequency of the stationary telegraph signal we obtain a new pulse random process, whose distribution of intervals between the fronts of pulses differs from Poisson distribution and is given by Eq. (2.33):

$$W_+(\tau,0) = \frac{\tau}{4\tau_0^2} * \exp\left(-\frac{\tau}{2\tau_0}\right)$$

The substitution of expressions for $m_{2+}(\tau,0)$, $B_+(\tau, 0)$, and $\lambda_+(0) = 1/4\tau_0$, which were derived earlier in Section 2.2, in inequalities (3.4) and (3.5) gives:

$$W_+(\tau,0) \ge \frac{1}{2\tau_0} * \exp\left(-\frac{\tau}{2\tau_0}\right) * \sin\left(\frac{\tau}{2\tau_0}\right) \tag{3.7}$$

$$W_+(\tau,0) \le \frac{1}{4\tau_0} * \exp\left(-\frac{\tau}{2\tau_0}\right) * \sin\left(\frac{\tau}{2\tau_0}\right) + \frac{[1 - \exp(-(\tau/\tau_0))]}{8\tau_0} \tag{3.8}$$

In Figure 3.1, the lower (3.7) and upper (3.8) estimates of $W_+(\tau,0)$ are shown. The same figure also shows a curve representing the true distribution density (2.33) of intervals between the fronts in a given pulse random process (the curve is shown

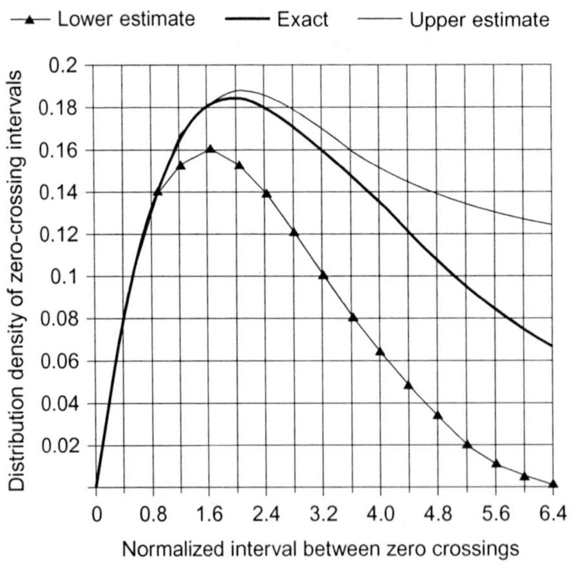

Figure 3.1 The distribution density of zero-crossing intervals between the fronts for a given pulse random process.

as a bold line). As we can see from the graphs, the first approximation (3.8) is pretty accurate, and for all τ lower than $\tau = 2\tau_0 = 0.5\tau_{av}$ the relative error does not exceed 0.83%.

In other words, the upper estimate remains sufficiently accurate until τ increases to values which correspond to the area of the maximum of true distribution density $W_+(\tau, 0)$ of intervals between zero crossings in a given process. Moreover, even for $\tau = 3\tau_0 = 0.75\tau_{av}$ the error is 4.2%, and it does not reach the level of 13.4% until $\tau = 4\tau_0 = \tau_{av}$.

In this case, the applicability range of the first approximation (3.8) is about twice as wide as it is in the case of a stationary telegraph signal. As long as the length of a time interval remains lower than $\tau = 1/\lambda_+(0) = \tau_{av}$, the first approximation $q_{1+}(\tau, 0)$ from Eq. (3.8) describes density $W_+(\tau, 0)$ (2.33) sufficiently accurately.

For τ lower than $\tau = 0.8\tau_0$ approximation (3.7), which, in general, is less accurate, practically coincides with Eqs. (2.33) and (3.8), but for $\tau = 2\tau_0 = 0.5\tau_{av}$ the relative error of the lower estimate already amounts to 18.8% and increases rapidly with τ from that point on.

Let us now verify the validity of approximations (3.4) and (3.5) by analyzing two broadband Gaussian random processes, one of which has a Gaussian and the other—a uniform power spectrum, respectively.

Figure 3.2 shows the lower and upper estimate curves, as well as a curve (it is shown as a bold line) representing the results of calculation of function $W(\tau, 0)$ of distribution of zero-crossing intervals in a Gaussian process with a Gaussian power spectrum. These results obtained in Ref. 9 by using the method of the so-called "two-tie" approximation fully coincide with the results of a thorough computer simulation of the process (the simulation results are presented in Table A1.1 of the Appendix 1).

We can see in Figure 3.2 that for $\tau \leq 0.75\tau_0$ the two approximation curves merge and coincide with the curve representing true density $W(\tau, 0)$. The applicability range of the first approximation $q_1(\tau, 0)$ from Eq. (3.5) is, as we expected,

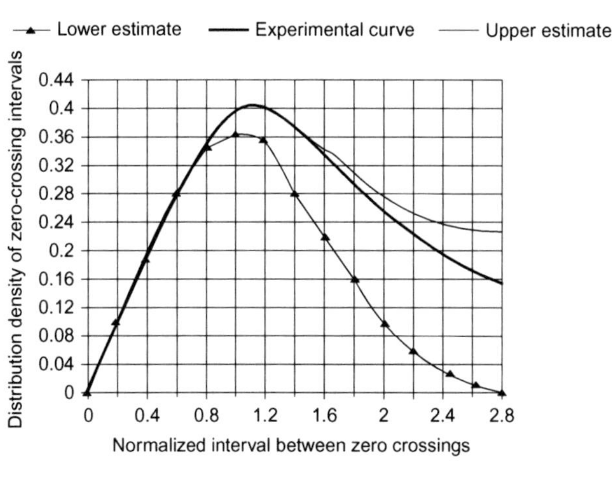

Figure 3.2 The distribution density of zero-crossing intervals for a Gaussian process with a Gaussian power spectrum.

twice as wide and extends to the value $\tau = 1.5\tau_0 \approx 0.7\tau_{av}$. In this area, the first approximation $q_1(\tau, 0)$ is very accurate and practically coincides with function $W(\tau, 0)$. Even for $\tau = \tau_{av} \approx 2.221\tau_0$ the first approximation $q_1(\tau, 0)$ remains acceptable, the relative error of the upper estimate being approximately 10%.

Figure 3.3 illustrates similar calculations of function $W(\tau, 0)$ of distribution of zero-crossing intervals in a Gaussian process with a uniform power spectrum. The lower and upper estimates obtained using formulae (3.4) and (3.5) are shown in thin lines, whereas the bold line represents the results of computer simulation of the process, (the simulation results are presented in Table A1.1 of the Appendix). The process has an limited spectrum; hence the wavy shape of curves $q_1(\tau, 0)$ and $W(\tau, 0)$.

As we can see in Figure 3.3, for τ lower than $\tau = 1.4\tau_0$ the two approximation curves merge and coincide with the curve representing true density $W(\tau, 0)$.

The applicability range of the first approximation $q_1(\tau, 0)$ from Eq. (3.5) in this example is also twice as wide and extends to the value $\tau = 2.8\tau_0 \approx 1.25\tau_{av}$, which coincides with the end of the first decline in the curve representing of function $W(\tau, 0)$. The reason why in this case the first approximation has such a wide applicability range is that the process with a rectangular spectrum has an limited frequency band and is characterized by very small probabilities of the occurrence of four or more zero crossings in an interval of length τ_{av}. A comparison of computer simulation results with the upper estimate $q_1(\tau, 0)$ shows that this approximation is very accurate: for values of τ lower than $\tau = 2.8\tau_0 \approx 1.25\tau_{av}$ the error does not exceed 5%.

Summarizing the above, we can draw the following conclusions:

1. Knowing correlation function $B(\tau, 0)$ and the first approximation $q_1(\tau, 0)$ of distribution density of values of length τ of intervals between successive zero crossings of the process

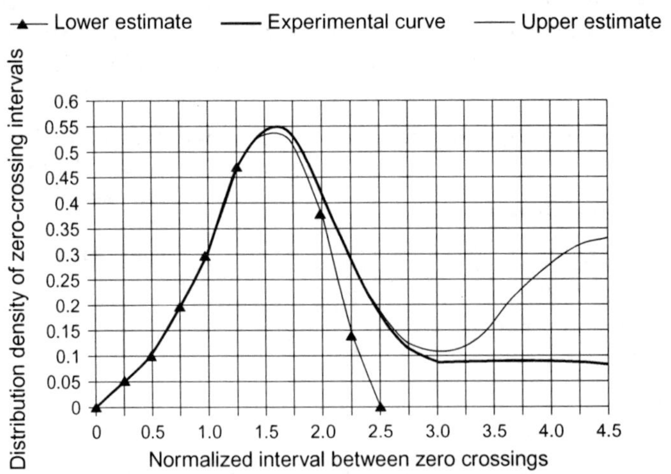

Figure 3.3 The distribution of zero-crossing intervals for a Gaussian process with a uniform power spectrum.

obtained at the output of an ideal limiter with zero threshold level, makes it possible to determine the boundaries of the area where the sought density $W(\tau, 0)$ of zero-crossing intervals belongs.

2. Function $q_1(\tau, 0)$ is the upper estimate which closely approximates density $W(\tau, 0)$ in a wide range of τ. In technically implementable random processes where distribution of momentary values lower than $\tau = \tau_{av}$ is symmetrical about zero the relative error of the upper estimate $q_1(\tau, 0)$ is comparatively small; it does not exceed 10%.
3. The first approximation functions $q_1(\tau, 0)$ obtained for processes with a limited power spectrum oscillate about value $\lambda_1(0) = 1/2\tau_{av}$. In this case, the applicability range of the first approximation $q_1(\tau, 0)$ extends beyond the limit $\tau = \tau_{av}$ to the end of the first decline in the curve representing oscillating function $q_1(\tau, 0)$.
4. Contrary to the widespread viewpoint,[7–9] the first approximation functions $q_1(\tau, 0)$ describe well not only the initial part but also an essential part of the curve representing probability density $W(\tau, 0)$ of duration of excursions above the zero level.
5. For Gaussian random processes with a given autocorrelation function the first approximation $q_1(\tau, 0)$ is not very difficult to calculate, especially when modern computers are used. In any case, it is much easier to calculate $q_1(\tau, 0)$ than to obtain, by using statistical computer simulation, probability density $W(\tau, 0)$ of excursions duration. In the cases where information on the shape of tails of distribution $W(\tau, 0)$ in the area $\tau \gg \tau_0$ is not relevant, this distribution function is almost ideally represented by the first approximation $q_1(\tau, 0)$.

Further in this monograph we will describe methods to determine the first approximation $q_1(\tau, 0)$ for a wide class of non-Gaussian random processes with a given distribution of momentary values and known autocorrelation function.

To conclude this section, let us estimate density of distribution $W_+(\tau, 0)$ of intervals between zero crossings which occur in the same direction (for instance, upward crossings).[29,45] Using the upper estimate (2.39) of distribution $P_{0+}(\tau)$ of time to the first upward zero crossing, after differentiating $P_{0+}(\tau)$ twice in terms of τ we obtain:

$$W_+(\tau, 0) = q_1(\tau, 0) - \frac{B''(\tau, 0)}{\lambda(0)} - \frac{1}{\lambda(0)} * \frac{d^2}{d\tau^2}(P_5 + 2P_6 + 4P_7 + \cdots)$$

The upper estimate, therefore, can be expressed as follows:

$$W_+(\tau, 0) \leq q_1(\tau, 0) - \frac{B''(\tau, 0)}{\lambda(0)} = q_{1+}(\tau, 0) \tag{3.9}$$

Function $q_{1+}(\tau, 0) = [q_1(\tau, 0) - (B''(\tau, 0)/\lambda(0))]$ is the first approximation of density of distribution of intervals between successive upward (or downward) crossings of the zero level by a random process. This approximation of true density $W_+(\tau, 0)$ has the form of differential probability $q_1(\tau, 0)$ that a process which crossed the zero level from below at moment $t = t_0$ will cross the same level from above at moment $t = (t_0 + \tau)$. Function $q_{1+}(\tau, 0)$ is the upper estimate of the sought density $W_+(\tau, 0)$, as it includes as summands differential probabilities of the occurrence of any given odd number of zero crossings within an interval of length τ. As long as the probabilities of the occurrence of three or more zero crossings during time τ are negligible, the upper estimate (3.9) remains very accurate.

To find the applicability range of the upper estimate (3.9) let us use the same example that we analyzed earlier. Let us consider a trigger flip-flop with a given stationary telegraph signal supplied to its input, and look at the pulse process obtained at the output of this flip-flop. In fact, in this case we are trying to determine the distribution density of intervals between successive zero crossings in a Poisson process whose frequency was divided by four. The formula of this density can be relatively easily obtained by using convolution of $W_+(\tau,0)$ from Eq. (2.33):

$$W_{++}(\tau,0) = \frac{\tau^3}{96\tau_0^4} * \exp\left(-\frac{\tau}{2\tau_0}\right) \tag{3.10}$$

At the same time, using expressions (3.4) and (3.5) obtained earlier for functions $q_1(\tau,0)$ and $B''(\tau,0)/\lambda(0)$ for this example, we can rewrite Eq. (3.9) as follows:

$$W_{++}(\tau,0) \le \frac{1-\exp(-(\tau/\tau_0))}{8\tau_0} - \frac{1}{4\tau_0} * \exp\left(-\frac{\tau}{2\tau_0}\right) * \sin\left(\frac{\tau}{2\tau_0}\right) \tag{3.11}$$

The curves shown in Figure 3.4 represent the upper estimate given by formula (3.11) and the true density of distribution of intervals between upward zero crossings given by formula (3.10).

From analyzing the curves in Figure 3.4 we can draw several conclusions about the applicability range of the first approximation $q_{1+}(\tau,0)$ and its behavior both in

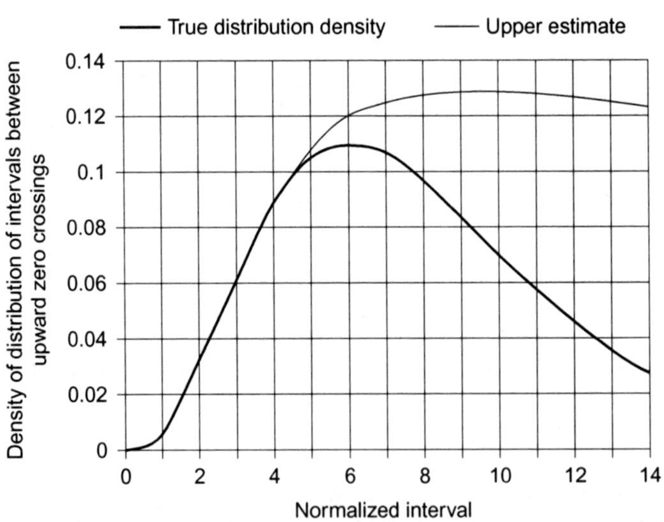

Figure 3.4 The distribution density of intervals between zero crossings which occur in the same direction in a pulse random process.

the area of short intervals ($\tau \ll \tau_{av}$), and in the area of long intervals ($\tau \gg \tau_{av}$), where τ_{av} is the average length of an interval between successive upward zero crossings (in a given example $\tau_{av} = 8\tau_0$). As it follows from the results of calculation presented in the form of graphs, the accuracy of approximation of the upper estimate (1.53) remains high for $\tau \leq 4\tau_0 = 0.5\tau_{av}$; in this area the relative error of function $q_{1+}(\tau, 0)$ in comparison with $W_{++}(\tau, 0)$ does not exceed 1.9%. For $\tau = 5\tau_0$ the error is still relatively small, it amounts to less than 4.7%; however, in the area of the maximum of the true density (3.9), that is, for $\tau = 6\tau_0 = 0.75\tau_{av}$ it reaches 9.74% and in the area of longer intervals increases rapidly.

Thus, for $\tau \leq 0.65\tau_{av}$ the first approximation $q_{1+}(\tau, 0)$ describes the sought density of intervals well, the error of the upper estimate does not exceed 5%. In the area of values close to the maximum of curve$W_{++}(\tau, 0)$, that is, for $\tau \leq 0.75\tau_{av}$ the accuracy of approximation is still acceptable, the relative error does not exceed 10%.

In the area of small values ($\tau \ll \tau_{av}$) the probability of the occurrence of short intervals between successive points at which the signal crosses zero level in the same direction is small and it increases proportionally to $C * (\tau/\tau_0)^3$, whereas the probability of the occurrence of short intervals between successive points at which the signal crosses zero level in the opposite directions increases much more rapidly, according to a linear law $K * (\tau/\tau_0)$. This is also true for the process in which the lengths of intervals between successive zero crossings are statistically independent. Thus, in the case of real processes we should expect smaller rates of increase of density $W_{++}(\tau, 0)$ in the area of small values of τ (for $\tau \ll \tau_{av}$). The reason for that is negative correlation between adjacent short intervals and small probability of the occurrence of two short intervals ($\tau \ll \tau_{av}$) in succession in a technically implementable process which has passed through a typical radio engineering section.

For $\tau > \tau_{av}$ the first approximation $q_{1+}(\tau, 0)$, according to formula (3.9) tends to function $q_1(\tau, 0)$ which, in turn, tends to intensity $\lambda_1(0)$ of excursions above the zero level. In this example $\lambda_1(0) = \lambda_{++}(0) = 1/8\tau_0$, as one could expect after analyzing the upper estimate (3.11).

Figure 3.5 shows the upper estimate $q_{1+}(\tau, 0)$ obtained on the basis of formula (3.9) for a stationary Gaussian random process with the zero mean and a Gaussian autocorrelation function. The same figure presents the results of computer simulation of density $W_{+}(\tau, 0)$ for this process, which are also shown in Table A1.1 in the Appendix 1 at the end of this book.

As we can see from Figure 3.5, in the area preceding the maximum of density (i.e., for values smaller than $\tau \approx 2.75\ \tau_0$) approximation $q_{1+}(\tau, 0)$ remains very accurate: the relative error does not exceed 5%. Even for $\tau = 0.75\tau_{av} \approx 3.3\tau_0$ the upper estimate is still acceptable and, as in the previous example, the error does not exceed 10%.

For $\tau < \tau_0$ the first approximation $q_{1+}(\tau, 0)$ which nearly coincides with true distribution density $W_{+}(\tau, 0)$ can be expressed as $(2.8/\tau_0) * 10^{-2} * (\tau/\tau_0)^4$.

This proves the validity of the earlier assumption that real processes are characterized by lower probability of the occurrence of short intervals ($\tau < \tau_0$) between points where the signal crosses zero level in a given direction.

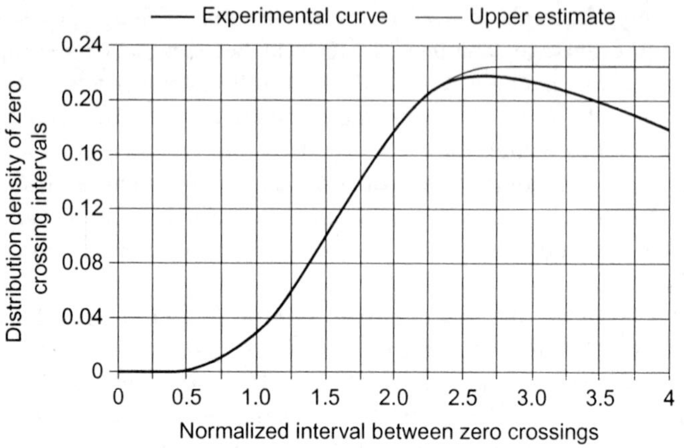

Figure 3.5 The distribution density of intervals between zero crossings which occur in the same direction in a Gaussian process with a Gaussian power spectrum.

In the area of $\tau > \tau_{av}$ function $q_{1+}(\tau,0)$ tends to $\lambda_1(0) = (1/\pi\sqrt{2} * \tau_0) = (1/\tau_{av}) \approx (0.225/\tau_0)$ and does not even give a vague idea of how the sought density $W_+(\tau,0)$ behaves.

Figure 3.6 shows the upper estimate $q_{1+}(\tau,0)$ obtained on the basis of formula (3.9) for a centered Gaussian stationary random process with a uniform power spectrum. The same figure presents the results of computer simulation of density $W_+(\tau,0)$ for this process, which are also shown in Table A1.2 in Appendix 1 at the end of this book.

From Figure 3.6, we can see that for $\tau \leq 4.4\tau_0 \approx \tau_{av}$ the first approximation $q_{1+}(\tau,0)$ and density $W_+(\tau,0)$ obtained by computer simulation are nearly identical. Like function $q_1(\tau,0)$ in Figure 3.3, $q_{1+}(\tau,0)$ has the form of a fading oscillation around limit $\lambda_1(0) \approx 0.225/\tau_0$. The applicability range of the upper estimate coincides with the end the first declining part of the estimate curve. Approximation $q_{1+}(\tau,0)$ obtained on the basis of formula (3.9) cover an essential part of the experimental curve representing density $W_+(\tau,0)$.

The fact that the applicability range of the approximation function extends to the value $\tau = \tau_{av}$ can be explained by the limited nature of power spectrum of the process and by a small probability of the occurrence of five or more zero crossings in an interval of length $\tau = \tau_{av}$. The initial part ($\tau \leq \tau_0$) of curves representing functions $q_{1+}(\tau,0)$ and $W_+(\tau,0)$ in Figure 3.6 is described well by expression $W_+(\tau,0) = (1,4/\tau_0) * 10^{-3} * (\tau/\tau_0)^4$, whereas density $W(\tau,0)$ in this case changes in a linear fashion $W(\tau,0) \approx (\tau/5 * \tau_0^2)$.

A comparison of the initial parts of curves representing function $W_+(\tau,0)$ in Figures 3.5 and 3.6 shows a 20-fold reduction in the probabilities of the occurrence of short intervals ($\tau \leq \tau_0$) for a process with a rectangular spectrum. Note that for both Gaussian processes the average length of an interval between successive

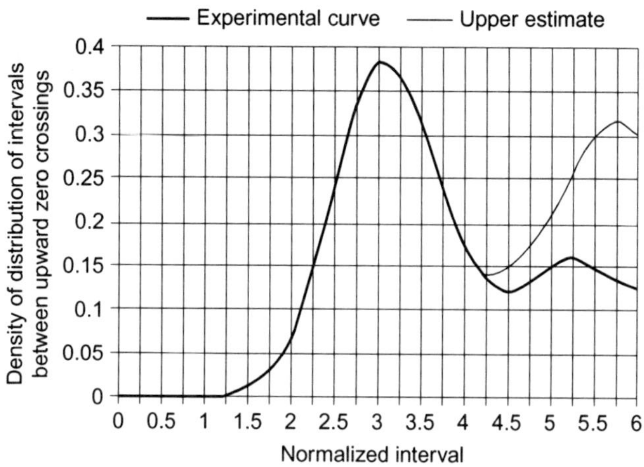

Figure 3.6 The distribution density of intervals between zero crossings which occur in the same direction in a Gaussian process with a uniform power spectrum.

excursions above the zero level is assumed to be the same: $\tau_{av} = (1/\lambda_1(0)) \approx 4.44\tau_0$.

Thus, knowing the first approximation $q_1(\tau, 0)$ of probability density of duration of intervals between successive zero crossings in a random process and knowing correlation function $B(\tau, 0)$ of the process obtained at the output of an ideal limiter with zero threshold level makes it possible to calculate, using formula (3.9) the upper estimate $q_{1+}(\tau, 0)$ of probability density $W_+(\tau, 0)$ of intervals between successive excursions above the zero level. For processes with an unlimited power spectrum the first approximation $q_{1+}(\tau, 0)$ remains very accurate for a wide range of values reaching as far as the maximum value of the sought one-mode curve of density $W_+(\tau, 0)$, whereas for processes with a limited power spectrum, the applicability range of approximation extends as far as the first minimum of the multimode density curve.

For Gaussian processes with a given autocorrelation function the calculation of the first approximation $q_{1+}(\tau, 0)$ does not cause any problems. The existing method of calculating functions $q_1(\tau, 0)$, $B(\tau, 0)$, and $q_{1+}(\tau, 0)$ for differentiable non-Gaussian random processes symmetrical about zero level with a given law of distribution of momentary values and the known autocorrelation function, is rather complicated as it involves calculation of multidimensional integrals.[2]

Further in this chapter a method of nonlinear noninertial transformation will be described which enables transformation of a wide class of non-Gaussian random processes into Gaussian ones. Relations between autocorrelation functions of the initial process and the transformed one will be established. Using the above-mentioned transformation method, we managed to cope with the difficult task of calculating functions $q_1(\tau, 0)$, $B(\tau, 0)$, and $q_{1+}(\tau, 0)$ for a number of non-Gaussian processes.

3.2 One Way to Increase the Accuracy of the First Approximation for the Distribution Density of Level-Crossing Time Intervals in a Stationary Random Process

Using the general formulae obtained for the first time in Ref. 2 and cited in Refs. 7–9, it is possible to calculate, accurately enough, unknown density $W(\tau, x_0)$ of distribution of duration τ of excursions above a given level x_0 as well as the distribution of intervals between such excursions in a random process. The formulae, however, are practically useless for engineers, since the calculations they involve are too complicated even for a powerful computer.

It is relatively easy to calculate the first approximation functions $q_1(\tau, x_0)$, which serve as upper estimates for true probability densities $W(\tau, x_0)$. In Section 3.1 we showed that the accuracy of approximation for $x_0 = 0$ remains acceptable for all values of τ that are comparable with the average duration of excursions τ_{av}. In this case relative error $q_1(\tau, 0)$ of the upper estimate for technically feasible Gaussian random processes does not exceed a few percentage points. However, for $\tau > \tau_{av}$ an increase in τ is accompanied by a rapid increase in the error of first approximation $q_1(\tau, 0)$.

Obviously, the range of duration values, for which a particular upper estimate is applicable, is not large enough; it can be extended by making corrections to functions $q_1(\tau, 0)$ of the first approximation.[40,46]

It is known that, for any even number, $q_1(\tau, 0)$ includes as a summand the differential probability that level x_0 will be crossed that number of times within an interval of length τ. Therefore, $q_1(\tau, x_0) > W(\tau, x_0)$ for all τ. The first correction $\Delta_1(\tau, x_0)$ must take into account the differential probability of occurrence of two level crossings within interval τ, that is, the probability that interval τ will be broken into three segments by the points where the process curve crosses level x_0.

Accordingly, the second correction $\Delta_2(\tau, x_0)$ is the differential probability of occurrence of four level x_0 crossings within interval τ, that is, the probability that interval τ will be broken into five segments by the points where the process curve crosses level x_0 as it goes up and down (which means that the values of the first derivative for any two successive crossings will have the opposite signs).

Generally speaking, the nth correction $\Delta_n(\tau, x_0)$ takes into account the differential probability of occurrence of $(2n)$ level crossings within interval τ, such that they break the interval into $(2n+1)$ segments. Probability density $W(\tau, x_0)$ which we are trying to determine can be treated as a correction of zero order or as the differential probability of the absence of level crossings within interval τ.

It is obvious that:

$$\sum_{n=0}^{\infty} \Delta_n(\tau, x_0) = q_1(\tau, x_0) \tag{3.12}$$

The first correction $\Delta_1(\tau, x_0)$ can be determined as a zero correction convolution or as a convolution of the sought function $W(u, x_0)$ and distribution $W_+(\tau - u, x_0)$

of intervals between successive upward or downward level crossings (i.e., successive crossings with the same—positive or negative—sign of the first derivative). Each interval under consideration, thus, comprises two intervals between successive level-crossing points.

In Refs. 35, 40, 47, and 55 it is shown that, in technically feasible stationary random processes, crossings of a specified level can be, as a first approximation, considered statistically independent. In view of the above, the first correction $\Delta_1(\tau, x_0)$ can be calculated using the formula for the convolution of the sum of two independent positive random variables:

$$\Delta_1(\tau, x_0) = \int_0^T W(u, x_0) * W_+(\tau - u, x_0)\mathrm{d}u \tag{3.13}$$

Similarly, the second correction $\Delta_2(\tau, x_0)$ can be obtained from the convolution of the first correction $\Delta_1(u, x_0)$ and distribution $W_+(\tau - u, x_0)$:

$$\Delta_2(\tau, x_0) = \int_0^T \Delta_1(u, x_0) * W_+(\tau - u, x_0)\mathrm{d}u \tag{3.14}$$

Generally speaking, the nth correction $\Delta_n(\tau, x_0)$ can be obtained from the convolution of correction number $(n - 1)$ and distribution $W_+(\tau - u, x_0)$:

$$\Delta_n(\tau, x_0) = \int_0^T \Delta_{n\text{-}1}(u, x_0) * W_+(\tau - u, x_0)\mathrm{d}u \tag{3.15}$$

In view of Eq. (3.12), by adding up Eqs. (3.13)–(3.15) for $n \rightarrow \infty$ we obtain:

$$W(\tau, x_0) = q_1(\tau, x_0) - \int_0^T q_1(u, x_0) * W_+(\tau - u, x_0)\mathrm{d}u \tag{3.16}$$

When $\tau \rightarrow \infty$, $q_1(\tau, x_0)$ tends to $\lambda_1(x_0)$, the intensity of excursions above level x_0, hence

$$lim\ W(\tau, x_0) = \lambda_1(x_0) - \lambda_1(x_0) * \int_0^\infty W_+(u)\mathrm{d}u = \lambda_1(x_0) - \lambda_1(x_0) = 0$$

Using the same logic, distribution $W_+(\tau, x_0)$ of interval lengths between neighboring excursions above level x_0 in a random process can be found from the following recurrent relation:

$$W_+(\tau, x_0) = q_{1+}(\tau, x_0) - \int_0^T q_{1+}(u, x_0) * W_+(\tau - u, x_0)\mathrm{d}u \tag{3.17}$$

Function $q_{1+}(\tau, x_0)$ is the first approximation of density $W_+(\tau, x_0)$ of distribution of intervals between successive upward or downward crossings of level x_0 in a random process. This function uses as an approximation of true probability density $W_+(\tau, x_0)$ the differential probability of the occurrence of level x_0 crossing point with a positive derivative at time moment $(t_0 + \tau)$ in a process which has crossed the same level at time moment t_0 with a positive derivative.

Function $q_{1+}(\tau, x_0)$ can serve as an upper estimate of probability density $W_+(\tau, x_0)$, since it includes differential probabilities of the occurrence of any odd number of level x_0 crossings within time interval τ. In this case, the first correction $\Delta_{1+}(\tau, x_0)$ must take into account the differential probability of the occurrence of three level crossings within interval τ, that is, the probability of breaking interval τ into four segments by the points where it the process curve crosses level x_0. The second correction $\Delta_{2+}(\tau, x_0)$ takes into account the differential probability of the occurrence of five level crossings (respectively, six segments), etc.

True probability density $W_+(\tau, x_0)$ may be treated as zero correction which includes the differential probability of the occurrence within interval τ of one crossing of level x_0 with the opposite sign of the derivative (a crossing which breaks the interval into two segments).

The errors of estimates (3.16) and (3.17) are determined by the strength of correlation between successive level-crossing points for which the first derivative of the process has the same or different sign. If the correlation is zero for all values of τ, then formulae (3.16) and (3.17) give accurate estimations (i.e., we have zero estimation errors). On the other hand, a strong correlation between successive level-crossing intervals causes big relative errors in the calculation of correction integrals in formulae (3.16) and (3.17).

It follows from the results obtained in Section 3.1 that for $\tau \le \tau_0$ there is a high correlation between duration values for successive level-crossing intervals, particularly zero-crossing ones. However, for $\tau \le \tau_0$ both absolute and relative values of correction are very small because of the proximity of functions $q_1(\tau, x_0)$ and $W(\tau, x_0)$, and functions $q_{1+}(\tau, x_0)$ and $W_+(\tau, x_0)$. This conclusion is borne out by the fact that, for a wide range of values of τ, the upper estimates calculated using formulae (3.5) and (3.9) practically coincide with functions $W(\tau, 0)$ and $W_+(\tau, 0)$.

For $\tau > \tau_0$ we see an increase in both absolute and relative values of convolution integrals in Eqs. (3.16) and (3.17) (speaking of relative values, we mean their values in comparison with density values); at the same time, relative errors of correction are expected to decrease because of the weakening of correlation between successive excursions above level x_0.

It follows from the results in Section 3.1 where distributions $W(\tau, 0)$ and $W_+(\tau, 0)$ were obtained for a Gaussian process with rectangular power spectrum by using a digital simulation, that those distributions fully coincide with the first approximation functions $q_1(\tau, x_0)$ and $q_{1+}(\tau, x_0)$ over a wide range of values of τ. Consequently, the integrals of convolutions in formulae (3.16) and (3.17) remain negligibly small for $\tau < 2.8\tau_0$ in the case of $W(\tau, 0)$, and for $\tau < 4.4\tau_0$ in the case of $W_+(\tau, 0)$.

If zero-crossing points were statistically independent for all τ, the distribution density of intervals between successive excursions above zero level could be obtained from the convolution as the distribution of the sum of two independent random variables distributed according to the same law $W(\tau, 0)$:

$$\overline{W_{+}}(\tau, 0) = \int_{0}^{\tau} W(u, 0) * W(\tau - u, 0)\mathrm{d}u \tag{3.18}$$

The differences between $W_{+}(\tau, 0)$ and $\overline{W_{+}}(\tau, 0)$, if any, indicate a correlation between neighboring intervals.

Table 3.1 shows values of $\overline{W_{+}}(\tau, 0)$, which were calculated using formula (3.18) in which experimental density values $W(\tau, 0)$ were obtained by means of a digital simulation. The experiment was carried out on a broadband centered Gaussian process with a uniform power spectrum. Table 3.1 also contains values of $W(\tau, 0)$ obtained by numerical integration using formula (3.16) and experimental density values $W(\tau, 0)$.

The analysis of the data in Table 3.1 reveals the fact that the assumption about the independence of successive zero-level crossings proves right only if $\tau \geq 3\tau_0$ where τ is the sum of duration values for two successive intervals between zero crossings. For small τ, neighboring intervals turn out to be highly correlated; hence, a considerable decrease in the differential probability of occurrence of two short intervals in a row (here "short" means that $\tau \leq \tau_0$). This explains why for $\tau \leq \tau_0$ the convolution from Eq. (3.18) exceeds by far density value $W_{+}(\tau, 0)$ obtained experimentally. At the same time, for $\tau \geq 3\tau_0$, the difference between $W_{+}(\tau, 0)$ and convolution (3.18) does not exceed 5%, that is, the assumption about statistical independence of duration values for adjacent intervals proves absolutely right.

We should also note that, because of correlation between duration values for adjacent intervals, the relative error of convolution (3.18) is higher than real errors in estimates (3.16) and (3.17), especially in (3.17).

Table 3.1 A Gaussian Process with a Uniform Power Spectrum

(τ/τ_0)	$q_1(\tau,0)$	$W(\tau,0)$	$q_{1+}(\tau,0)$	$W_{+}(\tau,0)$	$\overline{W_{+}}(\tau,0)$
0.5	0.1	0.1	0.0001	0.0001	0.004
1	0.3	0.3	0.0014	0.0014	0.032
1.5	0.545	0.545	0.015	0.015	0.06
2	0.435	0.435	0.075	0.075	0.115
2.5	0.175	0.17	0.255	0.255	0.292
3	0.085	0.078	0.39	0.39	0.399
3.5	0.16	0.082	0.305	0.305	0.32
4	0.28	0.08	0.164	0.163	0.165
4.5	0.33	0.065	0.145	0.12	0.126
5	0.3	0.069	0.215	0.145	0.15
5.5	–	0.04	0.3	0.15	0.145
6	–	–	0.31	0.13	0.125
7	–	–	–	0.05	0.05
8	–	–	–	0.04	0.04

Indeed, formula (3.16) is based on the assumption that the total duration of two successive intervals is independent from the duration of the preceding single interval (the first correction), the total duration of the three preceding intervals (the second correction), the five preceding intervals (the third correction), etc. Correlation between durations of successive interval groups weaken more rapidly than correlation between successive single intervals, especially for high levels x_0.

In the case of recurrent formula (3.17) the error will be even smaller, because correlation between in this case the total duration of two successive intervals and the total duration of two, four, etc. preceding ones weakens more rapidly.

Let us now consider a Gaussian random process with a Gaussian autocorrelation function. Earlier in Section 3.1 we obtained the first approximations (i.e., the upper estimates) $q_1(\tau,0)$ and $q_{1+}(\tau,0)$ of this type of process. Let us calculate, using recurrent relation (3.17), the estimate of density $W_+(\tau,0)$ of intervals between successive upward crossings of zero level and then, using formula (3.16) the estimate of density $W(\tau,0)$ of intervals between successive downward crossings of zero level. (In the two cases the first derivative of the process at zero-crossing points will be positive and negative, respectively).

Table 3.2 shows values of the first approximation functions $q_1(\tau,0)$ and $q_{1+}(\tau,0)$ obtained by calculation, values of density $W_+(\tau,0)$ obtained by numerical integration using formula (3.17), values of density $W(\tau,0)$ calculated using formula (3.16) and convolution values obtained using formula (3.18). Apart from that, Table 3.2 contains the values of the so-called "double-correlation" approximation $P_2(k\tau_0)$ of duration distribution for excursions above zero level in a given Gaussian process. These data were obtained in Ref. 9 for integer $k = 1, 2, 3, 4, 5, 6, 7$, and 8.

As we can see from Table 3.2, the assumption about the independence of successive zero crossings is only correct if $\tau \geq 1.5\tau_0$ where τ is the total duration of two adjacent intervals. For small values of τ adjacent intervals strongly correlate, and the occurrence of two short successive intervals (i.e., such that $\tau \leq 0.5\tau_0$) is much less probable. Therefore, for small τ, values of convolution $\overline{W_+}(\tau,0)$ obtained using Eq. (3.18) are significantly higher than density values $W_+(\tau,0)$ found by experimenting.

However, a process with a Gaussian spectrum is characterized by a rapid decrease in correlation between successive level-crossing intervals. Even for $\tau \geq 1.5\tau_0$ the difference between $W_+(\tau,0)$ and convolution (3.18) does not exceed 5%, that is, the assumption about statistical independence of adjacent intervals becomes completely valid.

Therefore, it is certain that for $\tau \geq 0.5\tau_{av}$ the estimates obtained using formulae (3.16) and (3.17) are applicable for any threshold level. Here τ_{av} is the average length of the interval between successive excursions above a given level; for both Gaussian processes it has the same value (in the case of zero level $\tau_{av} = 1/\lambda_1(0) \approx 4.44\tau_0$).

The accuracy of estimation of density $W(\tau,0)$ becomes evident from the comparison of calculated approximation $P_2(k\tau_0)$ of duration distribution in excursions above the zero level, and figures obtained for a Gaussian process with a Gaussian

Table 3.2 A Gaussian Process with a Gaussian Power Spectrum

$\frac{\tau}{\tau_0}$	$q_1(\tau,0)$	$W(\tau,0)$	$P_2(k\tau_0)$	$q_{1+}(\tau,0)$	$W_+(\tau,0)$	$\overline{W_+}(\tau,0)$
0.5	0.2395	0.2395	–	0.0018	0.0018	0.0052
1	0.393	0.3923	0.4	0.0258	0.0256	0.0373
1.11	0.4022	0.4012	–	0.04	0.04	0.048
1.5	0.3649	0.36	–	0.0988	0.0988	0.1035
2	0.2717	0.256	0.25	0.181	0.1804	0.172
2.22	0.25	0.2275	–	0.2015	0.2003	0.1933
2.5	0.2326	0.184	–	0.2173	0.2142	0.209
3	0.2258	0.1394	0.14	0.2243	0.2156	0.214
4	0.225	0.0769	0.08	0.225	0.1812	0.179
5	0.225	0.0426	0.045	0.225	0.1351	0.135
6	0.225	0.0236	0.025	0.225	0.0942	0.094
7	0.225	0.013	0.013	0.225	0.0624	0.064
8	0.225	0.0072	0.007	0.225	0.0403	0.043
9	0.225	0.004	–	0.225	0.0258	0.028
10	0.225	0.0022	–	0.225	0.01586	0.018

spectrum by numerical calculation using formula (3.16). The two methods of calculation produced the same results for the case where k is an integer.

In addition, there is a very good match between density $W(\tau,0)$ from Table 3.2 and the results obtained for a Gaussian process with power spectrum of the same name by using a PC simulation.

For shorter intervals $\tau < 1.5\tau_0$, as it follows from Table 3.2, absolute and relative errors of the first approximation $q_1(\tau,0)$ are very small compared to $W(\tau,0)$. Therefore, the error of convolution integral in Eq. (3.16), which is caused by the correlation of short excursions, is negligible.

Similar conclusions can be drawn regarding the estimate given by Eq. (3.17). In the case of $\tau < 0.5\tau_{\text{av}}$ correlation between the total duration of two successive intervals following the event and the total duration of two preceding level-crossing intervals can only cause an insignificant error, since the relative "weight" of correction integral in Eq. (3.17) does not exceed 0.6% of value $W_+(\tau,0)$. As τ decreases, the approximation of true density $W_+(\tau,0)$ given by function $q_{1+}(\tau,x_0)$ becomes more accurate, and the convolution integral in Eq. (3.17) rapidly tends to zero.

After coming up independently with formulae (3.16) and (3.17) the author of this monograph read about the integral equation obtained by McFadden.[40] This equation used for calculating the estimate values of distribution functions $W(\tau,x_0)$ is based on the assumption that durations of successive intervals between points where the process crosses level x_0 are quasi-independent. This assumption means that duration of an arbitrarily chosen interval does not depend on the total duration of $(2n+2)$ preceding intervals, where $n = 0, 1, 2, \ldots$

To be able to compare results obtained by McFadden with formula (3.16), let us rewrite equation[40] as follows:

$$W_M(\tau, x_0) = q_1(\tau, x_0) - \int_0^\tau W(u, x_0) * q_{1+}(\tau - u, x_0)\mathrm{d}u \tag{3.19}$$

Analysis of McFadden's integral Eq. (3.19) shows that, for small τ (i.e., for $\tau < \tau_0$), it bears close similarity to Eq. (3.16). In this domain $W(u, x_0) \approx q_1(u, x_0)$, $W_+(\tau - u, x_0) \approx q_{1+}(\tau - u, x_0)$, and, in spite of the difference between integrated functions, integrals in Eqs. (3.16) and (3.19) tend to zero at the same rate.

Using PC simulations of functions $W(\tau, 0)$ defined by formulae (3.16) and (3.19) for $x_0 = 0$, we obtained results which were similar for a wide range of τ. Nevertheless, judging from the estimated average duration τ_{av} of zero-crossing intervals in three broadband Gaussian processes with different forms of $R(\tau)$, formula (3.16) gives more accurate figures, which do not differ from theoretical values by more than a few tenths of a percent. The values of τ_{av} obtained using McFadden's Eq. (3.19) turned out to be somewhat overestimated in all cases, whereas the relative errors do not exceed a few percentage points.

One advantage of formula (3.19) is that it is quite simple in comparison with formula (3.16). McFadden's equation is a recurrent relation, which enables calculation of density $W(\tau, 0)$ in the case where the first approximations $q_1(\tau, x_0)$ and $q_{1+}(\tau, x_0)$ are known. Unlike Eq. (3.19), Eq. (3.16) which is more accurate and more complicated, suggests using function $W_+(\tau, 0)$ which makes part of Eq. (3.17). Solving the system of integral Eqs. (3.16) and (3.17) starts with recurrent relation (3.17).

Thus, knowing the first approximations $q_1(\tau, x_0)$ and $q_{1+}(\tau, x_0)$, we can estimate densities $W_+(\tau, x_0)$ and $W(\tau, x_0)$. The system of integral Eqs. (3.16) and (3.17) can, therefore, be considered as a development of McFadden's results.[40]

Apart from increasing the accuracy of calculating $W(\tau, x_0)$, formula (3.17) has one other important feature, which other formulae do not have: it allows for the estimation of distribution $W_+(\tau, x_0)$ of durations in intervals between successive upward or downward crossings of level x_0.

Several conclusions can be drawn from the above:

1. Obtained on the basis of the assumption that the points where a stationary random process crosses a given level x_0 are statistically independent, formulae (3.16) and (3.17) allow for more accurate estimates of distributions $W_+(\tau, x_0)$ and $W(\tau, x_0)$ of duration of intervals between successive zero crossings having the same or different signs of the first derivative, respectively, in the case where the first approximations $q_1(\tau, x_0)$ and $q_{1+}(\tau, x_0)$ are known.
2. Using as an example stationary Gaussian processes with rectangular and Gaussian spectrums we made observations which support conclusions drawn in Refs. 35, 40, 47, and 55 about practical independence of successive zero crossings for $\tau > 1.5\tau_{1\text{av}}$, where τ is the total duration of two adjacent intervals and $\tau_{1\text{av}}$ is the average duration of a single interval.

3. It is shown that relative errors of estimates obtained using formulae (3.16) and (3.17) do not exceed a few tenths of a percent in the case of short adjacent level-crossing intervals $\tau < \tau_{1\,\mathrm{av}}$, in spite of a sufficiently high correlation between interval lengths. The reason for that is that for small τ absolute values of convolutions integrals are also small.
4. Comparing estimates obtained using formulae (3.16) and (3.17) with results produced by PC simulation of the so-called "double-correlation" approximation[9] reveals that estimated density curves $W(\tau,0)$ and $W_{+}(\tau,0)$ closely match experimental data and theoretical findings obtained by methods described in Ref. 46, for a wide range of $\tau \gg \tau_0$.
5. In the case where specific calculations are required recurrent formula (3.17) becomes useful as a starter. Knowing in advance the first approximation (upper estimate $q_{1+}(\tau,x_0)$, it is possible to calculate, by gradually increasing τ, curve $W_{+}(\tau,x_0)$, and then substitute the result in formula (3.16) in order to obtain density $W(\tau,x_0)$ from the known value of the first approximation $q_1(\tau,x_0)$.
6. For Gaussian and Rayleigh stationary random processes with a given form of power spectrum or a known correlation function $R(\tau)$ it is not difficult to find the first approximations $q_1(\tau,x_0)$ and $q_{1+}(\tau,x_0)$ of duration distribution for excursions above level x_0 and for intervals between successive excursions.[8] Once you have written a computer program which makes calculations using formulae (3.16) and (3.17) it is possible to create a "data bank" to store information on the forms of distributions $W(\tau,x_0)$ and $W_{+}(\tau,x_0)$ for the most interesting and most common types of random processes. It also seems appropriate to establish relations between the form of $R(\tau)$ and the parameters of distributions $W(\tau,x_0)$ and $W_{+}(\tau,x_0)$ for a given type of a stationary random process.

3.3 Methods of Calculating Level-Crossing Parameters for Certain Classes of Non-Gaussian Stationary Random Processes

As noted above, the problems related to the study of parameters which characterize excursions of random processes have, for a long time, been the focus of attention of mathematicians and engineers. Up until now it has not been possible to obtain, even for Gaussian random processes, the closed form of a formula to accurately calculate the distribution function $W(\tau,x_0)$ of duration τ of excursions above a given level x_0. The general expression proposed by Rice[2] for the excursions duration distribution function has the form of a slowly converging infinite series of multiple integrals whose multiplicity grows without limitation. The first member of the series (for $x_0 = 0$) is called the first approximation $q_1(\tau,0)$ for the distribution function of the duration of excursions above (or below) the zero level in a centered random process. All attempts to find the consequent members of the series in Rice's formula have failed, since the calculation of multiple integrals presents an insurmountable difficulty even for state-of-the-art computers.

Earlier in this monograph, we showed that for broadband Gaussian random processes with several distinctly different energy spectra, the first approximation $q_1(\tau,0)$ for the distribution of excursions duration enables sufficiently accurate calculation of density $W(\tau, 0)$ for $\tau \leq \tau_0$; the relative error of approximation does not exceed 1%.

In Section 3.2, we also showed that knowing the first approximation $q_1(\tau, 0)$ makes possible to calculate the precise distribution function, which describes well the true density $W(\tau, 0)$ for a wide range of duration values $\tau > \tau_0$.

Thus, knowing the first approximation $q_1(\tau, 0)$ for the distribution of excursions duration can be very useful and, in some cases, sufficient for solving numerous applied problems. However, as it follows from Ref. 8, even for the calculation of the first term of Rice's infinite series, that is, function $q_1(\tau, 0)$, it is necessary to know the joint probability density of random process $X(t)$ and its first derivative $Z(t)$ at two moments of time, or four-dimensional density $W_4(x_1, z_1, x_2, z_2)$.

For a Gaussian random process, as long as you know its autocorrelation function $R(\tau)$, the value for $q_1(\tau, 0)$ can be relatively easily calculated using formulae.[8] In case of a non-Gaussian centered random function, symmetrical about the zero level, we are dealing with a much more complicated problem: in general, it is impossible to calculate $q_1(\tau, 0)$ knowing only its one-dimensional momentary values distribution function $W(x)$ and autocorrelation function $R(\tau)$.[76]

We are going to prove that there exists a rather wide class of non-Gaussian stationary random processes with a given one-dimensional distribution function $W(x)$ and autocorrelation function $R(\tau)$, for which it is possible to calculate the first approximation for the distribution of excursions duration, $q_1(\tau, 0)$, using known formulae valid for Gaussian processes with some other autocorrelation function $R_n(\tau)$ which depends on the form of $W(x)$ and $R(\tau)$.[71]

Let us consider a noninertial nonlinear element with characteristic $y = f(x)$ which converts Gaussian random process $X(t)$ into a random process $Y(t)$ with non-Gaussian distribution law $W(y)$ of momentary values. Autocorrelation function $R_y(\tau)$ of process $Y(t)$ will also differ from autocorrelation function $R_x(\tau)$ of the Gaussian process at the converter input.

Let $y = f(x)$ be an odd function, that is, $f(x) = -f(-x)$. Then the processes at the input and output of the converter will have the same zero crossing points, and conversion will not change distribution $W(\tau, 0)$ of intervals between zero crossings.

Let us consider a smoothed limiter with characteristic[8]:

$$y = f(x) = \sqrt{\frac{2}{\pi}} * \int_0^{x/l\sigma} e^{-t^2/2} dt \tag{3.20}$$

$$y = 2F\left(\frac{x}{l\sigma}\right) - 1 \tag{3.21}$$

where $F(x/l\sigma)$ is a tabulated function of a probability integral.

For the case where the input of such nonlinear noninertial converter is a Gaussian stationary random process with the zero mean, variance σ^2 and autocorrelation function $R(\tau)$ it is shown in Ref. 8 that correlation function of the process at the limiter output is described by the formula:

$$B(\tau) = \left(\frac{2}{\pi}\right) * \arcsin\left[\frac{R(\tau)}{1+l^2}\right] \tag{3.22}$$

When $l = 0$, the smoothed limiter degenerates into an ideal two-sided signal clipping limiter with characteristic:

$$y = f(x) = \begin{cases} 1, & x \geq 0 \\ -1, & x < 0 \end{cases}$$

At the other extreme, that is, when $l \to \infty$, there is no limitation and the distribution of momentary values at the output is a Gaussian process. For any l between these two extreme cases process $Y(t)$ at the output of a smoothed limiter with characteristic (3.21) is not Gaussian one and has a continuous distribution.

By making simple transformations it is possible to obtain the expression for the distribution of process $Y(t)$ at the output of a smoothed limiter:

$$W(y) = \frac{l}{2}\exp\left[-\left(\frac{x^2}{2\sigma^2}\right) * \left(1 - \frac{1}{l^2}\right)\right] \tag{3.23}$$

Here $x = \phi(y)$ and the relation between x and y is given by nonlinear Eq. (3.21).

Distribution (3.23) is symmetric about variable y for any value of parameter 1. As it can be seen from Eq. (3.23), with $l = 1$ distribution $W(y)$ becomes uniform:

$$W(y) = \frac{1}{2}, \quad -1 \leq y \leq 1$$

In the case where $l \gg 1$ distribution $W(y)$ tends to a Gaussian one, whereas in the case where $l < 1$ the power of exponent in Eq. (3.23) becomes positive. As l tends to zero, distribution $W(y)$ gets closer to a discrete one with two equiprobable levels ± 1.

Autocorrelation coefficient $R_y(\tau)$ of process $Y(t)$ at the output of a smoothed limiter can be expressed through autocorrelation coefficient $R_x(\tau)$ of Gaussian process $X(t)$ as follows:

$$R_y(\tau) = \frac{\arcsin(R_x(\tau)/1+l^2)}{\arcsin(1/1+l^2)} = \frac{\arcsin(R_x(\tau)/1+l^2)}{\phi} \tag{3.24}$$

where angle

$$\varphi = \arcsin\frac{1}{1+l^2} \tag{3.25}$$

As it follows from Eq. (3.25), when $l = 0$, angle $\varphi = \pi/2$. Such condition means ideal limitation on zero level where:

$$R_y(\tau) = \frac{2}{\pi} * \arcsin R_x(\tau)$$

When $l \to \infty$, angle $\varphi = 0$ and $R_y(\tau) = R_x(\tau)$. If $l = 1$, then angle $\varphi = \pi/6$ and from Eq. (3.25) we obtain

$$R_y(\tau) = \frac{6}{\pi} * \arcsin \frac{R_x(\tau)}{2}$$

For $\varphi = \pi/4$, $l = 0.6436$; for $\varphi = \pi/3$, $l = 0.3933$.

It is obvious that the less is value l, the stronger is the non-Gaussian character of the process at the smoothed limiter output and the more noticeable is the difference between $R_y(\tau)$ and $R_x(\tau)$.

The family of curves on Figure 3.7 represents amplitude characteristics of smoothed limiter $y = f(x/\sigma)$; the curves were built using formula (3.21).

Angle φ which takes a number of values, in particular, $\varphi = \pi/12$, $\varphi = \pi/6$, $\varphi = \pi/4$, $\varphi = \pi/3$, $\varphi = \pi/2$, was chosen as a parameter. The relation between l and φ is given by formula (3.25).

Figures 3.8–3.14 show distributions $W_\varphi(y)$ of momentary values of process $Y(t)$ at the limiter output, calculated for $\varphi = 0.01$, $\pi/12$, $\pi/6$, $\pi/4$, $\pi/3$, $0.45 * \pi$, and $\pi/2$. The calculations were performed using formula (3.23), as well as formulae (3.21) and (3.25). It was assumed that $-1 \le y \le 1$.

As you can see from the figures, even small deviations of angle φ may cause significant changes in the shape of distribution $W_\varphi(y)$.

Using formula (3.24) we calculated the family of autocorrelation functions $R_y(\tau)$ of process $Y(t)$ at the limiter output for $\varphi = 0$, $\pi/12$, $\pi/6$, $\pi/4$, $\pi/3$, and $\pi/2$ (see Figure 3.15). The calculations are based on the assumption that autocorrelation function $R_x(\tau)$ of normal process $X(t)$ at the limiter input has the form of a Gaussian curve:

$$R_x(\tau) = \exp\left(-\frac{\tau}{\tau_0}\right)^2 \tag{3.26}$$

where τ_0 is the correlate on interval of process $X(t)$.

The graphs clearly show that, as φ changes from $\varphi = 0$ to $\varphi = \pi/4$, the coefficient of autocorrelation for process $Y(t)$ at the limiter output does not change significantly and remains close to autocorrelation coefficient $R_x(\tau)$ of normal process $X(t)$ at the limiter input ($\varphi = 0$).

Therefore, the processes with very different distribution laws $W_\varphi(y)$ (Figures 3.8–3.11) correspond to autocorrelation functions of almost the same

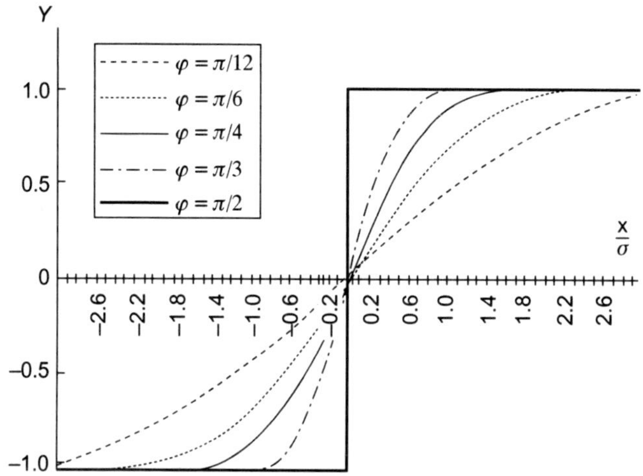

Figure 3.7 Amplitude characteristics of a smoothed limiter.

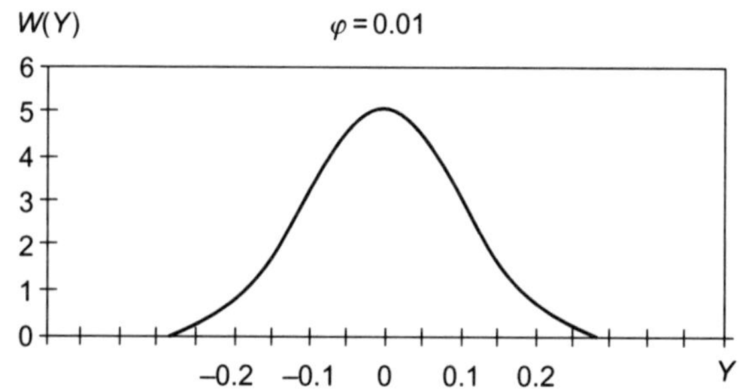

Figure 3.8 Distribution of momentary values of a process at the limiter output for $\varphi = 0.01$.

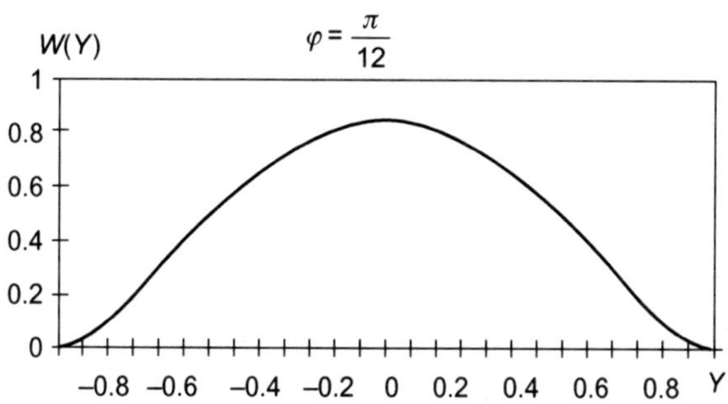

Figure 3.9 Distribution of momentary values of a process at the limiter output for $\varphi = \pi/12$.

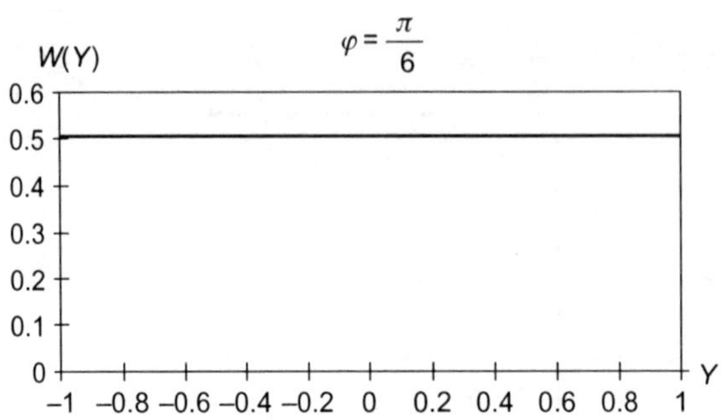

Figure 3.10 Distribution of momentary values of a process at the limiter output for $\varphi = \pi/6$.

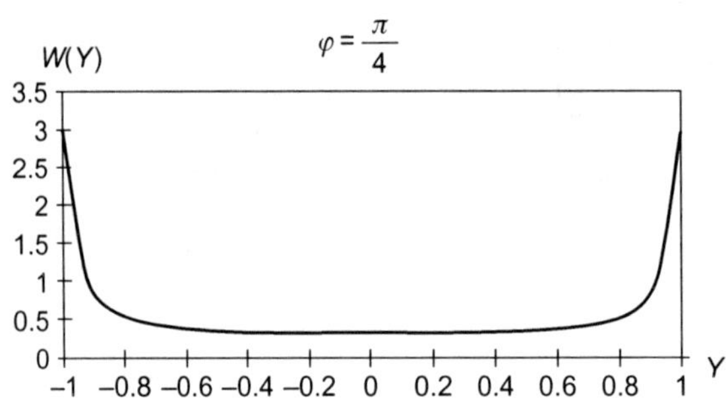

Figure 3.11 Distribution of momentary values of a process at the limiter output for $\varphi = \pi/4$.

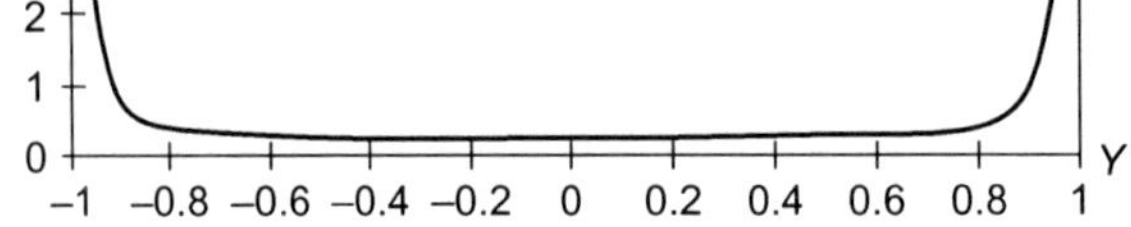

Figure 3.12 Distribution of momentary values of a process at the limiter output for $\varphi = \pi/3$.

shape, provided that the distribution of intervals between zeros $W(\tau, 0)$ is the same for all these processes.

If one-dimensional distribution $W(y)$ of momentary values of a non-Gaussian stationary random process $Y(t)$ which is symmetric about the zero level is described by expression (3.23) for arbitrary l, and its autocorrelation coefficient $R_y(\tau)$ is known, then a process of this class can undergo a nonlinear transformation, whose characteristic is the inverse of that of a smoothed limiter (3.21). In this case process $X(t)$ at such noninertial converter output will be a Gaussian one, its autocorrelation function $R_x(\tau)$ will be different from $R_y(\tau)$, but the zero level-crossing points will

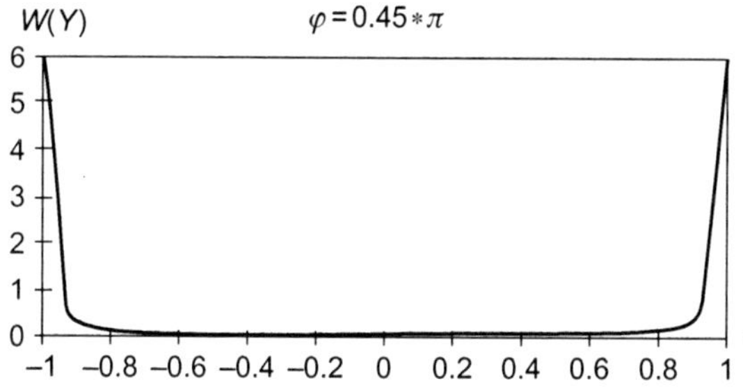

Figure 3.13 Distribution of momentary values of a process at the limiter output for $\varphi = 0.45 * \pi$.

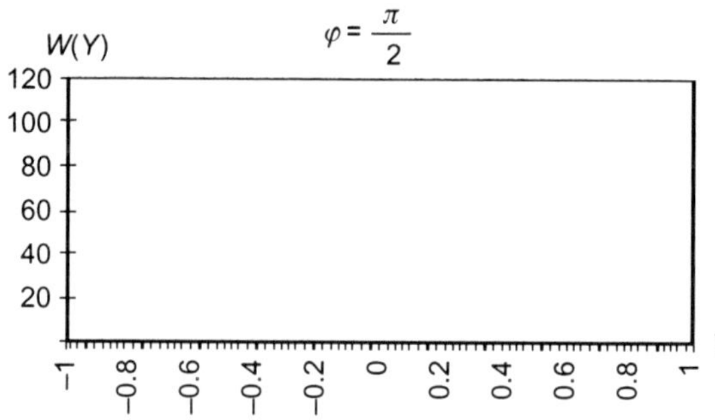

Figure 3.14 Distribution of momentary values of a process at the limiter output for $\varphi = \pi/2$.

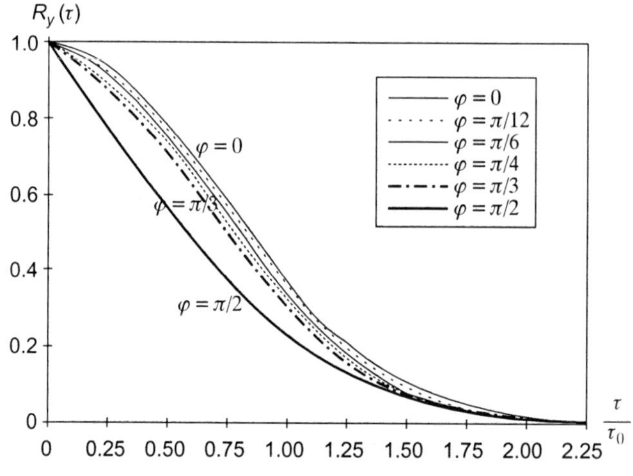

Figure 3.15 The family of autocorrelation functions of a process $Y(t)$ at the limiter output.

remain the same. Therefore, law $W(\tau, 0)$ of distribution of intervals between neighboring zeros of the process of interest will be the same.

Let us calculate autocorrelation function $R_x(\tau)$ of a normal random process at the output of a nonlinear transformer.

By using formula (3.24) we obtain:

$$R_x(\tau) = \frac{\sin[\varphi * R_y(\tau)]}{\sin \varphi} \tag{3.27}$$

It is known that for a normal random process the intensity of zero crossings $\lambda_x(0)$ can be expressed through its autocorrelation function $R_x(\tau)$ as follows:

$$\lambda_x(0) = \frac{1}{2\pi} * \sqrt{-R_x''(0)}$$

where $R_x''(0)$ is the second derivative of autocorrelation function $R_x(\tau)$ for $\tau = 0$.

From Eq. (3.27), we can obtain the following expression for the decrease K in the intensity of zero crossings of a given non-Gaussian random process $Y(t)$ compared to normal process $X(t)$, such that both processes have the same autocorrelation coefficient $R_y(\tau)$:

$$K = \frac{\lambda_x(0)}{\lambda_y(0)} = \sqrt{\frac{\varphi}{tg\varphi}} \tag{3.28}$$

where

$$\lambda_y(0) = \frac{1}{2\pi} * \sqrt{-R_y''(0)}.$$

Table 3.3 shows the values of coefficient K calculated for different values of angle φ.

As we can see from Table 3.3, for any given value $R_y(\tau)$ the intensity of zero crossings decreases as the non-Gaussian character of the random process becomes more obvious (i.e., with the increase in angle φ). At the same time, there is a wide range of values $\varphi \leq \pi/4$ for which the changes in the intensity of zero crossings are insignificant, in spite of a clearly non-Gaussian character of the distribution law.

It is, therefore, the autocorrelation function (power spectrum) of the process rather than the distribution of its momentary values that determines the distribution

Table 3.3 The Decrease K in the Intensity of Zero Crossings of a Given Non-Gaussian Random Process Compared to Gaussian Process

φ	0	$\pi/12$	$\pi/6$	$\pi/4$	$\pi/3$	$5\pi/12$	$11\pi/24$	$\pi/2$
K	1	0.9885	0.9523	0.8862	0.7776	0.5922	0.4354	0

density of intervals between zero crossings (we are talking here about a centered random process which is symmetric about zero).

Formula (3.27), in fact, provides a solution to our problem. Using this formula, we can find, for a given $R_y(\tau)$ and a given distribution law $W(y)$ of momentary values of a non-Gaussian process (3.23), the autocorrelation coefficient $R_{xy}(\tau)$ of such equivalent Gaussian process $X(t)$ that has the same distribution of intervals between zero crossings $W(\tau, 0)$ as the non-Gaussian process under consideration.

Let us consider, for example, a uniform distribution $W(y)$ of momentary values of process $Y(t)$ with a given autocorrelation coefficient $R_y(\tau)$. In this case $l = 1$, $\varphi = \pi/6$ and autocorrelation function $R_x(\tau)$ of equivalent Gaussian random process $X(t)$ at the nonlinear converter output, according to formula (3.27):

$$R_x(\tau) = 2 * \sin\left[\frac{\pi}{6} * R_y(\tau)\right] \tag{3.29}$$

For the case where autocorrelation coefficient $R_y(\tau)$ of the initial non-Gaussian process at the nonlinear converter input is given by Eq. (3.26), we calculated, using formula (3.29), the first approximation $q_{1x}(\tau, 0)$ of the distribution of excursions duration for Gaussian random process $X(t)$ with autocorrelation function $R_x(\tau)$.

The results are shown in Table 3.4 along with the first approximation $q_1(\tau, 0)$ of excursions duration distribution of a Gaussian random process with autocorrelation function expressed by Eq. (3.26).

Also shown in Table 3.4 taken from Section 3.2 (Table 3.2) density $W(\tau, 0)$ of a Gaussian random process with autocorrelation function given by Eq. (3.26). As we can see from Table 3.4, the difference in values between two functions having the same autocorrelation function (3.26)—function $q_{1x}(\tau, 0)$ for a process with a uniformly distributed momentary values and function $q_1(\tau, 0)$ for a Gaussian process—appears to be significant only in the area of small values, that is, for $\tau/\tau_0 \leq 0.8$.

In this area $q_{1x}(\tau, 0) < q_1(\tau, 0)$, that is, the probability of short excursions for a non-Gaussian process decreases, whereas the probability of long excursions above the zero level increases. This accounts for a decrease of about 5% (1.05 times decrease) shown in Table 3.3, in the intensity of zero crossings in a process with uniformly distributed momentary values compared with a Gaussian process with

Table 3.4 Calculation of the First Approximation of Excursions Duration Distribution for Gaussian and Non-Gaussian Random Processes

$\frac{\tau}{\tau_0}$	0	0.2	0.5	0.8	1
$q_{1x}(\tau, 0)$	0	0.061	0.1979	0.3189	0.3954
$q_1(\tau, 0)$	0	0.0994	0.2395	0.3503	0.3952
$W(\tau, 0)$	0	0.0994	0.2395	0.3502	0.3946
$\frac{\tau}{\tau_0}$	1.225	1.5	2	3	4
$q_{1x}(\tau, 0)$	0.4098	0.37	0.2659	0.2144	0.2143
$q_1(\tau, 0)$	0.4019	0.3649	0.2717	0.2255	0.2251
$W(\tau, 0)$	0.4017	0.36	0.2527	0.1408	0.0788

the same autocorrelation coefficient. If we compare figures presented in Table 3.4 for functions $q_1(\tau,0)$ and $W(\tau,0)$, we will see that $q_1(\tau,0) \geq W(\tau,0)$, which fully complies with the theory. However, the quality of the first approximation remains high not only for $\tau/\tau_0 \ll 1$, but also for a wide range of values $\tau/\tau_0 \leq 2$. As it follows from Table 1.4, for values no greater than $\tau/\tau_0 = 2$ the relative error of the first approximation does not exceed 7.1%.

Thus, the method proposed for calculating the first approximation of the distribution of intervals between zero crossings in the above-mentioned class of non-Gaussian processes with a given autocorrelation function and a given distribution of momentary values, makes it possible to identify, accurately enough for practical purposes, the character of distribution for the duration of excursions above the zero level.

The above example involving a smoothed limiter is just one particular case; however, a very important one, since it demonstrates the possibility of evaluating the distribution of intervals between zero crossings of a centered random process with uniformly distributed momentary values.

In more general terms, the problem can be formulated as follows. Supposing that we have a differentiable non-Gaussian stationary random process $Y(t)$ with the known variance $B(0)$ and autocorrelation function $R_y(\tau)$ and a given one-dimensional distribution $W(y)$ of momentary values, which is symmetric about the zero level, how can we find autocorrelation coefficient $R_x(\tau)$ of Gaussian process $X(t)$ which is similar to the above-mentioned non-Gaussian process in that they have the same distribution $W(\tau,0)$ of intervals between zero crossings? In order to answer this question, we have to find, for a given function $W(y)$, the characteristic $y-f(x)$ of noninertial nonlinear converter, which transforms a Gaussian process into a non-Gaussian one with a given one-dimensional function $W(y)$ of momentary values distribution. Knowing the relations[8] between $R_y(\tau)$ and $R_x(\tau)$, which are, respectively, the processes at the input and output of a noninertial nonlinear converter with a given characteristic $y=f(x)$, we can derive the unknown function $R_x(\tau)$ from $R_y(\tau)$. It is obvious that

$$W(x)\mathrm{d}x = W(y)\mathrm{d}y$$

$$\frac{\mathrm{d}y}{\mathrm{d}x} = \frac{\mathrm{d}f(x)}{\mathrm{d}x} = \frac{W(x)}{W(y)} \tag{3.30}$$

where $W(x)$ is a Gaussian distribution law for a process with a zero mean and a variance of 1, and $y=f(x)$ is a characteristic of a noninertial nonlinear converter which transforms a Gaussian process a variance of 1 into non-Gaussian process with a given one-dimensional function $W(y)$ of momentary values distribution.

$\mathrm{d}y/\mathrm{d}x$ in formula (3.30) can be presented as a Hermite sequence of orthogonal polynomials:

$$\frac{\mathrm{d}y}{\mathrm{d}x} = \sum_{n=0}^{\infty} c_{2n+1} * H_{2n}(x) \tag{3.31}$$

where $H_{2n}(x)$ is a Hermite polynomial of degree $(2n)$ which only contains even powers of argument x.

$$c_{2n+1} = \sqrt{\frac{2}{\pi}} * \frac{1}{(2n)!} * \int_0^{\infty} \frac{\mathrm{d}y}{\mathrm{d}x} * H_{2n}(x) * \exp\left(-\frac{x^2}{2}\right)\mathrm{d}x \tag{3.32}$$

As it follows from Eq. (3.30), the derivative of the nonlinear converter characteristic $y = f(x)$ is an even function of argument x, since distribution laws $W(x)$ and $W(y)$ are also even functions of respective arguments.

The integration of Eq. (3.31) will give us an odd function:

$$y = f(x) = \sum_{n=0}^{\infty} \frac{c_{2n+1}}{2n+1} * H_{2n+1}(x) \tag{3.33}$$

where $H_{2n+1}(x)$ is a Hermite polynomial of degree $(2n + 1)$, which only contains odd powers of x.

In Ref. 8, it is proved that the correlation function of the process at the output of a noninertial nonlinear element which has a stationary normal random process with autocorrelation function $R(\tau)$ at the input can be expressed as follows:

$$B(\tau) = \sum_{n=0}^{\infty} \frac{d_{2n+1}^2}{(2n+1)!} * R^{2n+1}(\tau) \tag{3.34}$$

where

$$d_{2n+1} = \sqrt{\frac{2}{\pi}} * \int_0^{\infty} \frac{\mathrm{d}y}{\mathrm{d}x} H_{2n}(x) * \exp\left(-\frac{x^2}{2}\right)\mathrm{d}x \tag{3.35}$$

Hermit polynomials form a system of orthogonal polynomials in relation to the weight of $\exp(-(x^2/2))$ on interval $(-\infty, \infty)$.

Considering Eqs. (3.32) and (3.35), we can write $d_{2n+1} = c_{2n+1} * (2n)!$.

Hence,

$$B(\tau) = \sum_{n=0}^{\infty} \frac{c_{2n+1}^2}{2n+1} * [(2n)!] * R^{2n+1}(\tau) \tag{3.36}$$

From Eqs. (3.30) and (3.31) we deduce that

$$W(y = f(x)) = \frac{W(x)}{\sum_{n=0}^{\infty} c_{2n+1} * H_{2n}(x)} \tag{3.37}$$

Thus, formulae (3.30), (3.32), (3.36), and (3.37) directly relate the nonlinear converter characteristic $y = f(x)$ and its first derivative dy/dx to distribution law $W(y = f(x))$ and correlation function $B(\tau)$ of non-Gaussian process $Y(t)$ at the output of the converter which has Gaussian process $X(t)$ with a given autocorrelation function $R(\tau)$ at the input. With these formulae the computation of excursion parameters of a non-Gaussian process becomes less complicated, as the parameters of a non-Gaussian process are expressed through those of the equivalent Gaussian process.

As an example, consider a nonlinear converter whose characteristic has the form of a hyperbolic sine: $y = f(x) = (\text{sh}\, ax/a)$, $a \geq 0$. The first derivative of such characteristic has the form of a hyperbolic cosine, $dy/dx = \text{ch}\, ax$. The distribution law of the non-Gaussian process at the converter output is:

$$W(y = f(x)) = \frac{W(x)}{\text{ch}\, ax} = \frac{1}{\sqrt{2\pi} * \text{ch}\, ax} * \exp\left(-\frac{x^2}{2}\right) \tag{3.38}$$

The function $\text{ch}\, ax$ can be expanded into a series in terms of Hermite polynomials:

$$\text{ch}\, ax = \exp\left(\frac{a^2}{2}\right) * \sum_{n=0}^{\infty} \frac{(a)^{2n}}{(2n)!} * H_{2n}(x)$$

Comparing it with formula (3.31) gives us the following expression:

$$c_{2n+1} = \frac{a^{2n} * \exp(a^2/2)}{(2n)!}$$

In view of formula (3.36), with a little algebra we get

$$B(\tau) = \frac{\exp(a^2)}{a^2} * \text{sh}\left[a^2 * R(\tau)\right]$$

For $\tau = 0$ the correlation function of the non-Gaussian process at the output of the converter, that is, the variance of the process, is $B(0) = (\exp(2a^2) - 1)/2a^2$.

Autocorrelation coefficient $R_y(\tau)$ of non-Gaussian process $Y(t)$ has the form:

$$R_y(\tau) = \frac{\text{sh}[(a^2) * R(\tau)]}{\text{sh}(a^2)} \tag{3.39}$$

It follows from the above that, for small values of parameter $a(a \to 0)$, the nonlinear transformation tends to a linear one.

At the same time, $y = f(x) \to x$, $dy/dx = \text{ch } ax \to 1$, the distribution of momentary values of non-Gaussian process $Y(t)$ at the converter output tends to the Gaussian one. The variance of the output process $B(0) = (\exp(2a^2) - 1)/2a^2 \to 1$, just as the variance of the input Gaussian process does.

The expression (3.38) for distribution law $W(y)$ of momentary values of the non-Gaussian process at the converter output can be rewritten in the explicit form, that is, the distribution can be expressed as a function of argument y as follows:

$$W(|y|) = \frac{1}{\sqrt{2\pi[1 + (ay)^2]}} * \exp\left\{ -\frac{1}{2a^2} \left[\ln\left(ay + \sqrt{1 + (ay)^2} \right) \right]^2 \right\} \tag{3.40}$$

The analysis of formula (3.40) shows that, in the case where $ay \geq 3$, the "tails" of distribution $W(|y|)$ are described accurately enough by the following function:

$$W(|y|) \approx \frac{1}{\sqrt{2\pi[1 + (ay)^2]}} * \exp\left(-\frac{|y|}{2a} \right) \tag{3.41}$$

where $|y|$ is a modulus of argument y.

In other words, distribution $W(|y|)$ from (3.40) tends to an exponential one when $|y|$ takes high values, and to a Gaussian one, when the modulus of argument y takes small values, that is, $|y| \ll (1/a)$.

It is worth mentioning that for $a \leq 1$ distribution (3.40) approximates well the distribution of momentary values of a speech signal. Normal distribution is typical for low-level speech signals such as pauses, voiceless consonants, sibilants, and noises. At the same time, according to experimental research, high-level signals representing vocalized speech are characterized by exponential distribution.

Table 3.5 shows values of distribution function $W_1(|y|) = W(|y|) * \sqrt{2\pi}$ describing the distribution of momentary values of the process; the figures presented in the table were calculated using formula (3.40) for three values of parameter a: $a = 0$ (Gaussian); $a = 0.707$, and $a = 1$.

As we can see from Table 3.5, the higher parameter a is, the heavier the distribution tails are, and for $|y| \geq 3$ the decrease in density $W(|y|)$ gets close to an exponential one.

From formula (3.39) it is possible to deduce the following relation between a given correlation function $R_y(\tau)$ of non-Gaussian process $Y(t)$ with distribution (3.40) and autocorrelation function $R(\tau)$ of equivalent Gaussian process $X(t)$ which has the same distribution of zero crossings:

$$R(\tau) = \frac{1}{a^2} * \text{arcsh}\left[R_y(\tau) * \text{sh}(a^2)\right] \tag{3.42}$$

Comparing the values of $R(\tau)$ obtained using formula (3.42) to those of $R_y(\tau)$ obtained for the same τ shows that for $a \leq 1$ the differences between the two autocorrelation functions are insignificant.

Using analogy with Eq. (3.28), we can now obtain from (3.42) the relation between the intensities of zero crossings in a Gaussian process and a non-Gaussian one with distribution (3.40) for the case where both processes have the same correlation function described by formula (3.26):

$$K = \frac{\lambda_x(0)}{\lambda_y(0)} = \sqrt{\frac{\mathrm{th}(a^2)}{(a^2)}} \tag{3.43}$$

The calculations of coefficient K are summarized in Table 3.6.

The comparison of Tables 3.5 and 3.6 shows that, in spite of a noticeable difference between distribution (3.40) and a Gaussian one, for $a \le 1$ the difference in the intensity of zero crossings between a non-Gaussian process and a Gaussian one with the same correlation function remains insignificant and increases only slightly with the increase in a. In other words, for a non-Gaussian process symmetric about the zero level, the distribution of intervals between zero crossings has little relation with the distribution law of the process momentary values, being mainly determined by the form of the correlation function (or power spectrum).

This conclusion is completely borne out by the data given in Tables 1.3 and 1.6. In spite of a great difference between them, distributions (3.23) and (3.40) whose correlation function is described by Eq. (3.26) have the same intensity of zero crossings for certain values of parameters φ and a. For instance, in the case of a process with uniformly distributed momentary values ($\varphi = \pi/6$), the intensity of zero crossings is not much different from that of a process whose distribution is

Table 3.5 The Distribution of Momentary values of a Non-Gaussian Random Process $Y(t)$ at the Converter Output

a = 0		a = 0.707		a = 1	
|y|	$W_1(\lvert y\rvert)$	|y|	$W_1(\lvert y\rvert)$	|y|	$W_1(\lvert y\rvert)$
0	1	0	1	0	1
0.4	0.9231	0.4	0.8901	0.4	0.8605
0.8	0.7261	0.8	0.6508	0.8	0.597
1	0.6065	1	0.5292	1	0.4795
1.2	0.4867	1.2	0.4214	1.2	0.3821
1.4	0.3753	1.4	0.3309	1.4	0.3042
1.6	0.278	1.6	0.2578	1.6	0.243
1.8	0.1979	1.8	0.2001	1.8	0.1951
2	0.1353	2	0.1552	2	0.1577
2.2	0.0889	2.2	0.1205	2.4	0.1053
2.4	0.0561	2.4	0.0938	3	0.0605
3	0.0111	3	0.0454	4	0.027
3.4	0.0031	3.4	0.0149	6	0.0074
4	0.0003	4	0.0055	8	0.0026

close to exponential (the case where $a = 0.707$), provided that both non-Gaussian processes have identical autocorrelation functions. Both processes, therefore, are likely to have similar distributions of intervals between zeros. It means that the first approximations $q_{1x}(\tau, 0)$ of distributions of excursion durations in the equivalent Gaussian processes with autocorrelation functions $R_x(\tau)$, which are calculated using formulae (3.29) and (3.42), are also likely to be close to each other.

Table 3.7 shows the values of the first approximation $q_{1x}(\tau, 0)$ of distribution of excursions duration for Gaussian random process $X(t)$ with autocorrelation function $R_x(\tau)$ given by formula (3.42) ($a^2 = 0.5$). The values were calculated for the case where autocorrelation coefficient $R_y(\tau)$ of the initial non-Gaussian process is given by expression (3.26).

For comparison, the same table shows values obtained for distribution law $W(\tau, 0)$ of intervals between zeros and the first approximation $q_1(\tau, 0)$ of distribution of excursions duration in a Gaussian random process whose autocorrelation function is given by formula (3.26).

The conclusion about essential proximity of the first approximations $q_{1x}(\tau, 0)$ of distribution of excursions duration for two non-Gaussian random processes with identical correlation functions and distinctly different distribution laws of momentary values is borne out by the comparison of data in Tables 3.4 and 3.7. This conclusion is also true for the case where we compare the data for two processes with the same power spectrum (or autocorrelation function), one of which is a non-Gaussian process and the other is a Gaussian one.

To conclude this section let us summarize its main points:

1. There is a large class of differentiable centered non-Gaussian random processes with a given distribution law of momentary values and a given correlation function, which can be converted, by way of a noninertial nonlinear transformation, to "equivalent" Gaussian random processes with a different correlation function. By equivalence here we mean that the processes at the input and output of the converter will have the same distribution law of intervals between zero crossings.
2. For certain types of centered processes (a Gaussian one, a uniformly distributed one, and a process whose distribution is close to exponential) it was shown that the distribution law of momentary values of a process has little impact on the distribution of intervals between zero crossings in it. The latter is mainly determined by the form of the autocorrelation function (or power spectrum) of the process. Therefore, the processes with same pattern of zero crossing points have nearly identical autocorrelation functions whose shape has little or no relation to the distribution law of momentary values.
3. The problem of zero crossings in the first approximation has been solved for a Gaussian process with a given autocorrelation function, but remains unsolved for non-Gaussian processes. The conclusions of this section form the basis for the extension of results

Table 3.6 The Decrease K in the Intensity of Zero Crossings of a Non-Gaussian Random Process

a	0	0.5	0.707	0.8	0.9	1	1.2	1.414
K	1	0.9898	0.9614	0.9395	0.9092	0.8727	0.7878	0.6943

Table 3.7 Calculation of the First Approximation of Excursion Duration Distribution of a Gaussian and for a Non-Gaussian Random Process

$\frac{\tau}{\tau_0}$	0	0.2	0.5	0.8	1
$q_{1x}(\tau,0)$	0	0.073	0.21	0.336	0.3905
$q_1(\tau,0)$	0	0.0994	0.2395	0.3503	0.3931
$W(\tau,0)$	0	0.0994	0.2395	0.3502	0.3929
$\frac{\tau}{\tau_0}$	1.225	1.5	2	3	4
$q_{1x}(\tau,0)$	0.408	0.37	0.272	0.22	0.2165
$q_1(\tau,0)$	0.4019	0.3649	0.2717	0.2258	0.2251
$W(\tau,0)$	0.4017	0.36	0.2537	0.14	0.0797

obtained earlier for Gaussian processes, to a large class of non-Gaussian processes with the same autocorrelation function.

4. The method proposed for calculating the correlation function of the equivalent Gaussian process makes it possible to reduce a complicated problem of finding parameters of excursions above a given level y_0 in a non-Gaussian process to the problem of finding parameters of excursions above a different given level x_0 in the equivalent Gaussian process. The relation between the two levels can easily be deduced from the characteristic $y = f(x)$ of a noninertial nonlinear converter.

4 Estimating Certain Informative Parameters of Random Process Excursions Above a Given Level

4.1 Estimating the Variance in Duration of Intervals Between Successive Excursions Above a Given Level in a Stationary Random Process

For some engineering applications it is sufficient to know the first two moments of distribution $W_+(\tau, x_0)$ of duration τ of intervals between successive excursions above level x_0 (or successive points on a time scale at which the given level is crossed in the same - for example, upward - direction). Knowing the average length of intervals between successive excursions and the variance in length of such intervals makes it possible to develop a good understanding of what distribution function $W_+(\tau, x_0)$ is like.

Calculating average length $\tau_{av}(x_0)$ of the above-mentioned intervals presents no difficulties, at least in the case of differentiable Gaussian random processes with known autocorrelation function $R(\tau)$; this parameter can be found as a reciprocal of intensity $\lambda_1(x_0)$ of excursions above level x_0.[8]

The situation is completely the opposite when we need to find variance $\sigma_+^2(x_0)$ in the length of intervals between successive crossings of level x_0 in a given direction. Even for Gaussian processes, estimating $\sigma_+^2(x_0)$ presents considerable difficulties as it usually involves computer calculations. In view of the above, obtaining simple formulae for the estimation of variance $\sigma_+^2(x_0)$ is, without any doubt, a very important issue.

The approximation we propose to use is based on the assumption made by some researchers[40,47] that moments at which a broadband signal crosses level x_0 in a given direction are statistically independent from each other. In the previous chapters of this book, some theoretical arguments were stated which also prove that this assumption is true. The results of a comprehensive computer simulation presented below lead us to a well-grounded conclusion about the absence of any noticeable correlation between combinations of two adjacent level-crossing intervals, even in the case where the given level is zero ($x_0 = 0$). The higher the given level with relation to variance ($|x_0/\sigma| > 0$) gets, the more rapidly correlation between moments when that level is crossed weakens.

Applications of Random Process Excursion Analysis. DOI: http://dx.doi.org/10.1016/B978-0-12-409501-4.00004-8

Let us try to express variance $\sigma_+^2(x_0)$ through the variance in the total duration of a large number of excursions above (or below) level x_0 occurring on an interval of length $\tau \gg \tau_0$. Let us assume that during a random interval of length τ which started at an arbitrary time moment the signal crosses level x_0, on average, 2 N_{av} times. That means that the number of excursions above level x_0 (which is the same as the number of upward level crossings) occurring within the same interval equals $N_{av.}$ Let us also assume that $N_{av} \gg 1$. Considering statistical independence of level-crossing moments, let us find variance $\sigma_+^2(x_0)$ by dividing variance $D(\tau)$ in length τ of intervals into average number N_{av} of excursions above level x_0 occurring in an interval of length τ. It was shown in Ref. 50 that in the case of a stationary differentiable random process a random number N of excursions above level x_0 occurring in an interval of length τ has an asymptotically normal distribution for $\tau \gg \tau_0$:

$$P(N,\tau) \approx \frac{1}{\sqrt{2\pi * \sigma_+^2(\tau, x_0)}} * \exp\left\{ -\frac{[N - \lambda_1(x_0) * \tau]^2}{2\sigma_+^2(\tau, x_0)} \right\} \tag{4.1}$$

Here, $P(N,\tau)$ is the probability that N excursions above level x_0 will occur in an interval of length τ which starts at an arbitrary time moment; $\sigma_+^2(\tau, x_0)$ is the variance in the number of excursions above level x_0 which occur in an interval of length τ; $\lambda_1(x_0)$ is the intensity of excursions above level x_0; $\lambda_1(x_0) * \tau = N_{av}$ is the average number of excursions above level x_0 which occur in an interval of length τ.

Formula (2.22) in Section 2.1 relates variance $\sigma_+^2(\tau, x_0)$ in the number of excursions to their average number N_{av} for $\tau \gg \tau_0$:

$$\sigma_+^2(\tau, x_0) = k_1(x_0) * \lambda_1(x_0) * \tau = k_1(x_0) * N_{av}$$

Coefficient $k_1(x_0)$ is obtained using the formula:

$$k_1(x_0) = \int_0^\infty [\Delta(\alpha, x_0) + \Delta(\alpha, \ -x_0)]\mathrm{d}\alpha$$

where $\Delta(\alpha, x_0) = q_1(\alpha, x_0) - \lambda_1(x_0)$, $\Delta(\alpha, \ -x_0) = q_1(\alpha, \ -x_0) - \lambda_1(x_0)$.

Constant coefficient $k_1(x_0)$ does not depend on τ; it is determined entirely by the form of the first approximation functions $q_1(\alpha, x_0)$ and $q_1(\alpha, \ -x_0)$. For $x_0 = 0$, the expression for $k_1(x_0)$ takes a less complicated form:

$$k_1(0) = 2 * \int_0^\infty [q_1(\alpha, 0) - \lambda_1(0)]\mathrm{d}\alpha$$

By using formula $k_1(0)$ and a computer simulation of a Gaussian process with a Gaussian autocorrelation function, we calculated the value $k_1(0) = 1/2\sqrt{3} \approx 0.2886$. Using the alternative calculation method described in Ref. 38 gave us a very close number: $k_1(0) \approx 0.286$.

After substituting $\sigma_+^2(\tau, x_0)$ in Eq. (4.1) we get:

$$P(N,\tau) \approx \frac{1}{\sqrt{2\pi * k_1(x_0) * N_{\text{av}}}} * \exp\left\{ -\frac{[N - N_{\text{av}}]^2}{2k_1(x_0) * N_{\text{av}}} \right\} \tag{4.2}$$

Random length τ of intervals characterized by the average number of excursions N_{av}, is distributed according to the normal law with variance $D(\tau)$ which can be estimated using formula (2.22). By analogy with formula (4.1) we can write:

$$P(N_{\text{av}},\tau) \approx \frac{1}{\sqrt{2\pi * D(\tau)}} * \exp\left\{ -\frac{[\tau - (N_{\text{av}}/\lambda(x_0))]^2}{2 * D(\tau)} \right\} \tag{4.3}$$

It is obvious that $D(\tau)/\mathrm{T}_{\text{av}}^2 = D(N)/N_{\text{av}}^2$, where $T_{\text{av}} = \tau_{\text{av}} * N_{\text{av}} = N_{\text{av}}/\lambda(x_0)$ is the average interval length ($\tau \gg \tau_0$).

We can now write a formula which gives us an estimate of variance $\sigma_+^2(x_0)$ in length of intervals between successive points where level x_0 is crossed in the same given direction[78]:

$$\sigma_+^2(x_0) = \frac{k_1(x_0)}{\lambda_1^2(x_0)} = k_1(x_0) * [\tau_{\text{av}}(x_0)]^2 \tag{4.4}$$

For high relative levels $|x_0/\sigma| \gg 1$ coefficient $k_1(x_0)$ tends to one and $\sigma_+^2(x_0) \to [\tau_{\text{av}}(x_0)]^2$, which indicates the exponential character of distribution of length values among intervals between successive excursions.

In centered processes, durations of excursions above zero level and durations of pauses between them are distributed equally. Assuming that in the first approximation there is no correlation between durations of successive excursions and pauses, we can obtain the following estimate of variance in the duration of excursions above zero level:

$$\sigma^2(0) = \frac{2k_1(0)}{\lambda^2(0)} \tag{4.5}$$

where $\lambda(0) = 2 * \lambda_1(0)$ is the zero-crossing rate of the process, which takes into account both upward and downward crossings.

Formulae (4.4) and (4.5) are only applicable for the case where the assumption about statistical independence of successive moments of level crossing occurring in the same direction (formula (4.4)) or in the opposite directions (formula (4.5)) is true. Estimates obtained using formula (4.5) are characterized by the biggest error, since at zero level intervals between successive excursions are, on average, shorter than they are at any other level, and correlation between successive moments of level crossing is the strongest.

Nevertheless, in real processes the average duration of excursions above zero level is much bigger than duration τ_0 of time correlation intervals.

Therefore, correlation between any two momentary values separated by a period of time shorter than the average duration of one excursion is much weaker compared to what it would be if the moments were separated by a longer period of time. Validity of formula (4.5) can be verified on a Gaussian process with a known distribution of lengths of intervals between successive zero crossings; for this kind of process there is a formula allowing to calculate the true variance in the duration of excursions above zero level. The process is studied in Ref. 42. Its autocorrelation function is

$$R(\tau) = \frac{3}{2} * \exp\left(-\frac{|\tau|}{\sqrt{3}}\right) * \left(1 - \frac{1}{3} * \exp\left(-\frac{2|\tau|}{\sqrt{3}}\right)\right)$$

The zero-crossing rate of this process is $\lambda(0) = 1/\pi$ and the average duration of a single excursion above zero level is $\tau_{av} = 1/\lambda(0) = \pi$. In Ref. 42, we find the formula for precise calculation of integral function $F(\tau)$ of distribution of length of intervals between successive zero crossings. Using this formula we managed to obtain the true value of variance:

$$\sigma^2(0) = 1.1567 * \pi^2$$

Coefficient $k_1(0)$ calculated for the Gaussian process under consideration equals $k_1(0) = \sqrt{3}/\pi \approx 0.55133$. Hence the approximate value of variance from Eq. (4.4):

$$\sigma^2(0) \approx 1.10266 * \pi^2$$

It is only a little less than the true value $\sigma^2(0)$ the proportion of the latter to the former is 1.049:1. The error, as we mentioned earlier, results from the assumption about statistical independence of zero-crossing moments. In real fact, in the above case with function $R(\tau)$ from formula (4.5) there is some correlation between successive excursions; thus, for $\tau_{av} = \pi$ we have $R(\pi) \approx 0.24$.

In real Gaussian processes correlation weakens more rapidly; as a result, the estimate given by formula (4.4) is more accurate and the estimation error, in any case, amounts to but a few percentage points. Thus, in a Gaussian random process with a Gaussian spectrum, correlation coefficient $R_1(\tau)$ for an interval of average duration $\tau_{av} \approx 2.221 * \tau_0$ equals to $R_1(\tau_{av}) \approx 0.0072$. With the same average duration of excursions in a process with a rectangular spectrum, correlation coefficient will be $R_2(\tau_{av}) \approx -0.137$.

Coefficients $k_1(0)$ for such processes were calculated in Ref. 38:

$$k_{11}(0) = \frac{1}{2\sqrt{3}} \approx 0.289, \quad k_{12}(0) \approx 0.241$$

Estimates of variance in duration of excursions above zero level, obtained using formula (4.5) have the following values:

$$\sigma_1^2(0) \approx 2.851 * \tau_0^2; \quad \sigma_2^2(0) \approx 2.379 * \tau_0^2$$

The respective estimates given by formula (4.4), that is, the estimates of variance in duration of intervals between successive upward (or downward) crossings of level $x_0 = 0$ in the above examples are twice as high:

$$\sigma_{1+}^2(0) \approx 5.755 * \tau_0^2; \quad \sigma_{2+}^2(0) \approx 4.757 * \tau_0^2$$

By experimenting with computer simulations we obtained variances in length of intervals between points where normalized level $x_0/\sigma = 0, 0.5, 1, 1.5, \ldots$ is crossed either in the same direction or in the opposite direction for processes with Gaussian and uniform (rectangular) spectra. According to our experimental findings, in the case of a Gaussian process with a Gaussian spectrum, the relative error of estimate $\sigma_{1+}^2(0)$ obtained using formula (4.4) for $x_0/\sigma = 0$ amounts to just a small figure of 0.9%. The relative error of estimate $\sigma_{2+}^2(0)$ obtained using formula (4.4) for zero level for a process with a uniform spectrum turns out to be a little higher but still remains under 1.6%. As we would expect, errors calculated for $x_0/\sigma > 0$ are even smaller; they are comparable with statistical errors of simulation.

Thus, the assumption about statistical independence of the total length of any two adjacent intervals from the total length of the preceding two intervals, measured at any level x_0/σ proves to be right, and a simple formula (4.4) gives fairly accurate estimates of variance $\sigma_+^2(x_0)$.

As for the estimate of variance $\sigma^2(0)$ in duration of excursions above zero level, which is obtained using formula (4.5), it turned out to be less accurate, because correlation between lengths of two adjacent zero-crossing intervals is too strong to be neglected. Nevertheless, even in this case calculations produce sufficiently accurate results. As it follows from the analysis of simulation results obtained for the two Gaussian processes, in both cases relative errors of estimates $\sigma_1^2(0)$ and $\sigma_2^2(0)$ given by formula (4.5) approach 3%.

Thus, knowing the correlation function of a Gaussian process makes it possible to calculate coefficients $k_1(x_0)$ and then, using formulae (4.4) and (4.5), estimate, accurately enough, variance in length of intervals between successive crossings of level x_0.

According to the findings of Section 3.3, this method of variance estimation can be applied to a number of non-Gaussian random processes with known correlation function. After a nonlinear noninertial transformation, a non-Gaussian process turns into an equivalent Gaussian process with a different correlation function; however, the level crossings distribution law will remain the same at the threshold levels corresponding to this nonlinear transformation. All the findings of the theory of excursions that were known prior to this publication, as well as the new findings of this chapter apply to the equivalent Gaussian process. That makes them applicable to a wide range of non-Gaussian processes.

4.2 Estimating Exponential Tail Parameters for Distribution of Excursions in Stationary Random Processes

The theory predicts that, for $\tau \gg \tau_0$, the tails of level-crossing intervals distribution functions $W_+(\tau, x_0)$ can be closely approximated by exponentials. Estimating parameters for exponential tails of distributions enables calculating the probability of super-long excursions above a given level x_0.

Estimating the small values that the distribution function $W_+(\tau, x_0)$ takes in the area of big values of τ is also important because experimental measurements of the function in this area have high statistical errors and can be time consuming. Attempts to use numerical computer simulation of a random process for studying the behavior of the function $W_+(\tau, x_0)$ in the entire range of its arguments, especially in the area of high standardized thresholds $(x_0/\sigma) \gg 1$, also face the necessity to solve a complex computational problem, which requires high speed and a lot of computer memory for analyzing long chunks of a realized random process.

Scarcity of experimental data[7,9,34,35,40,51,52] and a wide range of values obtained for the tails of distribution density functions $W_+(\tau, x_0)$ are yet more proof of the difficulty of using measurements and simulation.

In the area of maximum probability $P(N, \tau)$, for $N = N_{av}$, we obtain from formula (4.2):

$$P(N_{av}, \tau) \approx \frac{1}{\sqrt{2\pi k_1(x_0) * N_{av}}} \tag{4.6}$$

As we know, for $(x_0/\sigma) \gg 1$, the distribution of level crossings approaches a Poisson distribution:

$$P(N, \tau) = \frac{[\lambda_1(x_0) * \tau]^N}{(N)!} * \exp[-\lambda_1(x_0) * \tau] = \frac{(N_{av})^N}{(N)!} * \exp(-N_{av}) \tag{4.7}$$

In the case of a Poisson distribution, $k_1(x_0) = 1$ and variance in the number of excursions equals the mean. The Poisson distribution (4.7) also becomes normal for $N_{av} \gg 1$. Substituting Stirling's approximation for large factorials:

$$(N)! \approx \sqrt{2\pi * N} * (N)^N * \exp(-N) \tag{4.8}$$

in Eq. (4.6) gives us the following expression for near-maximum values of probability:

$$P(N_{av}, \tau) \approx \frac{1}{\sqrt{2\pi * N_{av}}}$$

It is known that, for $N \geq 4$, the relative error of approximation (Eq. (4.8)) does not exceed 2%; therefore, the Poisson distribution (Eq. (4.7)) gets noticeably close to normal for $\tau \geq 4/\lambda_1(x_0)$.

Given that successive moments of crossing a set level by a real process are statistically independent for $\tau \gg \tau_0$, with Eq. (4.6) in mind, we can write the following expression for $P(N, \tau)$, which is similar to Eq. (4.7):

$$P(N,\tau) \approx \frac{1}{k_1(x_0)} * \frac{[N_{av}/k_1(x_0)]^{N/k_1(x_0)}}{[N/k_1(x_0)]!} * \exp\left[-\frac{N_{av}}{k_1(x_0)}\right] \tag{4.9}$$

The distribution (4.9) also gets normal for $\tau \geq 4k_1(x_0)/\lambda_1(x_0)$ and reaches its maximum when $N = N_{av}$:

$$P(N_{av},\tau) \approx \frac{1}{\sqrt{2\pi k_1(x_0) * N_{av}}}.$$

This maximum coincides, in terms of both its value and location, with that of a Gaussian distribution (4.6). The same is true about the variances and means of distributions (4.6) and (4.9) for $\tau \geq 4k_1(x_0)/\lambda_1(x_0)$. As $k_1(x_0) \to 1$, the distribution (4.9) approaches the Poisson distribution (4.7), which fits in with our knowledge of random processes behavior at high levels $(x_0/\sigma) \gg 1$.

Assuming that $N = 0$ in Eq. (4.9) will give us for $\tau \gg \tau_0$:

$$P(0,\tau) \approx \frac{1}{k_1(x_0)} * \exp\left[-\frac{\lambda_1(x_0)}{k_1(x_0)} * \tau\right] \tag{4.10}$$

where $P(0, \tau)$ is the cumulative distribution function of time interval τ between a given moment of time and the first excursion above level x_0. The function $P(0, \tau)$ can also be interpreted as the probability that the random signal will make no excursions above level x_0 within time interval τ which starts at a given moment.

The author of Ref. 40 established the following relation between $P(0, \tau)$ and the density $W_+(\tau, x_0)$ of distribution for intervals between successive excursions of the process above level x_0:

$$W_+(\tau,x_0) = \frac{1}{\lambda_1(x_0)} * \frac{d^2}{d\tau^2} P(0,\tau)$$

This means that, for $\tau \geq 4k_1(x_0)/\lambda_1(x_0)$, parameters for the nearly exponential tails of the densities $W_+(\tau, x_0)$ should be the same as in Eq. (4.10).

Hence a simple formula for exponents $\Pi(x_0)$ which limit distributions of time intervals between excursions above level x_0:

$$\Pi(x_0) = \left[-\frac{\lambda_1(x_0)}{k_1(x_0)} * \tau\right], \quad \tau \geq \frac{4k_1(x_0)}{\lambda_1(x_0)} \tag{4.11}$$

For the above Gaussian process with a Gaussian autocorrelation function

$$\lambda_1(0) = \frac{1}{\sqrt{2} * \pi * \tau_0}, \quad k_1(0) = \frac{1}{2\sqrt{3}}$$

In the case where $x_0 = 0$, formula (4.11) gives us the inclination of the exponential curve:

$$\Pi(0) = -\frac{\sqrt{6}}{\pi * \tau_0} * \tau \approx -0.7797 * \frac{\tau}{\tau_0}$$

Note that the exponential tail parameter $\Pi(0)$ for the distribution $W_+(\tau, 0)$ of time intervals between above-zero excursions is expected to be the same as that for the density $W(\tau, 0)$ of a single excursion width distribution. The reason for this is that, for $\tau \gg \tau_0$, $W_+(\tau, 0)$ is the distribution density of the sum of two independent random variables, each distributed according to the law $W(\tau, 0)$. In this case the density $W_+(\tau, 0)$ can be obtained as a convolution of $W(\tau, 0)$; therefore, the two exponential curves will have the same inclination.[79,80]

Using proportionality factors $k_1(x_0)$ obtained in Refs. 7 and 38 for Gaussian processes with Gaussian and rectangular power spectra, it is easy to calculate the exponential tail parameters expressed by formula (4.11) for the distribution $W_+(\tau, 0)$ of time intervals between excursions above level x_0. The rate $\lambda_1(x_0)$ of excursions above level x_0 and the rate $\lambda_1(0)$ of above-zero excursions are related as follows:

$$\lambda_1(x_0) = \lambda_1(0) * \exp\left(-\frac{x_0^2}{2\sigma^2}\right)$$

The results of calculations are presented in Tables 4.1 and 4.2.

As we can see from Tables 4.1 and 4.2, for high standardized levels $(x_0/\sigma) \geq 2$, the distribution of level crossings in both processes is not different from the Poisson distribution and is practically independent of the spectral density; the only factor that determines its shape is the rate $\lambda_1(x_0)$ of excursions above the given level. To ensure a fair comparison, we assumed the same rate $\lambda_1(0)$ of above-zero excursions for both processes in calculating the data presented in Tables 4.1 and 4.2.

The Appendix 1 presents some results of statistical computer simulations of $W_+(\tau, x_0)$, the distribution density for time intervals between the excursions of the random process above level x_0.

Table 4.1 The Results of Calculations of the Constant $k_1(x_0)$

$\frac{x_0}{\sigma}$	$k_1(x_0)$ **Gaussian Spectrum**	$k_1(x_0)$ **Rectangular Spectrum**
0	0.289	0.241
0.5	0.374	0.335
1	0.576	0.556
1.5	0.765	0.801
2	0.915	0.938
3	0.9955	0.99923

Table 4.2 The Results of Calculations of the Exponents $\Pi(x_0)$

$\frac{x_0}{\sigma}$	$\Pi(x_0)$ **Gaussian Spectrum**	$\Pi(x_0)$ **Rectangular Spectrum**
0	−0.7797	−0.934
0.5	−0.5324	−0.593
1	−0.237	−0.246
1.5	−0.0953	−0.0912
2	−0.0333	−0.0324
3	−0.002512	−0.002502

Let us use gamma-distribution functions to approximate the experimentally obtained densities $W_+(\tau, x_0)$:

$$W_+(\tau, x_0) = \frac{b^m}{\Gamma(m)} * \tau^{m-1} * \exp(-b\tau) \tag{4.12}$$

where $b > 0$, $m > 0$. Both parameters can take integer as well as fractional values.

Let us express the distribution parameters b and m through the average time interval $\tau_{av}(x_0)$ between successive upward or downward crossings of level x_0/σ and variance $\sigma^2(x_0)$ in such intervals:

$$\tau_{av}(x_0) = \int_0^\infty \tau * W_+(\tau, x_0)d\tau = \frac{m}{b}$$

$$\sigma^2(x_0) = m_2(x_0) - [\tau_{av}(x_0)]^2 = \int_0^\infty \tau^2 * W_+(\tau, x_0)d\tau - \left(\frac{m}{b}\right)^2 = \frac{m}{b^2}$$

It follows from the above that

$$b = \frac{\tau_{av}(x_0)}{\sigma^2(x_0)}; \quad m = \frac{[\tau_{av}(x_0)]^2}{\sigma^2(x_0)} \tag{4.13}$$

In the previous section, we established the relation between the variance $\sigma^2(x_0)$ in the time interval between successive excursions of the signal above level x_0 and the mean value $\tau_{av}(x_0)$ of such interval:

$$\sigma^2(x_0) = \frac{k_1(x_0)}{\lambda_1^2(x_0)} = k_1(x_0) * [\tau_{av}(x_0)]^2$$

Given that $\tau_{av}(x_0) = 1/\lambda_1(x_0)$, after substituting Eq. (4.13) we get:

$$b = \frac{\lambda_1(x_0)}{k_1(x_0)}; \quad m = \frac{1}{k_1(x_0)} \tag{4.14}$$

Recall that the factor $k_1(x_0)$ which we introduced earlier established the relation between the variance in the number of excursions within the time interval τ and the mean of this number. At the same time, it follows from Eq. (4.13) that $k_1(x_0)$ is the square of the covariance factor ν for the density of distribution of time intervals between successive excursions:

$$\nu(x_0) = \frac{\sigma(x_0)}{\tau_{av}(x_0)}, \quad k_1(x_0) = \nu^2(x_0) \tag{4.15}$$

From Eqs. (4.12) and (4.14), we obtain:

$$W_+(\tau, x_0) = \frac{\lambda_1(x_0)/k_1(x_0)}{\Gamma[1/k_1(x_0)]} * \left[\frac{\lambda_1(x_0) * \tau}{k_1(x_0)}\right]^{(1/k_1(x_0))-1} * \exp\left[-\frac{\lambda_1(x_0)}{k_1(x_0)} * \tau\right] \tag{4.16}$$

Comparing formulae (4.16) and (4.10) shows that the two distribution functions have the same exponents. Thus, using two completely different approaches, we arrived at the same conclusions about the exponential tail parameters for the distribution of excursion durations.

The tails of experimentally measured distribution densities $W_+(\tau, x_0)$ and approximations (4.16) were analyzed for two broadband Gaussian processes and a number of standardized thresholds (x_0/σ ranged from 0 to 3 with 0.5 increments).

We assumed the lower boundary (i.e., the beginning) of the tail to equal $\tau = 4k_1(x_0)/\lambda_1(x_0)$, and the exponent in Eq. (4.16) to equal $\Pi(x_0) = -4$.

Table 4.1 shows the values of $k_1(x_0)$ calculated for processes with Gaussian and rectangular power spectrum densities.

Using the results of statistical simulation, we calculated, for each threshold x_0, the values of parameters $b = \tau_{av}(x_0)/\sigma^2(x_0)$ and $m = 1/\nu^2(x_0) = [\tau_{av}(x_0)]^2/\sigma^2(x_0)$.

They were compared with the values obtained from gamma-distribution approximations, and the comparison revealed a very close similarity of distribution densities in the area of $\tau > 4k_1(x_0)/\lambda_1(x_0)$.

The difference between the results obtained experimentally and the ones calculated proved to be commensurable with the statistical error which always increases as we move toward the tails of the distribution curve.

Statistical simulation proved that, in the area of distribution tails, time intervals between successive level crossings are practically not correlated, even if the threshold is set at level zero. This brings us to a conclusion that formulae (4.15) and (4.16), which were obtained on the basis of the assumption of independence of time intervals between excursions, are pretty accurate.

We can, therefore, claim that the problem of estimating the inclination of exponential tails for the distribution density has been solved. In the case of $\tau > 4k_1(x_0)/\lambda_1(x_0)$, parameter $b = \lambda_1(x_0)/k_1(x_0) = \tau_{av}(x_0)/D(x_0)$ in Eq. (4.16) fully determines the shape of the exponential tail of the distribution density curve.

Note that there is yet another experimental method of estimating parameters b and m in Eq. (4.14). It suggests using a simple computer program to calculate the average number $N_{av}(x_0) = \lambda(x_0) * \tau$ of excursions above level x_0 within time interval τ, as well as variance $\sigma_+^2(\tau, x_0)$ in the number of excursions of the random process above level x_0 time interval $\tau \gg 1/\lambda(x_0)$.

Statistical simulation enables finding the square of covariance factor $\nu^2(x_0) = k_1(x_0) = \sigma_+^2(x_0)/N_{av}(x_0)$ and then deriving all parameters for the distribution in Eq. (4.16) from the known rate $\lambda_1(x_0)$ of excursions.

In theory, the two methods should produce the same results; however, our experience showed that the second one is more preferable as it is more accurate and easier to implement.

In conclusion, we will give you an example of how the results we obtained can be used for solving problems of the theory of traffic flow and queuing. The article[32] analyzes non-Poisson models of teletraffic, described as the so-called heavy-tailed distributions of time intervals between events in a traffic flow. A random variable is believed to have a heavy-tailed distribution if its covariance factor is greater than one ($\nu > 1$).

One of the models analyzed in the article is a gamma distribution which meets this requirement. It was shown that the classical traffic flow theory, which is based on the Poisson model ($\nu = 1$), considerably underestimates both real delays of bursts in telecommunication networks, and the probabilities of their loss which increase with the covariance factor ν.

This brings us to the realization of the need for the analysis of a real flow in order to estimate the covariance factor ν and the related parameters such as the packet time delay and the optimum buffer memory requirement for the maximum traffic load and a given probability of packet loss.

Once we have calculated the average number of events in a flow and the variance in the number of events for sufficiently long time intervals, we will be able to estimate accurately enough the covariance factor ν, which can be later used in dynamic traffic control.

4.3 A Study into the Relation Between the Relative Root-Mean-Square Error of Measurement of the Cumulative Distribution Function of a Stationary Random Process and the Observation Time

Let us see if the results obtained so far can help us solve a vital problem of establishing the relation between the duration of the period during which the distribution law of a random process is analyzed for a given threshold level, and the statistical error of measurements.

The design of a system which makes it possible to determine experimentally the distribution of momentary values in stationary random processes cannot be improved without knowing exactly the minimum duration of the process observation period required for keeping the expected statistical error of measurements within the set limits. A similar problem arises when we estimate the minimum number of samples required for reaching a given accuracy level of computer simulation of a random process.

In Ref. 53, a formula for the accurate statistical error calculation is given for the case where the probability distribution of a Gaussian random process is measured using the method of establishing the proportion of the total time during which the process takes values above a given normalized level to the total observation time. In this formula, the error is calculated by integrating a slowly converging series; therefore, even in the case of zero normalized level ($x_0/\sigma = 0$), dozens of series terms have to be included in the sum in order to achieve the acceptable accuracy of calculations. With an increase in the threshold level the error of calculations increases too. Apart from cumbersome calculations, the formula has one other disadvantage; it cannot be used without knowing autocorrelation function $R(\tau)$ of the process.

In Ref. 53, we also find a formula which relates the mean square error to the duration of the period during which a stationary Gaussian process is measured. This formula does not require the knowledge of autocorrelation function $R(\tau)$ and provides the upper estimate of mean square error. However, this estimate is not very accurate, especially for high threshold levels. Thus, for a Gaussian process with correlation coefficient $R(\tau) = \exp(-\beta|\tau|)$ the relative error of estimate is 20% for $x_0/\sigma = 0$, and for $x_0/\sigma = 3$ the error is overestimated by order of 2.5.

Another approach to solving the problem is based on the use of certain results obtained by researchers into the applications of the theory of excursions in stationary random processes. Generally, cumulative probability distribution law $F(x)$ of ergodic stationary random process $X(t)$ is determined as the proportion of the total time during which the process takes values above normalized level x_0 to duration T of the entire observation period:

$$F(x \geq x_0) = \frac{\Sigma_{av}}{T} \tag{4.17}$$

where Σ_{av} is the average duration of a time period when the process is in excess of level x_0.

In experimental measurements (i.e., when the process is simulated on a PC) we actually have to replace Σ_{av} in formula (4.17) with random value Σ. It is variance $D(\Sigma)$ in Σ that determines mean square ε^2 for the relative statistical error of measurement:

$$\varepsilon^2 = \frac{D(\Sigma)}{(\Sigma_{av})^2} \tag{4.18}$$

It is obvious that, in the case where number N of level crossings x_0 tends to infinity, random total duration Σ of the process excursions above a given threshold is normalized in accordance with the central limit theorem of the probability theory. In this case, Gaussian distribution $W(\Sigma)$ is similar to the Gaussian distribution of random length τ of interval, during which level x_0 is crossed exactly N times ($N \gg 1$). The similarity of the two distributions implies that:

$$\frac{D(\Sigma)}{(\Sigma_{\text{av}})^2} = \frac{D(\tau)}{(\tau_{av})^2} \tag{4.19}$$

Here $\tau_{av} = N/\lambda(x_0)$ is the average length of an interval, on which level x_0 is crossed N times;

$\lambda(x_0) = 2\lambda_1(x_0)$ is the intensity (rate) of level x_0 crossings;
$\lambda_1(x_0)$ is the intensity of excursions above level x_0;
$1/\lambda(x_0)$ is the average length of an interval between two successive crossings of level x_0;
$D(\tau)$ is variance in random length $\tau$of an interval, on which N crossings of level x_0 have occurred.

Variance $D(\tau)$ in length τ of an interval on which $N/2 \gg 1$ excursions above a given level have taken place, can be derived from variance $\sigma_+^2(\tau, x_0)$ in the number of excursions which occur within time τ. Using formula (4.1) from the previous section, we obtain:

$$D(\tau) = \frac{\sigma_+^2(\tau, x_0)}{\lambda_1^2(x_0)} = \frac{k_1(x_0) * (N/2)}{\lambda_1^2(x_0)} \tag{4.20}$$

In view of fact that $\lambda(x_0) = 2\lambda_1(x_0)$, substituting Eq. (4.20) into Eq. (4.19) will give us the following expression:

$$\varepsilon^2 = \frac{k_1(x_0)}{(N_{\text{av}}/2)} = \frac{k_1(x_0)}{\lambda_1(x_0) * \tau_{\text{av}}} \tag{4.21}$$

Formula (4.21) provides a solution to the problem as it relates statistical error ε^2 of measurement either to average duration (τ_{av}) of the analyzed part of the process (for the case where number $N_{\text{av}}/2 = \lambda_1(x_0) * \tau_{\text{av}}$ of excursions above level x_0 is fixed) or to average number $N_{\text{av}}/2$ of excursions which occur within fixed time $\tau_{\text{av}} = \frac{N_{\text{av}}}{2*\lambda_1(x_0)}$.

For high relative levels ($x_0/\sigma \geq 2$) the distribution of the number of level crossings tends to a Poisson one, irrespective of the stationary random process distribution form. In this case, coefficient $k_1(x_0) \to 1$, that is, variance $\sigma_+^2(\tau, x_0)$ in the number of excursions above the given level tends to the average number of excursions occurring within time τ.

For relatively low thresholds $k_1(x_0) < 1$; therefore, in the general case, we can obtain from Eq. (4.21) the following upper estimate:

$$\varepsilon^2 \leq \frac{1}{\lambda_1(x_0) * \tau} \tag{4.22}$$

Unlike the upper estimate obtained in Ref. 53, the estimate in inequality (4.22) is the more accurate, the higher relative level x_0/σ chosen for analysis is. Thus, for $x_0/\sigma = 3$ the upper estimate in Ref. 53 is 2.5 times as high as the true error value, whereas the estimate in inequality (4.22) exceeds the true value by no more than a few tenths of a percentage point.

As an example, consider the exact relations between relative root mean square error ε^2 of measurements and the observation time ratio $\theta_k = \tau/\tau_k$, which were obtained using formula (4.21) for several values of relative level $x_0/\sigma = 0, 1, 2, 3$. Here τ_k is the time correlation interval of the process with autocorrelation function $R(\tau)$, which is determined by an expression similar to the one found in Ref. 8:

$$\tau_k = \int_0^\infty |R(\tau)| \mathrm{d}\tau$$

The calculations have been made for two stationary Gaussian random processes, one of which has a Gaussian power spectrum, and the other—a rectangular one. In both cases, the intensity of excursions above a given normalized level was chosen to be the same for all threshold levels. The relation between $\lambda_1(x_0)$ and $\lambda_1(0)$ is known:

$$\lambda_1(x_0) = \lambda_1(0) * \exp\left(-\frac{x_0^2}{2\sigma^2}\right)$$

For the process with a Gaussian power spectrum and correlation function $R(\tau) = \exp(-\tau^2/\tau_0^2)$ the correlation interval length is $\tau_{1k} = 1/2\sqrt{2\pi} * \lambda_1(0) = (\sqrt{\pi}/2) * \tau_0$, and for the process with a rectangular spectrum the value is $\tau_{2k} = (\pi/2\sqrt{6}) * \tau_0$.

The respective values calculated for the relative root-mean-square error ε of measurement are:

$$\varepsilon_1 = 2.239 * \sqrt{\frac{k_{11}(x_0)}{\theta_k}} * \exp\left(\frac{x_0^2}{4\sigma^2}\right)$$

$$\varepsilon_2 = 2.632 * \sqrt{\frac{k_{12}(x_0)}{\theta_k}} * \exp\left(\frac{x_0^2}{4\sigma^2}\right) \tag{4.23}$$

For zero threshold level, $k_{11}(0) = 1/2\sqrt{3} \approx 0.289$ in the case of the process with a Gaussian power spectrum, and $k_{12}(0) \approx 0.241$ in the case of the process with a rectangular spectrum. After substituting these values into formula (4.23), we obtain:

$$\varepsilon_1 \approx \frac{1.205}{\sqrt{\theta_k}}; \quad \varepsilon_2 \approx \frac{1.292}{\sqrt{\theta_k}}$$

It follows from the above that if we want to reduce statistical error to a few percentage points, then, even in the case where zero level is chosen as the threshold level, the number of noncorrelated samples of a random process should be increased to no less than $\theta_k = 10^3$ (in some cases to no less than 10^4).

A good estimate of the minimal measurement time required for achieving a given level of accuracy (i.e., obtaining a given statistical error ε), is given by the following formula from Ref. 8 which relates effective width $\Delta F_{\mathrm{eff.}}$ of the process power spectrum to time correlation interval length τ_k:

$$4\Delta F_{\mathrm{eff.}} * \tau_k = 1, \quad \text{where } \Delta F_{\mathrm{eff.}} = \frac{\int_0^\infty G(f)\mathrm{d}f}{G(0)}$$

Given that $\theta_k = 4\Delta F_{\mathrm{eff.}} * \tau$, from formula (4.23) we obtain:

$$\varepsilon_1 \approx 1.120\sqrt{\frac{k_{11}(x_0)}{\Delta F_{\mathrm{eff.}} * \tau}} * \exp\left(\frac{x_0^2}{4\sigma^2}\right)$$

$$\varepsilon_2 \approx 1.316\sqrt{\frac{k_{12}(x_0)}{\Delta F_{\mathrm{eff.}} * \tau}} * \exp\left(\frac{x_0^2}{4\sigma^2}\right) \tag{4.24}$$

Let us assume that we measure the cumulative distribution of momentary values of a Gaussian random process with a Gaussian spectrum and effective frequency bandwidth $\Delta F_{\mathrm{eff.}} = 100$ Hz. If the acceptable statistical error of measurement for all threshold levels chosen for analysis is limited by value $\varepsilon = 0.01$, then for the case of zero threshold the duration of the measurement period must not exceed $\tau = 36$ s. The number of uncorrelated samples of a random process must not exceed $\boldsymbol{\theta_k} = 1.45{*}10^4$. Similarly, for $|x_0/\sigma| = 1$ under the same conditions, the minimum requirement for the measurements period duration is $\tau \approx 121$ s, and the minimum requirement for the number of noncorrelated samples is $\theta_k \approx 4.9 * 10^4$. For $|x_0/\sigma| = 2$ the minimum requirements become much more stringent: measurement period duration $\tau \approx$ increases to 14.1 min, and number θ_k of noncorrelated samples increases to about $3.45 * 10^5$. Finally, in the case of the highest level chosen for analysis, $|x_0/\sigma| = 3$, the minimum requirements are $\tau \approx 189$ min = 3 h 9 min, and $\theta_k \approx 4.6 * 10^6$.

For the Gaussian process with a rectangular spectrum, to which the same requirements regarding measurement accuracy and intensity of excursions above the same given level were applied, we obtained values τ and θ_k of approximately the same order (only 1.07 times as high).

Weakening the minimum requirement for accuracy (e.g., increasing the acceptable error limit threefold to $\varepsilon = 0.03$) allows a much greater (in this example, ninefold) decrease in the measurement period duration and the number of samples. In other words, the more lenient the accuracy requirement is, the shorter will be the measurement time required for obtaining the distribution of momentary values in a random process.

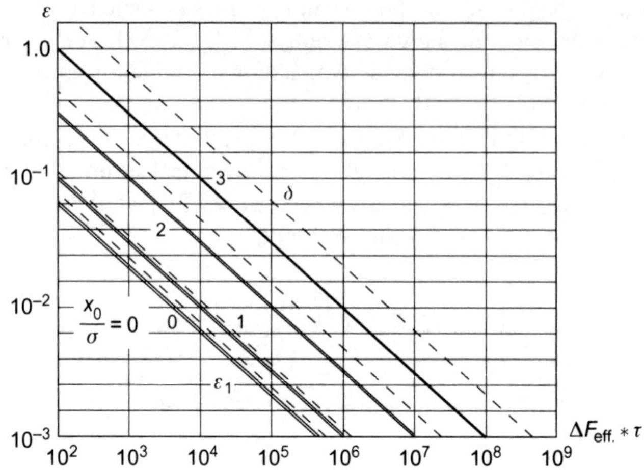

Figure 4.1 The relation between the root-mean-square error ε of measurement of a random process and the relative duration of the part analyzed.

Figure 4.1 shows families of curves $x = f(\Delta F_{\text{eff.}} * \tau)$ representing calculations of formula (4.24). Curves δ shown as broken lines represent upper estimates of ε taken from Ref. 53 and are shown for comparison. Calculations were made for $|x_0/\sigma| = 0, 1, 2,$ and 3.

As we can see from the graphs, the power spectrum shape (the correlation function form) has little, if any, impact on the statistical error ε of measurement, especially in the cases where the relative level $|x_0/\sigma|$ chosen for analysis is high.

This brings us to an important conclusion that the lack of information on the process correlation function form does not present a problem, since this information is practically irrelevant to us.

To choose the appropriate duration τ of the measurement period, we need to know either effective bandwidth $\Delta F_{\text{eff.}}$ of the process power spectrum, or intensity $\lambda_1(x_0)$ of excursions above level x_0 (i.e., the level chosen for analysis). The upper estimate obtained in Ref. 53 on the basis of the same kind of information is only acceptable when the chosen relative level $|x_0/\sigma|$ is low; at high levels it becomes too inaccurate.

5 Using a Family of Correlation Functions of a Clipped Random Process to Increase the Accuracy of Level-Crossing Parameters Estimation

5.1 One Method for Calculating Parameters of Zero Crossings in Broadband Centered Random Processes

Many problems related to the detection or recognition of random and deterministic signals against the background of noise with *a priori* unknown statistic properties are solved with the help of special devices that adapt to the most informative parameters of the excursions of received signal-plus-noise waveform. Examples of such parameters include the average number of points where the waveform crosses, within a given time limit, one or more relative levels chosen for analysis; the distribution of duration values for the waveform excursions above a given level and for intervals between excursions, and the distribution of durations for periods during which the signal remains within given boundaries. Other examples of data that may come useful include the data on the distribution law for the duration of intervals between a randomly chosen time moment and the first subsequent moment when the signal crosses zero level, the data on the average interval length for intervals between points of extremum, etc.

Such data may be used for solving various problems. Among them are the problem of signal-type recognition (speech, music, harmonics, pulses), the problem of detecting a given form of a signal against noise, the problem of detection and correction of "clicks" which are heard during the transmission of speech or music in a digital form using pulse-code modulation (PCM) method, and the problem of measuring the frequency of harmonic signals against noise background, to mention a few.[29,34,75]

In spite of the fact that for a long time the theory of excursions for random processes has been the focus of many researchers' attention and has a high potential for practical use, research into this theory is far from finishing. The reason for that is the difficulty of describing, in mathematical terms, the delicate structure of random processes. The well-studied parameters of excursions above a given level include the average duration of excursions, variance of duration, and the initial parts (first approximations) of excursions duration distribution density.

Applications of Random Process Excursion Analysis. DOI: http://dx.doi.org/10.1016/B978-0-12-409501-4.00005-X

Given that many problems that we deal with in the theory of excursions are difficult to solve if we talk about finding a perfect solution, solutions to such problems can be sought in the form of approximations, which are accurate enough for engineering applications.

At the same time, over the last few decades the researchers who study the properties of excursions have been widely using their personal computers for simulation of random processes.

This section offers a general description of a new method for the estimation of the distribution law for the length of interval between a randomly chosen moment and the first subsequent moment when the random process curve crosses zero level, regardless of whether it is an upward or downward crossing. The new method also enables the estimation of zero-crossings distribution density for several broadband Gaussian random processes. Computer simulations of such processes enable us to assess the accuracy of approximations. We showed in this section that the new method allows for an infinite reduction of the estimation error at the cost of making calculations more complicated. One other advantage of the new method is its applicability to centered random processes with an arbitrary distribution of momentary values and arbitrary autocorrelation function.[81]

In the calculation of zero-crossing parameters for centered stationary random process $X(t)$ correlation functions $R_k(\tau)$ of clipped sequences $B_k(t)$ can be used. Figure 5.1 shows an example of random process $X(t)$ and a family of clipped pulse sequences $B_k(t)$, where $k = 1,2,3,4$.

For $k = 1$ sequence $B_1(t)$ is a two-polar pulse process at the output of an ideal limiter with characteristic $f(x) = 1,\ x \geq 0,\ f(x) = -1,\ x < 0$, whose input is driven by random process $X(t)$. Sequences $B_1(t)$ where $k = 2,3,4$, etc. are derived from the generating sequence $B_1(t)$ by dividing the number of zero crossings by k (k intervals between successive zero-crossing points are combined and their lengths are added up).

It is easy to show that correlation functions $R_k(\tau)$ of clipped sequences $B_k(t)$ can be expressed through probabilities P_k of the occurrence of k zero crossings (upward or downward, does not matter) within time τ after a randomly chosen moment. Omitting a little algebra, we can write the following general expression:

$$\begin{aligned} R_k(\tau) = {} & \left[1 - \frac{2\lambda(0) * \tau}{k}\right] + \frac{4}{k} * [P_{k+1} + 2P_{k+2} + 3P_{k+3} + \cdots + (k-1)P_{2k-1}] \\ & + 4 * [P_{2k} + P_{2k+1} + P_{2k+2} + \cdots + P_{3k}] \\ & + \frac{4}{k} * [(k+1)P_{3k+1} + (k+2)P_{3k+2} + \cdots + (2k-1)P_{4k-1}] \\ & + 8 * [P_{4k} + P_{4k+1} + P_{4k+2} + \cdots + P_{5k}] \\ & + \frac{4}{k} * [(2k+1)P_{5k+1} + \cdots + (3k-1)P_{6k-1}] \\ & + 12 * [P_{6k} + P_{6k+1} + P_{6k+2} + \cdots + P_{7k}] + \cdots \end{aligned} \tag{5.1}$$

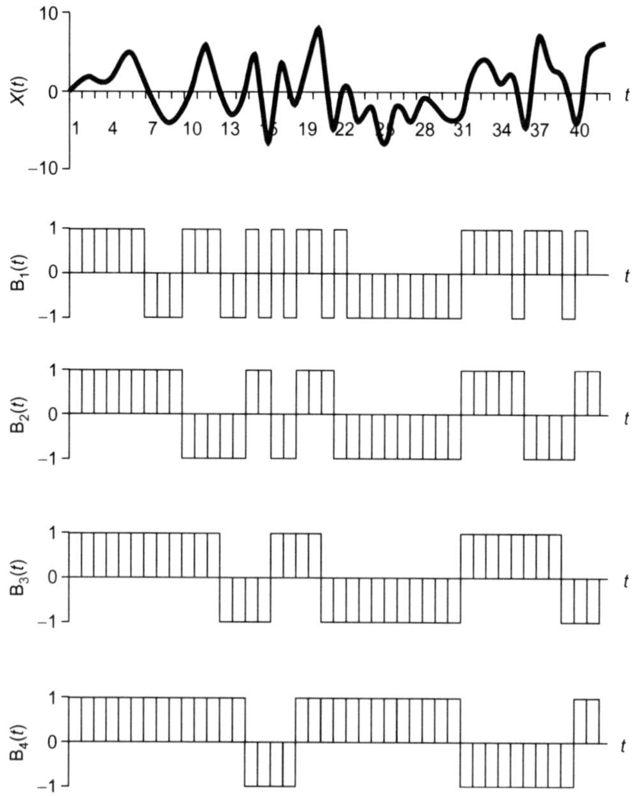

Figure 5.1 Clipped sequences $B_k(t)$, $k = 1,2,3,4$.

Here $\lambda(0)$ is the average number of zero crossings occurring in random process $X(t)$ per unit of time (zero-crossing intensity). Substitution of $k = 1$ into formula (5.1) will give us:

$$R_1(\tau) = [1 - 2\lambda(0) * \tau] + 4 * (P_2 + P_3) + 8 * (P_4 + P_5) + 12 * (P_6 + P_7) + 16 * (P_8 + P_9) + \cdots$$

In view of obvious relations, we have:

$$\sum_{k=0}^{\infty} P_k = 1, \quad \sum_{k=1}^{\infty} k * P_k = \lambda(0) * \tau \tag{5.2}$$

Using formula (5.1), we can express the sought function $P_0(\tau)$ in terms of correlation functions $R_k(\tau)$.

Function $P_0(\tau)$ is the probability that no zero crossings occur over time τ after an arbitrarily chosen moment. By substituting different values of parameter k from the range $k = 1,2,3,\ldots,8$ in formula (5.1) and Eq. (5.2), we obtain:

$$
\begin{aligned}
P_0(\tau) \quad &= \lambda(0) * \tau + \frac{R_1(\tau)}{4} + \frac{R_2(\tau)}{2} - \frac{3 * R_3(\tau)}{4} \\
&+ 2 * R_4(\tau) - \frac{5 * R_5(\tau)}{4} - \frac{3 * R_6(\tau)}{2} - \frac{7 * R_7(\tau)}{4} \\
&+ 8 * R_8(\tau) - \frac{9}{2} - P_{11} - 3 * P_{12} - 7 * P_{13} - 12 * P_{14} - 18 * P_{15} - \cdots
\end{aligned} \tag{5.3}
$$

Once we look at expression (5.3), it becomes obvious that it gives us the upper estimate of true function $P_0(\tau)$.

The estimate of $P_0(\tau)$ is expected to be accurate enough for those values of τ, for which the probability of the occurrence of 11 or more zero crossings is negligible. It is easy to show that if we double the maximum value $k_{\max}$ (increase it from $k_{\max} = 8$ to $k_{\max} = 16$), the error (Eq. (5.3)) will reduce from P_{11} to P_{21}. However, there is no need to increase $k_{\max}$, since an increase in length of series (Eq. (5.3)) and interval τ will inevitably cause an increase in error of calculation of $R_k(\tau)$ and consequently of function $P_0(\tau)$.

Formula (5.1) is quite cumbersome; therefore, it was used in combination with formula (5.2) only for the derivation of formula (5.3). The easiest way to calculate intensity $\lambda(0)$ of zero crossings for random process $X(t)$ and correlation functions $R_k(\tau)$ of clipped sequences $B_k(t)$ in (Eq. (5.3)) is to use a commercial computer application which calculates in the time domain.

To calculate the sought density $W(\tau)$ of distribution of lengths τ of intervals between successive zero crossings for random process $X(t)$, let us use the formula presented in Ref. 40:

$$W(\tau) = \frac{1}{\lambda(0)} * \frac{\mathrm{d}^2}{\mathrm{d}\tau^2} P_0(\tau) \tag{5.4}$$

After differentiating twice with respect to τ, we obtain from (Eq. (5.3)):

$$
\begin{aligned}
W(\tau) = \frac{1}{\lambda(0)} * \left[\frac{R_1''(\tau)}{4} + \frac{R_2''(\tau)}{2} - \frac{3 * R_3''(\tau)}{4} + 2 * R_4''(\tau) - \frac{5 * R_5''(\tau)}{4}\right. \\
\left. - \frac{3 * R_6''(\tau)}{2} - \frac{7 * R_7''(\tau)}{4} + 8 * R_8''(\tau)\right] - \frac{1}{\lambda(0)} * \{p_{11}''(\tau) + 3 * p_{12}''(\tau) + \cdots\}
\end{aligned} \tag{5.5}
$$

It is obvious that the expression in brackets in Eq. (5.5) gives us the upper estimate of true density $W(\tau)$. This estimate is expected to be accurate enough for those values of τ, for which the second derivatives of probabilities of the occurrence of 11 or more zero crossings are negligible.

Interestingly, the first term in Eq. (5.5) is the lower estimate of density $W(\tau)$.

As it is shown in Refs 8 and 29:

$$W(\tau) \geq \frac{1}{\lambda(0)} * \frac{R_1''(\tau)}{4} \tag{5.6}$$

The lower estimate in Eq. (5.6) is less accurate compared to the upper estimate in Eq. (5.5). It follows from Ref. 29 that

$$W(\tau) = \frac{1}{\lambda(0)} * \frac{R_1''(\tau)}{4} + \frac{1}{\lambda(0)} * \frac{d^2}{d\tau^2}\left[\sum_{k=1}^{\infty} k * (p_{2k+1} + p_{2k+2})\right] \tag{5.7}$$

As we can see from Eq. (5.7), the error in Eq. (5.6) is determined by the sum of the second derivatives of probabilities of the occurrence of three or more zero crossings in random process $X(t)$ within time limit τ. This error increases much more rapidly than the error in Eq. (5.5). It means that, for a wide range of values of τ, density $W(\tau)$ is expected to be very close to the upper estimate from Eq. (5.5).

To verify this conclusion, we suggest using well-known Poisson distribution expressions for $p_k(\tau)$:

$$p_k(\tau) = \frac{[\lambda(0) * \tau]^k}{(k)!} * \exp[-\lambda(0) * \tau]$$

Here $\lambda(0) = (1/2\tau_0)$ is the zero-crossing intensity of a Poisson process.

The average length of an interval between successive zero crossings is

$$\tau_{av} = (1/\lambda(0)) = 2\tau_0$$

For $k = 0$, $p_0(\tau) = \exp(-(\tau/2\tau_0))$. After differentiating twice $p_0(\tau)$, we obtain from Eq. (5.4):

$$W(\tau) = \frac{1}{2\tau_0} * \exp(-(\tau/2\tau_0)) \tag{5.8}$$

It is easy to calculate the absolute error of estimate in Eq. (5.5):

$$\Delta(\tau) = \frac{1}{\lambda(0)} * \{p_{11}''(\tau) + 3p_{12}''(\tau) + \cdots\}$$

and compare it with the true value of density in Eq. (5.8) for various values of τ. Calculations show that for $\tau \leq 6\tau_0 = 3 * \tau_{av}$ the relative error of the upper estimate obtained using Eq. (5.5) does not exceed 5%.

For $\tau = 6\tau_0 = 3\tau_{av}$ we see a 20-fold decrease in the value of $W(\tau)$ in Eq. (5.8) in comparison with its maximum, which is reached when $\tau = 0$. Thus, the upper estimate obtained using formula (5.5) turns out to be fairly accurate even for the "tails" of distribution $W(\tau)$.

Note that in real broadband processes the relative error of estimate (5.5) may get even smaller. Due to a certain correlation which exists between the lengths of adjacent intervals, the probability of the occurrence of several short intervals in succession in real processes is reduced. It causes a decrease, in comparison with a Poisson process, in the probability of high intensity of zero crossings over time τ, provided that intensity $\lambda(0)$ remains the same.

There have been extensive simulations of three broadband Gaussian random processes with very different power spectra. The following processes came into the focus of our attention:

1. A process with a uniform power spectrum.
2. A process with a Gaussian power spectrum.
3. A process obtained from the previous one by one-time differentiation.

To ensure a non biased comparison of the above processes, it was assumed that all of them have the same intensity $\lambda(0)$ of zero crossings.

For each process, statistical data comprised $N = 3 * 10^8$ signal samples, which were taken at the time interval $\Delta t = 0.1\tau_0$, where τ_0 is the correlation interval length for a particular process.

The average zero-crossing interval length was the same for all processes: $\tau_{av} = (1/\lambda(0)) = (\pi/\sqrt{2}) * \tau_0 \approx 2.221\ \tau_0$. An average interval between successive zero crossings, therefore, contained about 22 signal samples. Statistical data collected for each process covered $L = (N * \Delta t/\tau_{av}) \approx 1.35 * 10^7$ intervals.

Due to a large number of samples, the relative error of the estimation of distribution density $W(\tau)$ of intervals does not exceed 0.14% for maximum density values. At the same time, with a 100-fold decrease in density in distribution tails area the upper limit for the relative statistical error of measurement rises to 1.4%.

The power spectral density (PSD) of a process with a uniform power spectrum is represented by the following expression:

$$G_1(\omega) = \frac{\sqrt{\pi}}{6} * \tau_0 = \text{const}, \quad 0 \leq \omega \leq \frac{\sqrt{6}}{\tau_0} \tag{5.9}$$

The formula for the Gaussian PSD is given as:

$$G_2(\omega) = \sqrt{\pi} * \exp\left(-\frac{\omega^2\tau_0^2}{4}\right) * \tau_0 \tag{5.10}$$

The PSD of a process which represents the first derivative of the previous one has the following form:

$$G_3(\omega) = \frac{3\sqrt{3\pi}}{2} * (\omega\tau_0)^2 * \exp\left(-\frac{3\omega^2\tau_0^2}{4}\right) * \tau_0 \tag{5.11}$$

Note that zero crossings of the first derivative correspond to extreme points (extrema) of the initial process with a Gaussian PSD. It is known[8] that in a Gaussian process with a Gaussian PSD the intensity of extrema is $\sqrt{3}$ times higher than intensity $\lambda(0)$ of zero crossings in the initial process.

To enable an objective comparison of all three processes, PSD (Eq. (5.11)) was normalized so as to make square mean of spectrum frequency $\omega_1^2 = (2/\tau_0^2)$ and zero-crossing intensity $\lambda(0) = (\omega_1/\pi) = (\sqrt{2}/\pi\tau_0)$ the same for all three cases.

The first two processes are generated by passing white noise through an ideal filter of low frequencies and a Gaussian one, respectively.

The third process is formed at the output of a band-pass filter, whose frequency response is similar to that of an unloaded tuned circuit with a low Q-factor.

Each of the Figures 5.2–5.4 shows computer-simulated density $W(\tau)$ of distribution for lengths of intervals between successive zero crossings in one (the estimates were calculated using formula (5.5)).

Comparing $W(\tau)$ curves in pairs (Figures 5.2 and 5.5, Figures 5.3 and 5.6, and Figures 5.4 and 5.7) related to the same process shows their close coincidence for a wide range of values.

A small difference (temporary deviation from the trend) can only be seen in the area of exponential tails of density curves shown in Figures 5.5–5.7. In this case, we can see a 20-fold or even more serious decrease in the values of $W(\tau)$ in comparison with maximum values.

The difference in the area of tails is apparently caused by errors which occur in the calculation of the second derivatives of correlation functions $R_k(\tau)$ of clipped

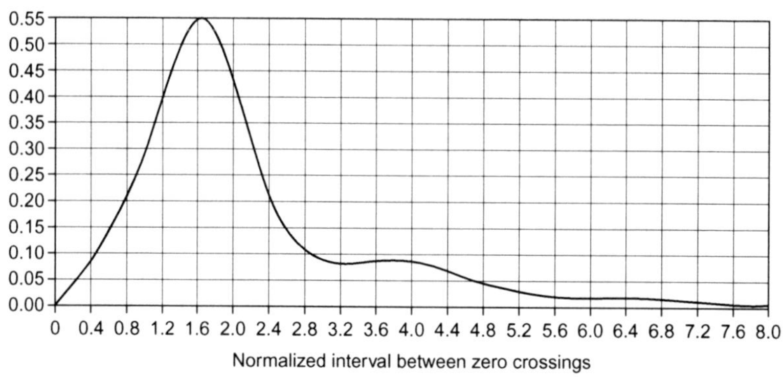

Figure 5.2 Experimental distribution density curve for intervals between zero crossings in a Gaussian process with a uniform power spectrum.

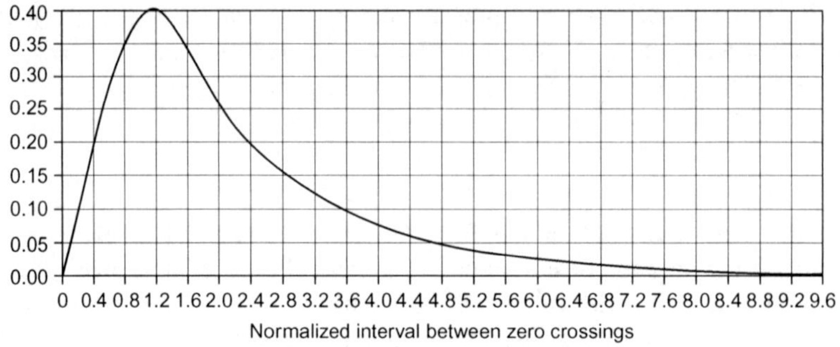

Figure 5.3 Experimental distribution density curve for intervals between zero crossings in a Gaussian process with a Gaussian power spectrum.

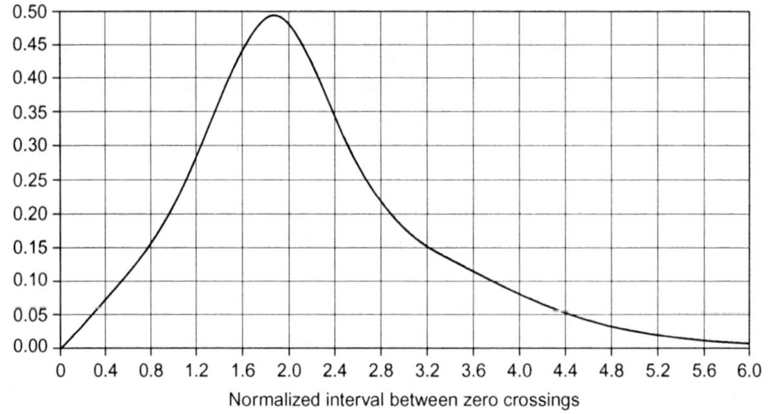

Figure 5.4 Experimental distribution density curve for intervals between zero crossings in the first derivative of a Gaussian process with a Gaussian power spectrum.

sequences (the functions are calculated using formula (5.5)). By shortening the interval between successive samples of a process, it is possible to reduce errors in calculation of the second derivative in Eq. (5.5) to arbitrarily small values.

5.1.1 Conclusions

1. It was shown that, in principle, it is possible to achieve any level of accuracy in the estimation of function $P_0(\tau)$ representing the distribution, for a particular random process, of time interval τ between an arbitrary moment and the first subsequent moment of upward or downward zero crossing.
2. Generating clipped sequence $B_1(t)$ is determined solely by its power spectrum (or by its autocorrelation function), which correlates with the power spectrum (or autocorrelation function) of initial random process $X(t)$. Moreover, such a sequence can be generated

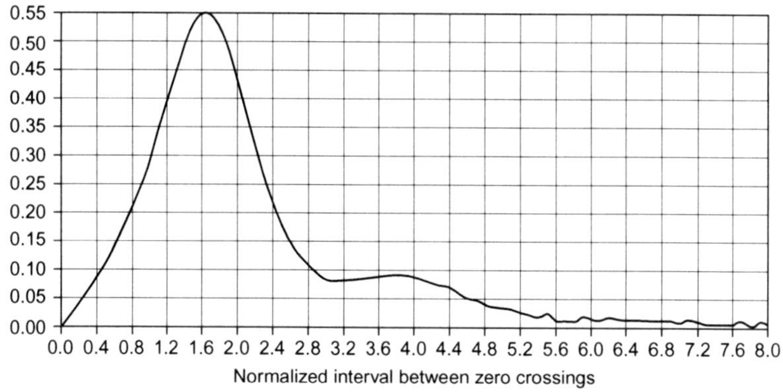

Figure 5.5 Calculated distribution density curve for intervals between zero crossings in a Gaussian process with a uniform power spectrum.

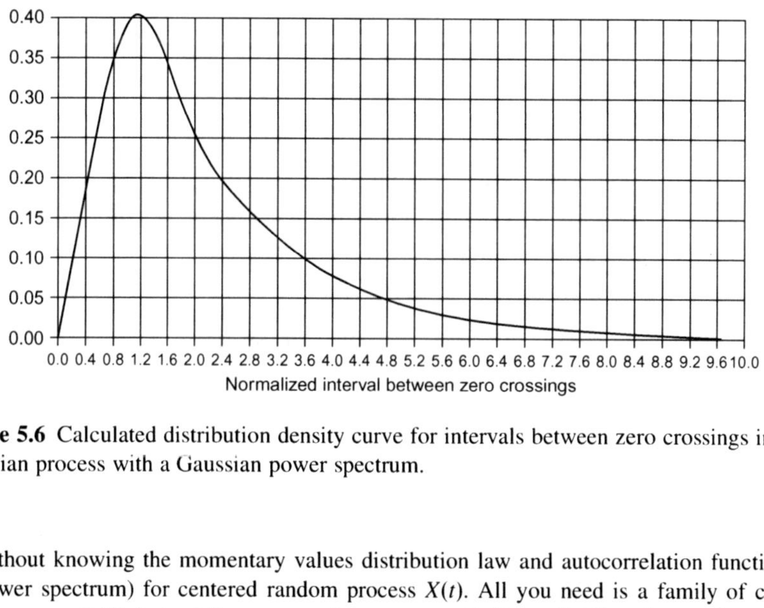

Figure 5.6 Calculated distribution density curve for intervals between zero crossings in a Gaussian process with a Gaussian power spectrum.

without knowing the momentary values distribution law and autocorrelation function (or power spectrum) for centered random process $X(t)$. All you need is a family of clipped sequences $B_k(t)$ derived from generating sequence $B_1(t)$ by dividing the number of zero crossings by k.

Therefore, the form of function $P_0(\tau)$ and the form of density $W(\tau)$ of interval length distribution are mainly determined by the form of power spectrum. As for the distribution of momentary values of centered random process $X(t)$, it has little impact on those forms.

3. Computer simulations of three broadband Gaussian random processes confirmed that, in the case where eight clipped sequences were used for calculation of correlation functions ($k = 8$), the error of calculating $W(\tau)$ remains negligible and comparable to the statistical error of measurements for a wide range of τ, which includes exponential tails of interval distribution density $W(\tau)$.

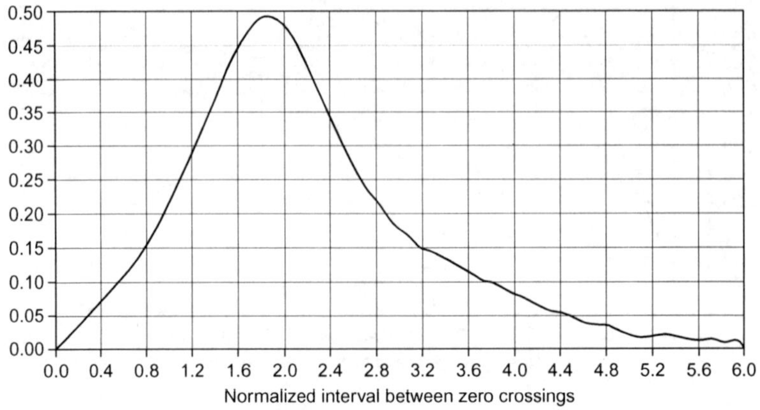

Figure 5.7 Calculated distribution density curve for intervals between zero crossings in the first derivative of a Gaussian process with a Gaussian power spectrum.

5.2 One Method for Calculating Parameters of Crossing a Given Standardized Threshold Level by a Random Process

In the previous chapter we analyzed the case of crossing level zero ($x_0 = 0$) by a random signal. This very important case can be extrapolated to crossing a nonzero standardized level ($(x_0/\sigma) \neq 0$), where σ is the effective value of random process $X(t)$.[82]

Figure 5.8 shows an example of a centered random process $X(t)$, which is symmetrical about level zero. In this case the standardized level was chosen to equal one ($(x_0/\sigma) = 1$).

Choosing a symmetrical process $X(t)$ gives us two options: (1) to double the number of excursions beyond the standardized thresholds with modulus one $|(x_0/\sigma)| \geq 1$ and reduce the statistical error of measurement and (2) to keep the same error rate and halve the length of the pulse sequence under analysis.

Figure 5.9 shows a sequence of rectangular pulses derived from excursions whose amplitude modulus is greater than the predetermined standardized threshold value. The width of pulses, all of which have amplitude of 1, is the same as that of corresponding excursions. The pulses are interleaved by pauses during which the process remains within the set amplitude limits. To shorten the time of analysis, we need to get rid of the pauses between unipolar pulses as shown in Figure 5.9 without losing information on the pulse width. This task can be accomplished by changing the sign of every other pulse in the flow before eliminating the pauses.

Figure 5.10 shows a bipolar sequence of rectangular pulses with amplitude of 1, which carries information on the length of intervals during which the random process makes excursions beyond the limits set for the standardized amplitude.

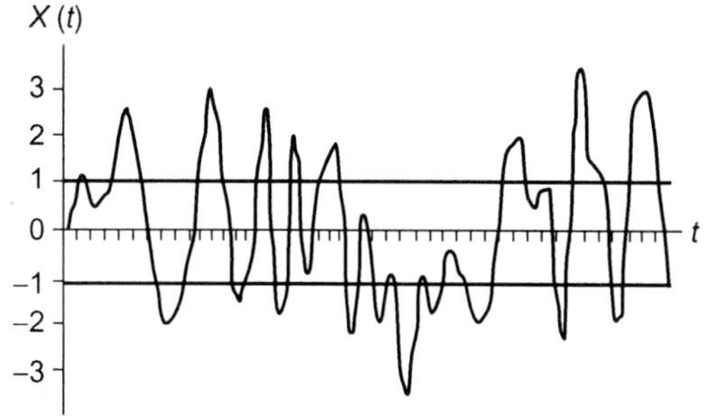

Figure 5.8 An example of a centered random process $X(t)$.

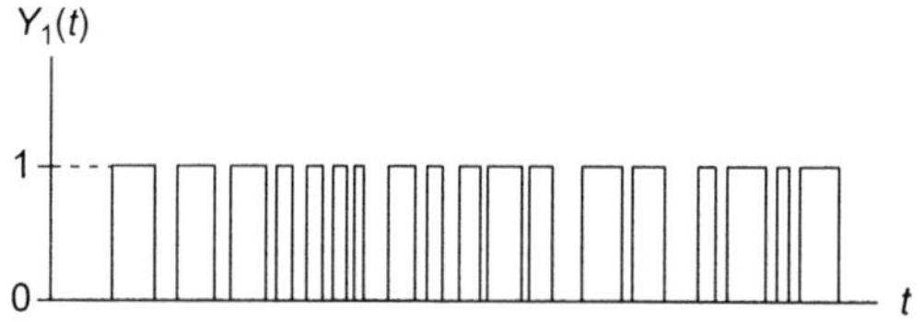

Figure 5.9 A sequence of rectangular pulses derived from excursions.

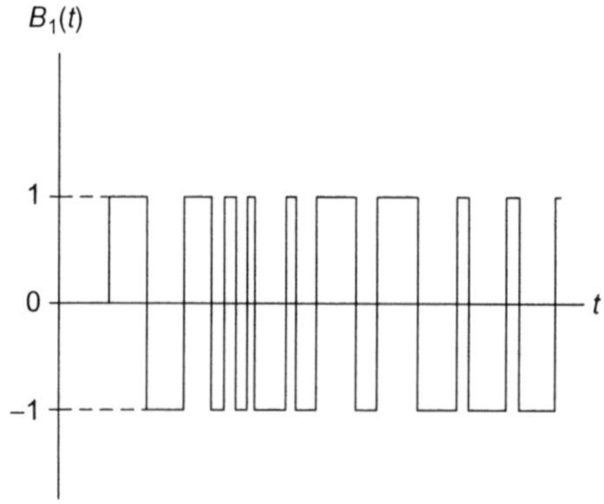

Figure 5.10 A bipolar sequence of rectangular pulses.

The pulse sequence shown in Figure 5.10 is shorter than that shown in Figure 5.9 or the signal in Figure 5.8, due to the fact that excursions above level $(x_0/\sigma) = 1$ occur less frequently than those above level zero.

Similarly, to the way we transformed the pulse sequence shown in Figure 5.9 into the one shown in Figure 5.10, we can transform it into a pulse sequence which carries information on the duration of intervals between excursions beyond the set limits (the so-called confinement intervals).

If we invert the sequence from Figure 5.9, the width of every positive pulse in the inverted sequence will equal that of a corresponding confinement interval (i.e., interval during which the process $X(t)$ remains within the set limits). The following transformation will again consist of the above-mentioned operations of inverting every other pulse and elimination of pauses between pulses.

Elimination of pauses from the sequence in Figure 5.9 requires storing the sequence in the computer memory and performing a few other simple procedures. The pulse sequence in Figure 5.10 can therefore be used as a new generating sequence $B_1(t,(x_0/\sigma))$, the same way the initial process $X(t)$ ideally limited by level zero was used to generate the sequence $B_1(t,0)$.

This means that the conclusions of the previous section are true not only for zero-threshold level but also for any standardized threshold $((x_0/\sigma) \neq 0)$.

In addition to that, our method can be used for analyzing the distribution of widths for both "spilling" and "confinement" intervals, and the upper and lower threshold values do not have to be symmetrical about zero. By setting, for example, two standardized thresholds (x_1/σ) and (x_2/σ) such that $(x_1/\sigma)<(x_2/\sigma)$, we can obtain two unipolar pulse sequences, one of which consists of pulses corresponding to excursions beyond the lower threshold, whereas the other consists of pulses representing intervals during which the process remains under the upper threshold limit. Applying the logical operation of conjunction (logical AND) to the two sequences will give us a sequence of pulses representing intervals during which the random process $X(t)$ is confined within the corridor limited by the two standardized threshold values. By inverting this sequence we can obtain information on the widths of "spilling" intervals (i.e., intervals during which the process "spills" over the set limits).

In other words, by setting two standardized thresholds (x_1/σ) and (x_2/σ) such that $(x_1/\sigma)<(x_2/\sigma)$, we can obtain information on width distribution laws for six different types of intervals, each type corresponding to a certain area to which the process is confined. These include the areas above and below the upper limit, above and below the lower limit, the area confined to the corridor formed by the two threshold lines, and the area outside that corridor. This kind of information may be required for the assessment of quality and reliability of products, product compliance with set standards, tolerance specifications, etc.

Note that the random process $X(t)$ does not have to be symmetrical about horizontal axis (level zero). Our method of calculating excursion parameters can be effectively applied for the analysis of any random processes—not just centered Gaussian ones but also Rayleigh, Rice, and other processes.

Knowing the standard deviation and range of momentary values of a process, we can reasonably choose the standardized threshold value (x_0/σ) and analyze level-crossing parameters for a wide range of such values. There are a number of cases where the average modulus of the process for the analyzed period can be substituted for the effective value σ.

Using formula (5.3), we can obtain the upper estimate of the function $P_0(\tau,(x_0/\sigma))$ which expresses the probability that the generating sequence $B_1(t,(x_0/\sigma))$ does not cross level zero during an randomly chosen time interval of length τ.

The function $P_0(\tau,(x_0/\sigma))$ can be expressed through a family of correlation functions $R_k(\tau,(x_0/\sigma))$ of high order pulse sequences $B_k(t,(x_0/\sigma))$.

Such sequences are obtained from the generating sequence $B_1(t,(x_0/\sigma))$ by adding up $k=2,3,4,\ldots$ successive intervals between zero crossings. The most effective way to calculate the correlation functions $R_k(\tau,(x_0/\sigma))$ of pulse sequences $B_k(t,(x_0/\sigma))$ is to use special software.

An alternative approach consists in finding the density $W(\tau,(x_0/\sigma))$ of width distribution among intervals between successive zero crossings by the generating sequence $B_1(t,(x_0/\sigma))$. This distribution density function can be obtained by processing the bipolar pulse sequence as shown in Figure 5.10 with a personal computer. As it was mentioned earlier, intervals between zeros of the sequence $B_1(t,(x_0/\sigma))$ have the width of corresponding excursions above threshold (x_0/σ).

As we can see from formula (5.4) above, the calculation of $W(\tau,(x_0/\sigma))$ boils down to finding the second derivative of the function $P_0(\tau,(x_0/\sigma))$ with respect to τ. In the process of calculating $W(\tau,(x_0/\sigma))$ we can also find the cumulative function $F(\tau,(x_0/\sigma))$ which expresses the distribution of above-threshold excursion durations:

$$F\left(\tau,\frac{x_0}{\sigma}\right) = -\frac{1}{\lambda(x_0/\sigma)} * \frac{\mathrm{d}}{\mathrm{d}\tau} P_0\left(\tau,\frac{x_0}{\sigma}\right) \tag{5.12}$$

Here $\lambda(x_0/\sigma)$ is the intensity of zeros in the generating sequence$B_1(t,(x_0/\sigma))$.

The function $P_0(\tau,(x_0/\sigma))$ can also be found experimentally with a simple computer program. Compared to calculation using formula (5.3), this alternative experimental method of calculating the density $W(\tau,(x_0/\sigma))$ of distribution has the advantage of using a less complicated and more accurate procedure for finding experimentally the function $P_0(\tau,(x_0/\sigma))$.

On the other hand, formula (5.3) suggests using a special computer application for calculating the correlation functions $R_k(\tau,(x_0/\sigma))$ of pulse sequences $B_k(t,(x_0/\sigma))$, which may, in some cases, be a better alternative.

5.3 Estimating the Distribution of Values for the Total Duration of Two or More Successive Excursions of a Random Process Above a Given Threshold

As it was mentioned in the previous section, the pulse sequence in Figure 5.10 can be regarded as the generating sequence $B_1(t,(x_0/\sigma))$ from which we can derive the sequences $B_k(t,(x_0/\sigma))$ for $k = 2,3,4\ldots$.

Let $p_{0k}(\tau)$ denote the probability that the pulse sequence $B_k(t,(x_0/\sigma))$ will not cross level zero over time τ which starts at the randomly chosen moment t_0. It is easy to prove that

$$p_{0k}(\tau) = \frac{1}{k} * \sum_{i=0}^{k-1}(k-i) * p_i(\tau), \quad \text{for } k \geq 1 \tag{5.13}$$

Here $p_i(\tau)$ is the probability that, over time τ, the generating sequence $B_1(t,(x_0/\sigma))$ will cross level zero i times, where $i = 0, 1, 2, \ldots(k-1)$.

It follows from Eq. (5.13) that,

$$\text{for } k=1,\quad p_{01}(\tau)=p_0(\tau);\quad \text{for } k=2,\quad p_{02}(\tau)=p_0(\tau)+\frac{1}{2}*p_1(\tau);$$

$$\text{for } k=3,\quad p_{03}(\tau)=p_0(\tau)+\frac{2}{3}*p_1(\tau)+\frac{1}{3}*p_2(\tau);$$

$$\text{for } k=4,\quad p_{04}(\tau)=p_0(\tau)+\frac{3}{4}*p_1(\tau)+\frac{1}{2}*p_2(\tau)+\frac{1}{4}*p_3(\tau);\ \text{etc.}$$

Let $p_{k+}(\tau)$ denote the probability that, over time τ which starts at the randomly chosen moment t_0, the generating sequence $B_1(t,(x_0/\sigma))$ will cross level zero k times in a particular direction (upward or downward).

$p_{k+}(\tau)$ can be derived from $p_k(\tau)$, the probability of k crossings of level zero in any direction over time τ. Given that $B_1(t,(x_0/\sigma))$ is symmetrical about level zero, we obtain:

$$p_{0+}(\tau)=p_0(\tau)+\frac{1}{2}*p_1(\tau),p_{1+}(\tau)=p_2(\tau)+\frac{1}{2}*[p_1(\tau)+p_3(\tau)],$$

$$p_{2+}(\tau)=p_4(\tau)+\frac{1}{2}*[p_3(\tau)+p_5(\tau)],p_{k+}(\tau)=p_{2k}(\tau)+\frac{1}{2}*[p_{2k-1}(\tau)+p_{2k+1}(\tau)] \tag{5.14}$$

Similarly, from Eq. (5.14) we can derive the following:

$$\begin{aligned}p_{k++}(\tau)&=p_{(2k)+}(\tau)+\frac{1}{2}*\left[p_{(2k-1)+}(\tau)+p_{(2k+1)+}(\tau)\right]=p_{(4k)}(\tau)\\&+\frac{3}{4}*\left[p_{(4k-1)}(\tau)+p_{(4k+1)}(\tau)\right]+\frac{1}{2}*\left[p_{(4k-2)}(\tau)+p_{(4k+2)}(\tau)\right]\\&+\frac{1}{4}*\left[p_{(4k-3)}(\tau)+p_{(4k+3)}(\tau)\right]\end{aligned} \tag{5.15}$$

Comparing Eqs. (5.13) and (5.14) shows that $P_{02}(\tau)=P_{0+}(\tau)$ expresses the probability of the absence of zero crossings in a particular direction over time τ.

In the case of $k=0$, formula (5.15) will look like this:

$$P_{0++}(\tau)=p_0(\tau)+\frac{3}{4}*p_1(\tau)+\frac{1}{2}*p_2(\tau)+\frac{1}{4}*p_3(\tau)=P_{04}(\tau)$$

As you can see, formula (5.13) gives us the same result for $k=4$.

Likewise, from Eqs. (5.14) and (5.15) we can obtain a common expression for calculating $p_{(k+++)}(\tau), p_{(k++++)}(\tau)$, etc., on the basis of probabilities of i crossings level zero over time τ by the generating sequence $B_1(t,(x_0/\sigma))$. Thus, knowing the expression for sequence $B_1(t,(x_0/\sigma))$, we can write a simple program which will allow us to obtain experimentally the probabilities $p_i(\tau)$ of i zero crossings over time τ which starts at an randomly chosen moment, where $i = 0, 1, 2, \ldots$.

Once we have calculated, using formula (5.13), the probability $p_{0k}(\tau)$ for a given k, we will be able to find the sought density $W_{0k}(\tau,(x_0/\sigma))$ of distribution of values for the total duration of k subsequent excursions of the random process above the randomly chosen standardized level (x_0/σ).

Using the approach described in the previous section, let us now express probabilities $p_{0k}(\tau)$ in Eq. (5.13) through a family of correlation functions $R_k(\tau)$ of higher-order pulse sequences $B_k(t)$. For that purpose, let us use the formulae obtained earlier, which establish relations between $R_k(\tau)$ and probability $p_i(\tau)$ that the generating sequence $B_1(t,(x_0/\sigma))$ will cross level zero i times over time τ.

For $k = 2$ a little algebra will give us

$$p_{02}(\tau) = p_{0+}(\tau) = \frac{1}{4} * [\lambda_{01}(0) * \tau + R_2(\tau) + 2 * R_4(\tau) - 3 * R_6(\tau) + 8 * R_8(\tau) - 4] \\ - \frac{p_{11}}{2} - p_{12} - 2 * p_{13} - 3 * p_{14} - \frac{9 * p_{15}}{2} - 6 * p_{16} - \cdots$$

For $k = 3$, we obtain

$$p_{03}(\tau) = \frac{1}{3} * \begin{bmatrix} \dfrac{\lambda_{01}(0) * \tau}{4} + \dfrac{R_2(\tau)}{4} + R_4(\tau) - \dfrac{3 * R_6(\tau)}{4} + 3 * R_8(\tau) - \dfrac{1}{2} - \\ - \dfrac{p_{11}}{2} - p_{12} - \dfrac{5 * p_{13}}{2} - 4 * p_{14} - \dfrac{5 * p_{15}}{2} - p_{16} - \cdots \end{bmatrix}$$

In the case of $k = 4$

$$p_{04}(\tau) = \frac{1}{4} * [R_4(\tau) + 2 * R_8(\tau) + 1] - \frac{p_{13}}{4} - \frac{p_{14}}{2} - \frac{3 * p_{15}}{4} - p_{16} - \cdots$$

As we know, the relation between the density $W_{0k}(\tau,(x_0/\sigma))$ and probability $p_{0k}(\tau,(x_0/\sigma))$ is established by the following equation:

$$W_{0k}(\tau,(x_0/\sigma)) = \frac{1}{\lambda_{0k}(x_0/\sigma)} * \frac{d^2}{d\tau^2} p_{0k}(\tau,(x_0/\sigma)) \tag{5.16}$$

where $\lambda_{0k}(x_0/\sigma) = \frac{1}{k} * \lambda_{01}(x_0/\sigma)$ is the frequency of zero crossings for the sequence $B_k(t,(x_0/\sigma))$; $\lambda_{01}(x_0/\sigma)$ is, therefore, the frequency of zero crossings for the sequence $B_1(t,(x_0/\sigma))$.

Let us see how the distribution density $W_{0k}(\tau, (x_0/\sigma))$ changes with the change of k, the number of intervals between subsequent zero crossings, whose durations are added up, in the case of a Poisson distribution of zero crossings.

The probability of the occurrence of i zero crossings over time τ is expressed as follows:

$$p_i(\tau) = \frac{[\lambda(0) * \tau]^i}{(i)!} * \exp[-\lambda(0) * \tau]$$

where $\lambda(0) = (1/2\tau_0)$ is the frequency of zero crossings, and $2\tau_0$ is the average total duration of a time interval between successive zero crossings in a Poisson sequence. As we know, durations of subsequent excursions in such a sequence are statistically independent.

Using formulae (5.13) and (5.16), we can easily obtain the sought density:

$$W_{0k}(\tau) = \frac{1}{2\tau_0} * \left(\frac{\tau}{2\tau_0}\right)^{k-1} * \frac{1}{(k-1)!} * \exp\left(-\frac{\tau}{2\tau_0}\right) \tag{5.17}$$

In Eq. (5.17) we are dealing with a chi-square (χ^2) distribution with ($2k$) degrees of freedom, also known as gamma distribution.

As one would expect, the average total duration of k intervals $\tau_{av}(k)$ and duration variance $D_k(\tau)$ in Eq. (5.17) increase in direct proportion to k, being $\tau_{av}(k) = k * (2\tau_0)$ and $D_k(\tau) = k * (2\tau_0)^2$, respectively.

The rate of increase in the mean $\tau_{av}(k)$ is $\sqrt{k}$ times higher that of increase in the root-mean-square value of the interval duration $\sqrt{D_k} = \sqrt{k} * (2\tau_0)$.

As k increases, the distribution density $W_{0k}(\tau)$ gets closer to normal, reaching its maximum at

$$\tau_{max} = \tau_{av}(k) * \left(1 - \frac{1}{k}\right) = (2\tau_0) * (k-1)$$

The maximum distribution density can be expressed as follows:

$$W_{0k}(\tau_{max}) = \frac{1}{2\tau_0} * (k-1)^{k-1} * \frac{1}{(k-1)!} * \exp[-(k-1)]$$

For $k \gg 1$ we can used the following approximation:

$$W_{0k}(\tau_{max}) \approx \frac{1}{(2\tau_0) * \sqrt{2\pi * (k-1)}}$$

This value reduces as k grows. The curve of distribution density $W_{0k}(\tau)$ shifts to the right along τ axis in proportion to $k * (2\tau_0)$, and we see a $\sqrt{k}$-fold increase in the standard deviation.

Obviously, in case of real random processes, the change in distribution density $W_{0k}(\tau,(x_0/\sigma))$ with the increase in the number k of intervals between successive zero crossings in the generating sequence $B_1(t,(x_0/\sigma))$ follows the same pattern. Naturally, the change in the mean duration and variation in duration of a single excursion will depend on the standardized threshold (x_0/σ). As for the change in distribution density $W_{0k}(\tau,(x_0/\sigma))$ with the increase in k, it will have the same key features such as getting closer to normal, shifting to the right along the τ axis, and $\sqrt{k}$-fold increase in the standard deviation.[35,54–58]

A number of publications dedicated to the theory of random process excursions prove that there is little correlation between durations of successive excursions, even at level zero.[7,47,55]

As the standardized level (x_0/σ) gets high enough, this correlation almost disappears. It is, therefore, safe to assume that, in a real process, the distribution of above-threshold excursion durations will be close to a Poisson one for $(x_0/\sigma)\geq 2$.[7,8,53] It means that real random processes have densities $W_{0k}(\tau,(x_0/\sigma))$ similar to the density in Eq. (5.17), with the only difference that the mean duration and variance in duration of a single excursion above level (x_0/σ) will decline and no longer remain the same if we raise the threshold value.

In the case of centered Gaussian processes, for example, there is a simple approximation[29] which allows calculating, accurately enough, the mean duration $\tau_{av}(x_0/\sigma)$ of excursions above level (x_0/σ):

$$\tau_{av}(x_0/\sigma)\approx\frac{\tau_{av}(0)}{1+(x_0/\sigma)}$$

where $\tau_{av}(0)$ is the mean duration of above-zero excursions.

The above conclusions and estimates are fully confirmed by numerous statistical simulations of random processes, including Gaussian and Rayleigh processes. Some results obtained through simulation are presented in the Appendices.

We have analyzed broadband Gaussian processes with a wide variety of power spectra. These include a limited homogeneous spectrum, a Gaussian spectrum, the spectrum of the first derivative of a process with a Gaussian spectrum, the spectrum of the first derivative of a process with a limited homogeneous spectrum, and spectra of processes obtained at the outputs of Butterworth filters (of orders 2 through 10).

All Gaussian processes under analysis have been transformed into Rayleigh processes which have also been analyzed.

Standardized thresholds (x_0/σ) varied widely (their values ranged from 0 to 3) for Gaussian processes and from 0.5 to 4 for Rayleigh processes, with 0.5 increments.

The number of digital samples in all cases was somewhere between $3*10^8$ and 10^{10}, which allowed us to keep the statistical error of measurement sufficiently low (it ranged from 10^{-3} to 10^{-2}), regardless of the interval width τ and any threshold (x_0/σ) chosen for obtaining the sought density $W(\tau,(x_0/\sigma))$ of excursion durations distribution.

5.3.1 Conclusions

1. A method has been proposed which allows deriving from a random process $X(t)$ the generating sequence $B_1(t,(x_0/\sigma))$ which carries all information on durations of excursions above the predetermined standardized threshold (x_0/σ).

   Using $B_1(t,(x_0/\sigma))$ allows to estimate, with any degree of accuracy, the distribution $P_0(\tau,(x_0/\sigma))$ of durations τ of intervals between a randomly chosen moment and the nearest point where the random process crosses level (x_0/σ) in any particular direction (upward or downward).

   Likewise, knowing $P_0(\tau,(x_0/\sigma))$, we can estimate the density $W(\tau,(x_0/\sigma))$ of distribution for duration of excursions above level (x_0/σ).
2. We have demonstrated that the aforementioned method can be used for calculating the distribution of durations for two kinds of intervals: the ones during which the process $X(t)$ stays within the set boundaries and the ones during which it "spills" over across such boundaries (in the general case, the two boundaries do not have to be symmetrical about zero level). By setting two different standardized thresholds we can obtain information on width distribution laws for six different types of intervals, each type corresponding to a certain area to which the process is confined. These include the area above and below the upper limit, above and below the lower limit, the area confined to the corridor formed by the two threshold lines, and the area outside that corridor.

   The random process $X(t)$ does not have to be symmetrical about horizontal axis (level zero). The proposed method of calculating excursion parameters can be effectively applied for the analysis of any random processes—not just centered Gaussian ones but also Rayleigh, Rice, and other processes.
3. As an alternative, we have analyzed another method for calculating the density $W(\tau,(x_0/\sigma))$ of width distribution for intervals between successive crossings of level zero by the generating sequence $B_1(t,(x_0/\sigma))$.

   This method suggests using a special PC program to process the generating pulse sequence. Computation of $W(\tau,(x_0/\sigma))$ boils down to finding the second derivative of the function $P_0(\tau,(x_0/\sigma))$ with respect to τ. In the process of calculating $W(\tau,(x_0/\sigma))$, we can also obtain the cumulative function $F(\tau,(x_0/\sigma))$ which describes the distribution of above-threshold excursion durations and is directly proportionate to the first derivative $P_0(\tau,(x_0/\sigma))$ with respect to τ.
4. Formulae have been provided for estimating the distribution of the total duration of k (two or more) successive excursions of a random process above a randomly chosen standardized threshold (x_0/σ). For those values of interval duration τ, which allow neglecting the second derivatives of functions expressing the probability of the occurrence of 11 or more zero crossings, the relative error of the estimation of $W_{0k}(\tau,(x_0/\sigma))$ remains small and commensurate with statistical error of measurement.
5. The results of computation and statistical simulation confirm our earlier conclusions that, for any given root-mean-square frequency of spectrum, there is very little correlation between the parameters of excursions above high standardized thresholds and the power spectrum shape.

   It is safe to assume that, in a real process, the durations of excursions above a threshold which is set high enough ($(x_0/\sigma)\geq 2$) will be distributed in a pattern which is close to the Poisson law.

   The key factor that determines the pattern of crossing any particular level (x_0/σ) by the random process is the parameter $\lambda(x_0/\sigma)$, which is the average number per time unit (i.e., frequency or intensity) of zeros in the generating sequence $B_1(t,(x_0/\sigma))$. In its turn, the parameter $\lambda(x_0/\sigma)$ fully depends on the root-mean-square value of the random process power spectrum frequency. To ensure that the comparison is fair, the power spectra of all Gaussian process under analysis were chosen to have the same root-mean-square frequency.

6 Estimates Obtained Through the Study of Certain Less-Known Parameters of Excursions in Differentiable Random Processes

6.1 Distribution Density of Time Before the Next Point Where the Set Upper or Lower Boundary Is Reached by a Differentiable Random Process

Probability characteristics of interval between a randomly chosen time moment and the next moment when the random process reaches the predetermined boundary constitute valuable information for the users of many important applications. The authors of Refs 7 and 8, for example, pointed out that the problem of finding the probability distribution function (or at least the mean and variance) which would accurately describe the distribution of the above-mentioned interval duration has only been solved for continuous Markovian processes. For stationary random processes, no accurate methods of solving this problem have been proposed so far.[77,86,88,90]

Let us consider the results obtained in Section 5.1, particularly formula (5.3) which relates the function $P_0(\tau)$ with correlation functions $R_i(\tau)$ of bipolar pulse sequences. Earlier, we defined the function $P_0(\tau)$ as the probability that the random process will not cross the predetermined boundary during time τ counted from a randomly chosen moment.

In Section 5.1, we analyzed the case of the first crossing of level zero by a centered process. In Section 5.2, we provided formula (5.3) which extends the case to any standardized level, (x_0/σ). We also noted that the process should not necessarily be symmetrical about the horizontal axis. The function $P_0(\tau)$ and the integral function $Q(\tau)$, which gives the distribution of time τ before the next crossing of the predetermined boundary by a random process, are obviously related as follows:

$$P_0(\tau) = 1 - Q(\tau) \tag{6.1}$$

Applications of Random Process Excursion Analysis. DOI: http://dx.doi.org/10.1016/B978-0-12-409501-4.00006-1

Hence, the formula for the sought density $W_2(\tau,(x_0/\sigma))$ of distribution for the time before the next crossing of boundary (x_0/σ) is given by

$$W_2\left(\tau,\frac{x_0}{\sigma}\right)=\frac{\mathrm{d}}{\mathrm{d}\tau}Q\left(\tau,\frac{x_0}{\sigma}\right)=-\frac{\mathrm{d}}{\mathrm{d}\tau}P_0\left(\tau,\frac{x_0}{\sigma}\right) \tag{6.2}$$

For small $\tau \ll (1/\lambda(x_0/\sigma))$ function $P_0(\tau,(x_0/\sigma))$, it is true that $P_0(\tau,(x_0/\sigma)) \approx 1-\lambda(x_0/\sigma)*\tau$. It follows from Eq. (5.12) that for $\tau \to 0$, $W_2(\tau,(x_0/\sigma)) \to \lambda(x_0/\sigma)$. Knowing the relation between the integral function $F(\tau,(x_0/\sigma))$, which describes the distribution of durations of excursions beyond level (x_0/σ), and the function $P_0(\tau,(x_0/\sigma))$,

$$F\left(\tau,\frac{x_0}{\sigma}\right)=-\frac{1}{\lambda(x_0/\sigma)}*\frac{\mathrm{d}}{\mathrm{d}\tau}P_0\left(\tau,\frac{x_0}{\sigma}\right)$$

From Eq. (6.2) we can obtain

$$W_2\left(\tau,\frac{x_0}{\sigma}\right)=\lambda\left(\frac{x_0}{\sigma}\right)*F\left(\tau,\frac{x_0}{\sigma}\right) \tag{6.3}$$

where $\lambda(x_0/\sigma)$ is the intensity of zero crossings by the generating sequence, or the inverse of the average duration of an excursion above the standardized level (x_0/σ) (see Section 5.2), and $F(\tau,(x_0/\sigma))=1-\int_0^{\tau} W_1(\tau,(x_0/\sigma))\mathrm{d}\tau$. $W_1(\tau,(x_0/\sigma))$ is the above-level excursion duration distribution density.

It follows from Eq. (6.3) that the distribution density $W_2(\tau,(x_0/\sigma))$ of time before the next crossing of boundary (x_0/σ) strongly depends on the distribution density $W_1(\tau,(x_0/\sigma))$ of excursion duration. Knowing the function $W_1(\tau,(x_0/\sigma))$, we can find the integral function $F(\tau,(x_0/\sigma))$ and then determine the sought density $W_2(\tau,(x_0/\sigma))$. In the absence of information on the function $W_1(\tau,(x_0/\sigma))$, the information on the function $P_0(\tau,(x_0/\sigma))$ can be used; once we have found its first derivative, we can use formulae (5.3) and (6.2) to devise the formula for the density, $W_2(\tau,(x_0/\sigma))$:

$$\begin{aligned}W_2\left(\tau,\frac{x_0}{\sigma}\right)=&-\lambda\left(\frac{x_0}{\sigma}\right)-\frac{R_1'(\tau)}{4}-\frac{R_2'(\tau)}{2}+\frac{3*R_3'(\tau)}{4}\\&-2*R_4'(\tau)+\frac{5*R_5'(\tau)}{4}+\frac{3*R_6'(\tau)}{2}+\frac{7*R_7'(\tau)}{4}\\&-8*R_8'(\tau)+P_{11}'+3*P_{12}'+7*P_{13}'\\&+12*P_{14}'+18*P_{15}'+\cdots\end{aligned} \tag{6.4}$$

Here, $\lambda(x_0/\sigma)$ is the intensity of changes from positive to negative and from negative to positive in the generating bipolar pulse sequence, $R_1'(\tau)$ is the first derivative of the correlation function for that sequence, and $R_i'(\tau)$ is the first derivative of the correlation function for the ith sequence.

Let us now look at two examples of calculating the density $W_2(\tau,(x_0/\sigma))$ of distribution of time before the next crossing of the boundary $(x_0/\sigma)=0$, assuming that the distribution of intervals, $W_1(\tau,(x_0/\sigma))$, between zero crossings is known. Let us start with the Poisson distribution $W_1(\tau)=(1/2\tau_0)*\exp(-(\tau/2\tau_0))$.

In this case, the value of the function $P_0(\tau)=\exp(-(\tau/2\tau_0))$ is known. From Eq. (6.2), we obtain $W_2(\tau)=(1/2\tau_0)*\exp(-(\tau/2\tau_0))=W_1(\tau)$. This relation is a consequence of the absence of any correlation between the moments when the Poisson pulse sequence switches from positive to negative or from negative to positive.

Now, let us consider a sequence in which each interval between subsequent zero crossings is a combination of two adjacent intervals between zeros in a Poisson pulse sequence. It was shown in Section 2.2 that in this case,

$$\mathrm{P}_{0+}(\tau)=\exp\left(-\frac{\tau}{2\tau_0}\right)*\left(1+\frac{\tau}{4\tau_0}\right)$$

After one-time differentiation of the function $[-P_{0+}(\tau)]$, we obtain

$$W_{2+}(\tau)=\frac{1}{4\tau_0}*\left(1+\frac{\tau}{2\tau_0}\right)*\exp\left(-\frac{\tau}{2\tau_0}\right) \tag{6.5}$$

The formula for the distribution $W_{1+}(\tau)$ density of the total duration of two noncorrelated intervals between zero crossings in a Poisson sequence was obtained by convolution:

$$W_{1+}(\tau)=\frac{\tau}{4\tau_0^2}*\exp\left(-\frac{\tau}{2\tau_0}\right) \tag{6.6}$$

In this case, the comparison of Eq. (6.5) and Eq. (6.6) shows that the densities are no longer the same. The difference between them arises from the fact that, in the second case, intervals between zero crossings are correlated. In the new sequence, we see a 50% reduction in the rate (intensity) $\lambda_{1+}(0)=(1/4\tau_0)$ of zero crossings compared to the Poisson sequence. Accordingly, the average duration of intervals between zero crossings $\tau_{1+\mathrm{av}}$ doubles:

$$\tau_{1+\mathrm{av}}=\frac{1}{\lambda_+(0)}=4\tau_0$$

The variance $D_{1+}=8\tau_0^2$ in the duration of intervals in sequence (6.6) also doubles. It no longer equals the square mean, which was the case with the Poisson distribution.

Let us now determine the mean and variance for the distribution $W_{2+}(\tau)$ of time interval between a randomly chosen moment and the next zero crossing. It is easy to show that the mean can be found by integrating the function $P_{0+}(\tau)$:

$$\tau_{2+\mathrm{av}} = \int_0^{\infty} P_{0+}(\tau)\mathrm{d}\tau = 3\tau_0 = \frac{3}{4} * \tau_{1+\mathrm{av}}$$

The second moment of the distribution $W_{2+}(\tau)$ can also be derived from the function $P_{0+}(\tau)$:

$$(M2)_{2+} = 2 * \int_0^{\infty} \tau * P_{0+}(\tau)\mathrm{d}\tau = 16\tau_0^2$$

Hence, the formula for the sought variance is

$$D_{2+} = (M2)_{2+} - (M1)_{2+}^2 = 7\tau_0^2$$

It is only natural that, in the case of correlated time intervals between a randomly chosen moment and the nearest subsequent crossing of x-axis, both the mean interval duration and the variance in interval width turned out to be smaller than those obtained for the case of noncorrelated intervals between subsequent zero crossings.

Let us now compare the squares of variance ratios $\nu^2 = \left(D/(M1)^2\right)$ for the two distributions. We have $\nu_1^2 = (1/2)$ for $W_{1+}(\tau)$ and $\nu_2^2 = (7/9)$ for $W_{2+}(\tau)$, respectively.

The relation between the parameters of the two distributions can be easily established. Knowing the mean time interval $\tau_{1+\mathrm{av}}$ and variance D_{1+}, we can devise the formula for the mean interval $\tau_{2+\mathrm{av}}$:

$$\tau_{2+\mathrm{av}} = \frac{(M2)_{1+}}{2 * (M1)_{1+}} = \frac{(D_{1+}) + (\tau_{1+\mathrm{av}})^2}{2 * (\tau_{1+\mathrm{av}})} = \frac{(\nu_1^2 + 1)}{2} * (\tau_{1+\mathrm{av}}) \tag{6.7}$$

In the case of the Poisson distribution $\nu_1^2 = 1$, we have, and it follows from Eq. (6.7), that both distributions have the same mean value.

Knowing the first two moments of the distribution density $W_{1+}(\tau)$ is, however, insufficient for the calculation of variance or the second moment of the distribution $W_{2+}(\tau)$. We also need to know the third moment $(M3)_{1+}$ in order to find the second moment $(M2)_{2+}$:

$$(M2)_{2+} = \frac{(M3)_{1+}}{3 * (M1)_{1+}} \tag{6.8}$$

In this example, the third moment is easy to find:

$$(M3)_{1+} = \int_0^{\infty} \tau^3 * W_{1+}(\tau)\mathrm{d}\tau = 192\tau_0^3$$

From Eq. (6.8) we obtain $(M2)_{2+} = 16\tau_0^2$ and $D_{2+} = 7\tau_0^2$. The same results were obtained earlier using the known function.

All of the above can be summarized as follows:

1. The distribution density $W_2(\tau, (x_0/\sigma))$ of time before the next boundary crossing strongly depends on the distribution density $W_1(\tau, (x_0/\sigma))$ of excursions beyond the standardized level (x_0/σ).
2. The function $P_0(\tau, (x_0/\sigma))$ also provides full information on the density $W_2(\tau, (x_0/\sigma))$. This function shows the probability that the random process will not cross the set boundary during the time τ counted from a randomly chosen moment.
3. There are a number of reasons why we should give preference to using the function $P_0(\tau, (x_0/\sigma))$ for calculating the density $W_2(\tau, (x_0/\sigma))$.

First, when statistical simulations are used, this function is easier to find and can be calculated more accurately than the density $W_1(\tau, (x_0/\sigma))$.

Second, the proposed formula (5.3) allows us to calculate $P_0(\tau, (x_0/\sigma))$ with high accuracy for a wide range of values of τ.

Finally, it is much easier to derive the moments of the density $W_2(\tau, (x_0/\sigma))$ (the mean, the variance, and the moments of higher orders) from the function $P_0(\tau, (x_0/\sigma))$ than from the density $W_1(\tau, (x_0/\sigma))$.

6.2 On the Distribution of Random Process Excursion Areas for Excursions Above a Given Level

A number of practical problems cannot be solved without calculating random process excursion areas (i.e., the areas of shapes formed by a random process curve and a given horizontal line it crosses). A problem of this kind arises, for example, in the calculation of transmission channel quality or reliability of communication for multipath channels transmitting signals in shortwave or ultra-shortwave range, where the signal fades at the receiver due to the interference of radio waves. This calculation involves estimating both the duration of the signal fading periods and the depth of fading troughs, i.e., the duration and depth of signal excursions below a certain acceptable level. Knowing the distribution of areas of shapes formed by the radio signal curve and the horizontal line representing a set threshold below which the curve "drops" from time to time gives you a good idea about the reliability of communication in the channel.

The need to know the distribution of excursion areas may arise, e.g., when we evaluate the surface quality of finished products, the roughness of the sea, the effect of overburden load on buildings, and other structures.

Whenever both the load duration and the amount of overburden pressure are relevant, we have to evaluate the area under the pressure curve, which is proportional to the product of the average pressure and the load duration.

Problems arising from the need to find the distribution law for level excursion areas are usually very complicated and the least studied. In fact, the author is only aware of two sources[4,53] which describe methods to estimate excursion areas for short excursions above high thresholds. In Ref. 4 it was assumed for simplicity that such excursions have the shape of a square parabola.

The author(s) of Ref. 53 went further and suggested, for calculation and measurement purposes, that short excursions are triangular in shape. The area of a triangular excursion can be calculated as the product of its duration and half its amplitude. It should be noted that in a real random process, excursions cannot possibly be triangular in shape as triangular shape of excursions would imply a jump in the first derivative of the process; in other words, the process becomes nondifferentiable.

The proposed method for calculating the distribution of areas for random process excursions above or below a given standardized level is based on the results presented in Sections 5.1 and 5.2. After sampling and amplitude quantization of a random signal with a standardized level, the stream of randomly shaped excursions is converted into a stream of bipolar rectangular pulses with a unit amplitude. The duration of each pulse equals to the area of the original sampled excursion (this area is calculated as the product of the average amplitude and the real duration of the excursion). The signs of pulses in the sequence alternate. The average amplitude is obtained as a simple mean of sampled values of the signal taken while the excursion remains above a given level.

The bipolar rectangular pulse sequence so obtained is the generating sequence defined in Sections 5.1 and 5.2. The data on the size of intervals between successive zeros of the generating sequence is fully sufficient to calculate excursion areas. The areas of excursions calculated for a normalized random process with a unit variance (they are measured in volt-seconds) have the same numerical values as the lengths of pulses in the generating sequence. It means that the formulae given in Section 5.2 can also be used for calculating the distribution of areas for random process curve excursions.

If you are interested in finding the distribution of areas for excursions of a normalized random process below a given level, you can take the same approach as we suggested for finding the distribution of areas for excursions above a given level. If you are dealing with downward excursions, all you have to do is invert (i.e., change signs of) all samples of the process and add them to the given threshold value. All downward excursions will then become upward, and the problem will take a familiar shape.

Let us now estimate the average areas of excursions above or below various given standardized levels for Gaussian and Rayleigh processes. For our calculations, we will use the formula

$$S_{av}\left(\frac{x_0}{\sigma}\right) = A_{av}\left(\frac{x_0}{\sigma}\right) * \tau_{av}\left(\frac{x_0}{\sigma}\right) \tag{6.9}$$

where $S_{av}(x_0/\sigma)$ is the average area of excursions above the standardized level (x_0/σ), $A_{av}(x_0/\sigma)$ is the average amplitude of excursions, and $\tau_{av}(x_0/\sigma)$ is the average excursion duration.

The average duration of excursions above or below the level (x_0/σ) is known for both Gaussian and Rayleigh processes [8].

As far as the average amplitude is concerned, in the case of a Gaussian process, after doing a little algebra we have,

$$A_{\text{av}}\left(\frac{x_0}{\sigma}\right) = \frac{\exp(-(x_0^2/2\sigma^2))}{\sqrt{2\pi} * [1 - F(x_0/\sigma)]} - \left(\frac{x_0}{\sigma}\right) \tag{6.10}$$

Here, $F(x_0/\sigma)$ is a probability integral and a known tabulated function.

The average amplitude in Eq. (6.10) is measured in volts.

By substituting Eq. (6.10) and the known expression for the average excursion duration into Eq. (6.9) we obtain

$$S_{\text{av}}\left(\frac{x_0}{\sigma}\right) = \sqrt{\pi} * \left\{1 - \sqrt{2\pi} * \left(\frac{x_0}{\sigma}\right) * \exp\left(\frac{x_0^2}{2\sigma^2}\right) * \left[1 - F\left(\frac{x_0}{\sigma}\right)\right]\right\} * \tau_0 \tag{6.11}$$

where τ_0 is the duration of the correlation time interval for the random process.

The average area of excursions above the level (x_0/σ) in Eq. (6.11) is measured in volt seconds.

Formulae (6.10) and (6.11) were used for computing the data presented in Table 6.1

As one should expect, the average area of excursions above a given standardized threshold (x_0/σ) goes down significantly as the threshold rises. The reason for such reduction becomes obvious if you take into consideration the simultaneous reduction of the average duration and the average amplitude of excursions.

Following the same way we obtained formula (6.10), we can easily deduce a formula for computing the average amplitude of excursions below the given level (x_0/σ) (downward excursions). All it takes is inverting the sign of the expression for the standardized level (x_0/σ) and replacing $[1 - F(x_0/\sigma)]$ with $F(x_0/\sigma)$:

$$A_{\text{av}}\left(-\frac{x_0}{\sigma}\right) = \frac{\exp(-(x_0^2/2\sigma^2))}{\sqrt{2\pi} * F(x_0/\sigma)} + \left(\frac{x_0}{\sigma}\right) \tag{6.12}$$

The average amplitude of a downward excursion in Eq. (6.12) is measured in volts.

Table 6.1 The Average Area of Excursions Above a Given Threshold (x_0/σ) for a Gaussian Process

(x_0/σ)	0	0.5	1	1.5	2	2.5	3
$\tau_{\text{av}}(x_0/\sigma)$ (mcs)	$(\pi/\sqrt{2}) = 2.2214$	1.5533	1.1623	0.9143	0.7467	0.6280	0.5399
$A_{\text{av}}(x_0/\sigma)$ (V)	$\sqrt{(2/\pi)} = 0.7979$	0.6411	0.5251	0.4386	0.3732	0.3226	0.2828
$S_{\text{av}}(x_0/\sigma)$ (V mcs)	$\sqrt{\pi} = 1.77245$	0.9958	0.6103	0.4010	0.2787	0.2026	0.1527
$S_{\text{av}}\exp(x_0/\sigma)$ (V mcs)	1.77240	0.9974	0.6126	0.4034	0.2811	0.2050	0.1551

By substituting Eq. (6.12) and the known expression for the average excursion duration into Eq. (6.9) we obtain

$$S_{av}\left(-\frac{x_0}{\sigma}\right) = \sqrt{\pi} * \left\{1 + \sqrt{2\pi} * \left(\frac{x_0}{\sigma}\right) * \exp\left(\frac{x_0^2}{2\sigma^2}\right) * \left[F\left(\frac{x_0}{\sigma}\right)\right]\right\} * \tau_0 \qquad (6.13)$$

Formulae (6.12) and (6.13) were used for computing the data presented in Table 6.2.

It follows from Table 6.2 that the average area of downward excursions grows rapidly as the standardized threshold (x_0/σ) rises. This happens because of a sharp increase in the average size of intervals between short successive excursions above high thresholds. As far as the average amplitude of downward excursions is concerned, it follows from formula (6.12) that the higher the value of the standardized level (x_0/σ) is, the closer this amplitude gets to this value.

For a normalized Rayleigh process (i.e., a process with standard deviation $\sigma = 1$) with the duration τ_0 of the correlation time interval, the average duration of excursions above the level (r_0/σ) can be obtained using the known formula $\tau_{av} = (\sqrt{\pi}/(r_0/\sigma)) * \tau_0$.

After doing a little algebra, we can easily get the formula for computing the average amplitude of excursions above the level (r_0/σ):

$$A_{av}\left(\frac{r_0}{\sigma}\right) = \sqrt{2\pi} * \exp\left(\frac{r_0^2}{2\sigma^2}\right) * \left[1 - F\left(\frac{r_0}{\sigma}\right)\right] \qquad (6.14)$$

Substituting Eq. (6.14) into Eq. (6.9) will give us

$$S_{av}\left(\frac{r_0}{\sigma}\right) = \frac{\pi * \sqrt{2} * \exp(r_0^2/2\sigma^2) * [1 - F(r_0/\sigma)]}{(r_0/\sigma)} * \tau_0 \qquad (6.15)$$

It follows from Eq. (6.15) that for low thresholds $((r_0/\sigma) \ll 1)$

$$S_{av}\left(\frac{r_0}{\sigma}\right) \approx \frac{\pi}{\sqrt{2}} * \frac{1}{(r_0/\sigma)}$$

Table 6.2 The Average Area of Excursions Below a Given Threshold (x_0/σ) for a Gaussian Process

(x_0/σ)	0.1	0.5	1	1.5	2	2.5	3
$\tau_{av}(x_0/\sigma)$ (mcs)	2.4104	3.4811	6.1628	12.771	32.082	100.49	399.40
$A_{av}(x_0/\sigma)$ (V)	0.83533	1.0092	1.2876	1.6388	2.0553	2.51764	3.00444
$S_{av}(x_0/\sigma)$ (V mcs)	2.01350	3.51301	7.9352	20.9286	65.936	253.001	1199.96
S_{av} exp(x_0/σ) (V mcs)	2.01709	3.51537	7.9340	20.9068	65.759	250.767	1155.24

In the case of high thresholds $((r_0/\sigma) \gg 1)$, we have

$$S_{av}\left(\frac{r_0}{\sigma}\right) \approx \sqrt{\pi} * \frac{1}{(r_0/\sigma)^2}$$

Formulae (6.14) and (6.15) were used for computing the data presented in Table 6.3.

Comparing the figures in Table 6.3 with those in Table 6.1 shows a striking similarity between them in the area of high thresholds $((r_0/\sigma) \geq 2)$.

Because of Rayleigh distribution density curves having longer tails compared to Gaussian ones, the average areas obtained for excursions above the same high thresholds are somewhat bigger in the case of a normalized Rayleigh process than in the case of a Gaussian process.

The average duration of excursions below the level (r_0/σ) can be obtained using the known formula:

$$\tau_{av} = \frac{\sqrt{\pi}}{(r_0/\sigma)} * \left[\exp(r_0^2/2\sigma^2) - 1\right] * \tau_0$$

For low thresholds $((r_0/\sigma) \ll 1)$, $\tau_{av} \approx (r_0/\sigma) * (\sqrt{\pi}/2) * \tau_0$.

With a little algebra, we obtain the formula for computing the average amplitude of excursions below the level (r_0/σ):

$$A_{av}\left(\frac{r_0}{\sigma}\right) = \left(\frac{r_0}{\sigma}\right) + \frac{1}{1 - \exp(-(r_0^2/2\sigma^2))} * \left\{\left(\frac{r_0}{\sigma}\right) * \exp\left(-\frac{r_0^2}{2\sigma^2}\right) - \sqrt{2\pi} * \left[F\left(\frac{r_0}{\sigma}\right) - \frac{1}{2}\right]\right\} \quad (6.16)$$

Table 6.3 The Average Area of Excursions Above a Given Threshold (r_0/σ) for a Rayleigh Process

(r_0/σ)	0.1	0.2	0.3	0.4	0.5	1.0	1.3
$\tau_{av}(r_0/\sigma)$ (mcs)	17.7245	8.86227	5.90818	4.43113	3.54491	1.77245	1.36343
$A_{av}(r_0/\sigma)$(V)	1.1593	1.07594	1.00184	0.93567	0.87637	0.65570	0.56486
$S_{av}(r_0/\sigma)$(V mcs)	20.5473	9.53530	5.91906	4.14609	3.10666	1.16220	0.77015
S_{av} exp(r_0/σ)(V mcs)	20.3597	9.56083	5.91066	4.13875	3.10149	1.16767	0.77515
(r_0/σ)	1.5	2.0	2.5	3.0	3.5	4.0	5
$\tau_{av}(r_0/\sigma)$ (mcs)	1.18164	0.88623	0.70898	0.59082	0.50642	0.44311	0.35449
$A_{av}(r_0/\sigma)$(V)	0.51584	0.42137	0.35428	0.30461	0.26356	0.24271	0.19704
$S_{av}(r_0/\sigma)$(V mcs)	0.60953	0.37343	0.25118	0.17997	0.13347	0.10755	0.06985
S_{av} exp(r_0/σ)(V mcs)	0.61439	0.37823	0.25259	0.18136	0.13526	0.10893	0.07089

The average area of excursions below the level (r_0/σ) is

$$S_{av}\left(\frac{x_0}{\sigma}\right) = \frac{\sqrt{\pi}}{(r_0/\sigma)} * \left[\exp\left(\frac{r_0^2}{2\sigma^2}\right) - 1\right] * \left[\left(\frac{r_0}{\sigma}\right) + \frac{1}{1-\exp(-(r_0^2/2\sigma^2))}\right.$$

$$\left. * \left\{\left(\frac{r_0}{\sigma}\right) * \exp\left(-\frac{r_0^2}{2\sigma^2}\right) - \sqrt{2\pi} * \left[F\left(\frac{r_0}{\sigma}\right) - \frac{1}{2}\right]\right\}\right] * \tau_0 \quad (6.17)$$

It follows from Eq. (6.17) that for low thresholds $((r_0/\sigma) \ll 1)$

$$S_{av}\left(\frac{r_0}{\sigma}\right) \approx \frac{\sqrt{\pi}}{6} * \left(\frac{r_0}{\sigma}\right)^2 * \tau_0$$

Formulae (6.16) and (6.17) were used to calculate the data presented in Table 6.4. As we can see, the average area of excursions below the level (r_0/σ) grows rapidly as this level rises. This happens because of a sharp increase in the average size of intervals between short successive excursions above high thresholds.

Using the data on the average areas of random process excursions, one can estimate the gradient of exponential tails of distribution density curves for big values of excursion duration $(\tau \gg \tau_0)$. It is known that in a real random process, the distribution of excursion durations tends to exponential. This is especially true for excursions above low thresholds and excursions below high thresholds, as in these cases, the distribution law for excursion durations can be described as the exponential for a wide range of duration values $(\tau \geq \tau_{av})$.

In Section 4.2, a method of finding exponential tail parameters for excursion duration distribution is described. We will now demonstrate that the same method is applicable to the distribution of excursion areas in the case of long excursions $(\tau \gg \tau_0)$. The amplitude of a long excursion can be calculated as the simple mean

Table 6.4 The Average Area of Excursions Below a Given Threshold (r_0/σ) for a Rayleigh Process

(r_0/σ)	0.1	0.2	0.3	0.4	0.5	0.8	1.0
$\tau_{av}(r_0/\sigma)$ (mcs)	0.08884	0.17903	0.27194	0.36906	0.47200	0.83555	1.14983
$A_{av}(r_0/\sigma)$(V)	0.03228	0.06690	0.10098	0.13553	0.17090	0.28388	0.36696
$S_{av}(r_0/\sigma)$(V mcs)	0.00287	0.01198	0.02746	0.05002	0.08066	0.23719	0.42194
S_{av} exp(r_0/σ) (V mcs)	0.00298	0.01237	0.02798	0.05094	0.08169	0.23845	0.42372
(r_0/σ)	1.3	1.5	1.8	2.0	2.5	3.0	3.5
$\tau_{av}(r_0/\sigma)$ (mcs)	1.81060	2.45806	3.99106	5.66215	15.4274	52.5929	230.999
$A_{av}(r_0/\sigma)$(V)	0.50720	0.61325	0.79385	0.92951	1.32026	1.76973	2.25219
$S_{av}(r_0/\sigma)$(V mcs)	0.91833	1.50739	3.16831	5.26301	20.3681	93.0753	520.253
S_{av}exp(r_0/σ) (V mcs)	0.91973	1.51208	3.17377	5.27095	20.3805	92.8778	515.751

of a large number of noncorrelated samples of a signal all of which exceed a given threshold. Once excursions get long enough ($\tau \gg \tau_0$), their amplitude tends to a constant average value which, for any given process, depends on the threshold level, but not on the excursion duration.

It follows from the above that, for large duration values, the random area changes in direct proportion to the random duration of an excursion (the area is calculated as the product of the average amplitude and duration). The proportionality factor is determined by the average amplitude of excursions and depends solely on the distribution law for momentary values of the random signal and the height of the standardized threshold.

Random excursion duration values are known to be distributed exponentially for $\tau \gg \tau_0$; the same is true about the random values of excursion areas. The gradient of the exponential curve which represents the distribution density of excursion area values exceeding the average $S_{av}(x_0/\sigma)$ can be easily found as the product of the known average amplitude of above-level excursions and the gradient of the exponential curve which represents the distribution of excursion durations for $\tau \gg \tau_0$.

This statement is based on numerous data provided by computer simulations. The data were collected for the area distribution density for both upward and downward excursions in Gaussian and Rayleigh processes with Gaussian and rectangular power spectra.

Appendix 3 shows excursion area distribution densities which were obtained using computer simulations for a number of standardized thresholds (x_0/σ). From the data obtained by simulation, the average excursion areas were found for all of the above-mentioned cases. Simulation results closely match the figures shown in Tables 6.1–6.4: the errors in all cases proved to be similar to statistical errors of simulation which never exceeded 3%.

Other calculations performed include cumulative distribution function and distribution density of excursion areas. Those data were computed using the formulae presented in Section 5.2. Computing excursion durations on the basis of the data on excursion areas enabled us to obtain the so-called generating sequence of bipolar rectangular pulses. From that sequence, one can easily obtain a family of sequences and calculate the related correlation functions, using software available on the market. The data computed using the proposed formulae are very close to simulation results even in the case of exponential tails.

7 Design Methodology of Adaptable Analyzers Used to Measure the Parameters of Excursions in Stationary Random Processes

7.1 Principal Features of Random Process Parameter Analyzers

The design of instruments to measure parameters of random processes presents has always been one of the most challenging tasks of instrumental design. Parameters of random processes have to be determined experimentally whenever theory does not provide a solution to the problem. Due to the lack of *a priori* information about the process passing through a typical radio engineering section it will be hardly possible to make a good mathematical description of a process without preliminary experimental investigation. We cannot rely on any theoretical conclusions about parameters of random processes unless those conclusions are confirmed by experiments.

Compared to deterministic signal measurements, the analysis of random processes is much more complicated. The *a priori* uncertainty of a number of parameters of the random signal at the input makes it difficult to work out a rational design of the random process analyzer. For instance, in the absence of information about the amplitude range and frequency range of the random process the gain and bandpass of the analyzer cannot be chosen appropriately; the same is true about choosing the right time integrator constants. Without information about the invariable component of the process it is difficult to set the threshold level, parameters of separation circuits, etc. Without knowing in advance the random process amplitude and frequency ranges, it is also impossible to optimize the setting of time and level sampling intervals in digital analyzers. When the distribution of values of the parameter measured is unknown, the chosen time sampling interval may turn out to be too short to allow the minimization of the total relative error of distribution function measurement for a given duration of the observation period.

It is known[53] that choosing the "wrong" duration for time sampling intervals increases either the statistical error of measurement (in the case where the sampling interval is too short) or the implementation error (in the case where the sampling

Applications of Random Process Excursion Analysis. DOI: http://dx.doi.org/10.1016/B978-0-12-409501-4.00007-3

interval is too long). The optimal length of the time sampling interval depends on the *a priori* unknown distribution of parameter values, the threshold level and the measurement session duration.

In the absence of information about the width of the random process power spectrum, such a critical parameter as the measurement session duration is remains completely uncertain. It is known that a process must be observed long enough for its statistical properties to show on the segment chosen for observation. Otherwise the data will be insufficient (i.e., the number of noncorrelated samples taken during the observation period will be too low), and the mean square error of statistical measurement will grow beyond the acceptable limit (usually this error increases in inverse proportion to statistical data volume). There is no need to explain that the same measurement session may prove too long or too short, depending on the width of the frequency passband. In the first case (i.e., in the case of a wide frequency band), the measurement process will be inefficient, whereas in the second case (i.e., in the case of a narrow frequency band) statistical error will increase.

Hence two approaches to the design of random process analyzers. The first one, which is based on the use of measurement tools with constant parameters (i.e., such parameters that can only be changed manually), suggests creating complicated multipurpose equipment. This equipment must comply with stringent requirements: it must be able to operate within wide amplitude and frequency ranges, have low level of interior noise, provide ample time for observation, have small nonlinear distortions, a large number of measurement channels, etc. In spite of its obvious irrationality, this approach was applied to the design of most random process analyzers known to the author.[36,52,53,59–61,70]

Attempts to compromise technical requirements to measuring equipment in order to make this equipment more simple lead to a loss of measurement accuracy and versatility of the equipment, and, in the long run, to an increase in the measurement session duration.

The other approach to the design of random process analyzers uses the principle of adaptability. In accordance with this principle, a small portion of the total period of time assigned for observation is spent on preliminary adaptation ("training") of the measuring instrument. In the process of adaptation the settings of the device are automatically adjusted to match the characteristics of a given stationary random process. Once the adaptation is complete, the analyzer will have the best settings in terms of ensuring a given accuracy of measurement during the entire time assigned for measuring the sought process characteristics. As in most cases the adaptation does not have to be very accurate, it does not take a long time in proportion to the total measurement session duration (the adaptation time accounts for no more than a few percentage points of the total measurement time). Adaptable analyzers contain sections with variable parameters which change their values automatically in the process of adaptation.

The most promising kind of adaptable analyzers of random processes are digital ones. They include pulse counters with variable frequency division ratios which can be set by entering codes, DAC and ADC, automatic switches for selecting amplitude range, threshold level, measurement session duration, sampling frequency, etc. All of the above parameters must be set to their optimal values (optimal in relation to the random process chosen) during the adaptation stage to make sure that all the critical measurements are carried out in the fixed parameter values mode.

In spite of their obvious advantages, adaptable analyzers of random processes have not yet started to be widely used. One possible reason for that is that the self-training mode is relatively difficult to implement, especially in the case where the old component base is used. However, a rapid development of microelectronics, the advent of special-purpose large integrated circuits such as RAM, ROM, programmed logical matrices, and microprocessors, and the possibility of using a PC to simulate adaptation processes open up good prospects for the design of digital adaptable analyzers of random processes. Connecting analyzers to a computer enables high-speed real-time processing of random signals, as well as display, accumulation and storage of measurement data.

In this chapter, we will discuss several approaches to the design of adaptable analyzers of stationary random process characteristics. The processes analyzed have to be stationary throughout the entire observation interval to ensure the consistency and reliability of measurement results. Real random processes usually remain stationary during limited intervals. The duration of such intervals depends on specific physical phenomena which caused a given random process.

For instance, the form of a shortwave signal depends on the condition of reflecting ionospheric layers. The intensity of solar activity, the altitude and thickness of ionization layer, dimensions of inhomogeneities in relation to wave length are subject to cyclical variations. Variation cycles may last a few minutes, hours or days; they may be seasonal, and one variation cycle matches the 11-year cycle of solar activity. A multipath shortwave signal reflected from ionosphere can be described as a random process caused by certain ionospheric phenomena. Characteristics of this process are of interest not only to experts in short-wave radio communication but also to those who study the ionosphere itself. Experiments show that minimum duration T_{st} of intervals during which this process remains stationary varies from 5 to 15 min. The interval of time correlation τ_0 for quick level oscillations in a fading shortwave signal is much shorter; its duration τ_0 varies from 0.1 to 10 s.

Due to the fact that $(T_{st}/\tau_0) \gg 1$ we can be sure of our ability to obtain a representative selection of samples within the interval where the signal is stationary, and reaches the acceptable accuracy of measurement for characteristics of a fading shortwave signal.

7.2 The Analyzer of Duration Values Distribution Density for Above-Threshold Excursions in Random Processes

Widely known digital analyzers of excursion duration distribution described in Ref. 53, in which the duration of above-threshold excursions is converted into a code, are, for the most part, characterized, by inappropriately chosen sampling frequency. All analyzers of this type use the same time sampling interval; its duration equals one counting pulse period and remains unchanged, regardless of the random process frequency range and threshold level. As we mentioned earlier, there exists an optimal time sampling interval length which minimizes the total relative error of measurement when we measure the probability density of excursion duration distribution.[53] This optimal length depends on the distribution of instant values in the random process, the threshold level, and variance in excursion durations. Using a count pulse generator with a relatively high constant frequency of pulses ($f_p \gg F_{max}$ where F_{max} is the highest spectrum frequency of processes analyzed) increases the number of channels in a multichannel analyzer and makes its structure too complicated.

On the other hand, attempts to simplify the measuring instrument structure by reducing the number of channels either narrow down the process frequency range, or, in the case where a constant frequency range is used, cause an increase in statistical and implementation error of measurement. Some analyzers such as the ones described in Ref. 61 cover a wider frequency range at the expense of a longer measurement session. The analyzers in which count pulse frequency f_p is relatively high and the number of channels is relatively small determine the probability density of excursion duration distribution in several stages, or cycles. In comparison with the one-cycle solution, the N-cycle approach enables an N-fold increase in the frequency range of processes being analyzed, but at the same time the measurement session becomes N times as long. However, due to the fact that for some random processes the time of being stationary is limited, using this approach does not always allow for real-time data processing.

In an ideal analyzer of the distribution of values of excursion duration τ, the length T of the measurement session must be set automatically in inverse proportion to the intensity $\lambda_1(x_0)$ of excursions above threshold level x_0, and time sampling interval $\Delta\tau$ must be adjusted automatically in direct proportion to the mean square $\sigma_\tau(x_0)$ of duration τ of random excursions above level x_0.

Meeting the first requirement, that is, $\lambda_1(x_0) * T = E = \text{const}$, where E is constant capacity of the device counting the number of excursions above the threshold level, enables accumulation of any amount of statistical data, regardless of the threshold level x_0.

As it was shown earlier in this book (see Section 4.3), relative mean square error ξ of measurement of total duration $\sum_{i=1}^{E} \tau_i$ of excursions above level x_0 can be determined from the following upper estimate:

$$\xi \leq \frac{1}{\sqrt{\lambda_1(x_0) * T}} = \frac{1}{\sqrt{E}}$$

If E is chosen within the range from 10^3 to 10^4, then statistical error of measuring the average duration of one excursion will not be higher than a few percentage points (because duration T of observation interval is limited). As we have assumed that our analyzer is a multichannel one, the interval for which we estimate density $W(\tau, x_0)$ of duration values for excursions above level x_0 is divided into N equal parts ("channels"); thus, the approximate number of excursions in one channel equals E/N. Hence, the upper estimate of statistical error of measurement, which arises from the fact that the observation time is limited:

$$\xi \leq \sqrt{\frac{N}{E}} \tag{7.1}$$

and the number of samples E must be multiplied by N.

Every two adjacent channels in the analyzer are separated from each other by a time interval of length $\Delta\tau$. As it follows from Ref. 53, the optimal value $(\Delta\tau)_{opt}$ is determined primarily by the mean square deviation $\sigma_\tau(x_0)$ in duration τ of random excursions above level x_0.

Given that the mean square deviation of a positive random value is usually comparable with its mean (e.g., for exponential distribution $\sigma_\tau = \tau_{av}$, for Rayleigh distribution $\sigma_\tau \approx 0.5\tau_{av}$), the average duration τ_{av} of an excursion is also an informative parameter. The values of σ_τ obtained earlier, either by calculation or by using PC simulations, for normal processes with zero-threshold level are $\sigma_{1\tau} \approx 0.76\tau_{av}$ and $\sigma_{2\tau} \approx 0.7\tau_{av}$ for processes with a Gaussian spectrum and a rectangular spectrum, respectively.

In any case, knowing the average duration $\tau_{av}(x_0)$ of excursions above a given threshold level makes it possible to determine accurately enough the boundaries of the range of values where the sought optimal value $(\Delta\tau)_{opt}$, which is proportional to $\tau_{av}(x_0)$, must belong.

Finding $\tau_{av}(x_0)$ experimentally is much easier than measuring $\sigma_\tau(x_0)$. As both parameters are equally informative, the author of this book proposed an adaptable analyzer of excursion duration distribution[62] which automatically maintains $(\Delta\tau)_{opt}$ in direct proportion to the average duration $\tau_{av}(x_0)$ of excursions. This feature enhanced by the capability of the analyzer to automatically set the length of the measurement period $T = E/\lambda_1(x_0)$ to any value which is allowed by excursions number counter with capacity E, makes it possible to expand the frequency range of processes being analyzed (it can be expanded by moving its lower boundary, without changing the upper one) and do so without any loss of accuracy or increase in the number of "channels."

Adaptability to changes in the intensity and average duration of excursions allows for a less complicated design of the analyzer, which can be made simpler by removing a few switches used for setting some of the parameters such as the frequency bandwidth (or observation time or the pulse generator frequency. The equipment required for the analysis will then take less space compared to a nonadaptable solution in which the same accuracy of measurements can only be achieved by increasing the number of channels.

In the process of adaptation, the analyzer determines the optimal number n_{opt} dividing the frequency f_{max} of the count pulse generator which sets the minimum duration $T_{min} = (1/f_{max})$ of the time sampling interval for the device which converts excursion duration values into a code. After dividing the frequency f_{max} into n_{opt}

the counting pulse period $T_{opt} = n_{opt} * T_{min}$ must have the same length as the optimal time sampling interval:

$$n_{opt} * T_{min} = (\Delta\tau)_{opt} \tag{7.2}$$

As it follows from Ref. 53, the limits of the range of values of $(\Delta\tau)_{opt}$ which have to be chosen to minimize the total error in the estimation of probability density for the distribution of excursion durations, are somewhat blurred. The optimum value $(\Delta\tau)_{opt}$ is found somewhere between 0.15 $\sigma_\tau(x_0)$ and 0.25 $\sigma_\tau(x_0)$; the choice depends on the number of samples E, the relative threshold level and the distribution of excursion duration values in excursions above that level. Considering that $\sigma_\tau(x_0) = (0.5/1) * \tau_{av}(x_0)$, we obtain $(\Delta\tau)_{opt} \approx (0.07/0.25) * \tau_{av}(x_0)$. From Eq. (7.2) it follows that:

$$n_{opt} * T_{min} \approx (0.07/0.25) * \tau_{av}(x_0) = k * \tau_{av}(x_0)$$

$$n_{opt} = k * f_{max} * \tau_{av}(x_0) \tag{7.3}$$

From Eq. (7.3) it follows that the problem of finding the optimal value of n_{opt} boils down to finding the average excursion duration $\tau_{av}(x_0)$ at the adaptation stage. Given that we do not know the range of excursions duration values until the completion of that stage, $\tau_{av}(x_0)$ must be measured for maximum frequency f_{max} of count pulse generator with the highest possible accuracy. The instrument that calculates $\tau_{av}(x_0)$ counts the total number Σ of pulses generated over the adaptation period of length T_a in which L excursions occur. In order to reduce the ratio between adaptation time and the total measurement time T, we must choose L much less than E ($L \ll E$). On the other hand, if we reduce adaptation time T_a too much, that will cause an increase in the statistical error of estimation of $\tau_{av}(x_0)$. As no clear boundaries are set for $(\Delta\tau)_{opt}$, high accuracy of adaptation is not required. For instance, if $L = 100$, then the mean square error in estimation of $\tau_{av}(x_0)$ will be in the order of 10%, which is acceptable. Over time T_a the counter measuring the average duration of excursions will "accumulate" Σ pulses of frequency f_{max}:

$$\Sigma = f_{max} * \sum_{i=1}^{L} \tau_i = f_{max} * L * \tau_{av}(x_0) \tag{7.4}$$

After substituting the expression for $\tau_{av}(x_0)$ derived from Eq. (7.3) into formula (7.4), we obtain

$$\Sigma = \frac{L}{k} * n_{opt} \tag{7.5}$$

As we can see from Eq. (7.5), the counter of the average duration of excursions must be implemented as a serial connection of two counters of which the first one (the scaling counter) counts (L/k) pulses and the second one provides the optimal code n_{opt} which is useful information. Considering that coefficient k in Eq. (7.3) varies between 0.07 and 0.25 (approximately), it would be appropriate to use as a base value the average value $k_{av} = 1/6$. Thus, from Eq. (7.5) we obtain:

$$\Sigma = (6L) * n_{opt} \tag{7.6}$$

During the measurement stage, after accumulating certain information about the relation between values $\sigma_\tau(x_0)$ and $\tau_{av}(x_0)$, it is possible to estimate more accurately the value of coefficient k_{av} and adjust scaling ratio (L/k_{av}) by feeding information back to the first pulse counter.

Thus, by the time L excursions are covered, the adaptation stage will be completed and random code n_{opt} will be found, which is proportional to the average duration of above-threshold excursions. This code sets the optimal frequency division ratio for the count pulse generator: from that moment on through the entire time of measurement the optimal frequency of count pulses is determined by the relation:

$$f_{opt} = \frac{f_{max}}{n_{opt}}$$

The time sampling interval length $T_{opt} = 1/f_{opt} = k_{av} * \tau_{av}(x_0)$ is proportional to an average excursion duration. It means that, one excursion covers approximately $(1/k_{av})$ periods of frequency f_{opt}.

All we have to do now is choose appropriately the number of channels in the analyzer. To a large extent, this number depends on the time sampling interval duration $(\Delta\tau)_{opt}$. Thus, considering that the maximum duration τ_{max} of an excursion usually varies between $3\sigma_\tau(x_0) \approx 2\tau_{av}(x_0)$ and $4\sigma_\tau(x_0) \approx 3\tau_{av}(x_0)$, for $k_{av} = 1/6$ the optimal number of channels in the analyzer equals $N_{opt} = \tau_{max}/(\Delta\tau)_{opt} = (12/18)$.

Choosing the number of channels greater than N_{opt} makes the analyzer too complicated; in addition to that, according to formula (7.1), it either increases statistical error, or lengthens the measurement time period. Choosing the number of channels smaller than the optimal value is not good either, as it leads to a loss in measurement accuracy due to the shift in probability density estimate.

For digital analyzers we recommend using $N_{opt} = 16$; in this case the maximum duration of an excursion will be

$$\tau_{max} = k_{av} * N_{opt} * \tau_{av}(x_0) \approx 2.67 * \tau_{av}(x_0)$$

that is, it will vary approximately between $2.67*\sigma_\tau(x_0)$ and $5.33*\sigma_\tau(x_0)$.

When the time sampling interval duration is chosen correctly, the number of ultra-long excursions (i.e., such that $\tau > \tau_{max}$) must be relatively small (no more than a few percentage points) compared to the total number E of excursions which occur during the measurement session. Ultra-long excursions represent the "tail" of distribution for probability density $W(\tau, x_0)$. Their number can be found experimentally in the end of the measurement session as the difference between the total number E of excursions registered and the total number of excursions which occurred in all 16 channels of the analyzer.

For exponentially distributed excursion duration values the probability that $\tau > 2.67 * \tau_{av}(x_0)$ is less than 7%, whereas for Rayleigh distribution it constitutes only 0.38%. According to the calculation and simulation results presented in Chapter 3 for the density $W(\tau, 0)$ of excursion duration values in a normal process with a Gaussian spectrum, the probability that $\tau > 2.67 * \tau_{av}(0)$ is less than 5%. For density $W_{+}(\tau, 0)$ of length distribution in intervals between zero level upcrossings the probability that $\tau > 2.67 * \tau_{+av}(0)$ equals 2%. Thus, in most cases the probability that excursion duration values will be found in the area of the truncated distribution tail ($\tau > \tau_{max}$) is comparable with the probability of their occurrence in the remaining $N_{opt} = 16$ time sampling intervals. According to the upper estimate (7.1), the choice of $N_{opt} = 16$ implies that the sampling volume E (the number of excursions during the measurement session) should be within the range from 10^4 to 10^5. In this case statistical error of measurement will be, on average, within the range from 1.25 to 4%. For $L \approx 10^3$ the adaptation time ratio L/E will range from 0.01 to 0.1, which means that it will not be high and will hardly affect the total duration of the measurement session.

The frequency range of random processes under investigation is limited from below by the maximum duration of the measurements period achievable for the highest relative threshold levels x_0/σ without a loss in accuracy (here σ is standard deviation of instant values of the process). The only factor that limits the frequency range from above is the speed of the analyzer components.

Let us now use as an example broadband Gaussian random processes with Gaussian and rectangular power spectra to show how we estimate the frequency range and measurement session duration for any particular type of random process. When the autocorrelation function $R(\tau)$ is known, we can take advantage of the fact that the effective power spectrum width ΔF_{eff} of the process is related to the time correlation interval length τ_k, and easily establish the relation between intensity $\lambda_1(0)$ of excursions above zero level and ΔF_{eff}.

Thus, for the process with a Gaussian spectrum $\lambda_1(0) = \sqrt{(2/\pi)} * \Delta F_{eff.}$, and for the process with rectangular spectrum $\lambda_1(0) = \Delta F_{eff.}/\sqrt{3}$.

The analyzer sets the measurement time T (measurement session duration) automatically; measurements are carried out until the moment when the device that counts the number of excursions above level x_0/σ reaches its capacity E.

For a Gaussian process:

$$T = \frac{E}{\lambda_1(x_0)} = \frac{E}{\lambda_1(0)} * \exp\left(\frac{x_0^2}{2\sigma^2}\right) = \frac{E}{\gamma * \Delta F_{eff.}} * \exp\left(\frac{x_0^2}{2\sigma^2}\right) \tag{7.7}$$

where $\gamma_1 = \sqrt{2/\pi} \approx 0.8$ for a Gaussian spectrum, and $\gamma_2 = 1/\sqrt{3} \approx 0.577$ for a rectangular one.

As we can see from Eq. (7.7), the length of the measurement period is set in inverse proportion to the effective width of the process power spectrum. For a given power spectrum width ΔF_{eff}, the measurement time is the greater, the higher the standardized threshold level x_0/σ is. As long as the effective width of the power spectrum is kept the same throughout the measurement session, the spectrum shape does not have any noticeable effect on the measurement time T.

Thus, a transition from the Gaussian spectrum to the rectangular one results in a $\sqrt{6/\pi} \approx 1.38$-fold increase in the measurement session duration. The average excursion duration $\tau_{\text{av}}(x_0)$ can be found from the standardized:

$$\tau_{\text{av}}(x_0) = \frac{F(x_0/\sigma)}{\lambda_1(x_0)} = \frac{F(x_0/\sigma)}{\gamma * \Delta F_{\text{eff.}}} * \exp\left(\frac{x_0^2}{2\sigma^2}\right) \tag{7.8}$$

where $F(x_0/\sigma) = 1/\sqrt{2\pi} * \int_{x_0/\sigma}^{\infty} \exp(-y^2/2)dy$ is the probability that the Gaussian process will go above the relative level x_0/σ.

Calculations made using formula (7.8) show that we can keep the error under 5% over the entire range of x_0/σ by using a simple approximation:

$$\tau_{\text{av}}\left(\frac{x_0}{\sigma}\right) \approx \frac{1}{2\gamma * \Delta F_{\text{eff.}}(1 + (x_0/\sigma))} \tag{7.9}$$

It follows from Eqs. (7.8) and (7.9) that in a broadband process the average above-threshold excursion duration is in inverse proportion to the power spectrum width, regardless of the threshold level. This conclusion becomes obvious when you look at the physics of the process.

A lack of interdependence between $\tau_{\text{av}}(x_0/\sigma)$ and the standardized threshold level x_0/σ is less obvious. With an increase in the threshold level the average excursion duration decreases more slowly than you would expect. With an increase in x_0/σ two opposite factors come into play: on the one hand, the total amount of time during which the given threshold level is exceeded decreases; on the other hand, the number of excursions occurring per unit of time decreases too. The first factor dominates somewhat over the second one; as a result, although $\tau_{\text{av}}(x_0)$ decreases with an increase in x_0/σ, the decrease is relatively slow.

Excursion parameters related to duration are, therefore, usually determined by the process power spectrum width or by the intensity of above-zero excursions in the process, which is essentially the same thing.

Remember that we are dealing with broadband processes whose spectrum width ranges from 0 to $\Delta F_{\text{eff.}}$. In narrowband processes the mean frequency of the spectrum is much higher than the upper boundary of this range ($f_0 \gg \Delta F_{\text{eff.}}$); therefore, the average excursion duration is primarily determined by f_0 being almost independent from $\Delta F_{\text{eff.}}$.

The measurement session duration expected and the average excursion duration can be easily calculated as functions of x_0/σ using formulae (7.7)–(7.9). Those calculations make it possible to estimate the analyzer frequency range, decide what speed is required, and find the time sampling interval duration for a given level x_0/σ.

The results of calculations made for two types of Gaussian processes with zero mean are shown in Table 7.1.

From analyzing figures presented in Table 7.1 we can see, for instance, that in the case of a process with the minimum width $\Delta F_{\text{eff.}} = 100$ of the spectrum the measurements on the zero-threshold level do not take longer than 3 min; in the case where $x_0/\sigma = 1$ they take less than 5 min, which is still a relatively short period of time; however, this time gets considerably longer as x_0/σ increases; thus, the measurements may take 15–20 min when $x_0/\sigma = 2$, and 3–4 h when $x_0/\sigma = 3$.

For high threshold levels ($x_0/\sigma \geq 2$) too long measurement periods are often unacceptable because real processes only remain stationary within certain time limits. The problem can be solved by reducing the number of samples E, which results in less strict requirements to the accuracy of measurements. Another solution is to narrow the frequency range of the process under consideration by increasing the minimum width $\Delta F_{\text{eff.}}$ of its spectrum.

When the threshold level is high, in some cases it is enough to measure the average duration of excursions at the adaptation stage, as it will reduce the measurement time requirements by order of 10–100. On the basis of the information presented in Table 7.1, we can estimate the frequency range width and speed requirements for the analyzer.

Considering that $\tau_{\text{av}}(3\sigma) = 0.157/\Delta F_{\text{eff.}}$ (a Gaussian spectrum) we obtain from formula (7.3) for $n_{\text{opt}} = 1$ and $k = 1/6$:

$$f_{\max} = \frac{(\Delta F_{\text{eff.}})_{\max}}{k * 0.157} \approx 40 * (\Delta F_{\text{eff.}})_{\max}$$

It means, for instance, that in the case where $(\Delta F_{\text{eff.}})_{\max} = 1$ MHz the counting pulse generator frequency $f_{\max}$ roughly equals 40 MHz and high-speed digital chips with TTL structure must be included in the analyzer scheme. Reducing $(\Delta F_{\text{eff.}})_{\max}$ by order of magnitude allows for the use of economical CMOS-chips in the analyzer design.

The minimum duration of the time sampling interval which coincides with the period $T_{\min}$ of counting pulse generator output will be set automatically for the highest values of threshold level x_0/σ in a process with the largest bandwidth:

$$(\Delta\tau)_{\min} \approx \frac{1}{40 * (\Delta F_{\text{eff.}})_{\max}}$$

Table 7.1 The Measurement Session Duration and the Average Excursion Duration

Gaussian spectrum $E = 10^4$	$\frac{x_0}{\sigma}$	0	1	2	3
	$\Delta F_{\text{eff}} * T * 10^{-4}$	1.253	2.066	9.256	112.8
	$\tau_{\text{av}}(x_0) * \Delta F_{\text{eff}}$	0.627	0.313	0.209	0.157
Uniform spectrum $E = 10^4$	$\frac{x_0}{\sigma}$	0	1	2	3
	$\Delta F_{\text{eff}} * T * 10^{-4}$	1.732	2.856	12.76	155.5
	$\tau_{\text{av}}(x_0) * \Delta F_{\text{eff}}$	0.866	0.433	0.289	0.216

The study of Gaussian processes shows that the frequency range of an adaptable analyzer[62] designed for measuring distribution density of excursion duration values in excursions above a given level can extend from 10 Hz to hundreds of kilohertz, covering three to four decades. In that case the division ratio n_{opt} of a code-controlled frequency divider varies from 1 to 10^3 and in some cases may reach 10^4. The total time taken by measurements is automatically set to a value which is derived from the frequency range of the process and a given threshold level. It varies from a few hundredths of a second to several hours. The total relative error of measurement which includes the deviation of density estimate and the mean square error for $E = 10^4$ and $N = 16$ does not exceed 5%, which is acceptable in a random process analysis.

We can always increase the accuracy of measurement to a given value by increasing the counter capacity E and the number N of channels in the analyzer, but then the time T required for taking measurements will also increase.

In conclusion, we would like to point out that the need to analyze the distribution of excursion duration values in random processes arises quite often in radio engineering, medicine, meteorology, construction, and many other fields. Such analysis may be necessary, for example, in speech recognition, cardiography, and encephalography, study of atmospheric phenomena, the measurement of excessive stress in building structures etc.

The data in Ref. 52, for instance, represent the distribution of excursion durations obtained experimentally for above-threshold excursions of a broadcast signal envelope. In the absence of a special device which analyzes the distribution of excursion duration values, the experimenters used a device which converts pulse duration into amplitude, in combination with a multichannel amplitude analyzer of AI-256 series made in Russia. The same measurement period duration was chosen for all threshold levels; as a result, the number of samples E varied from 5000 for low levels to 100 for high levels. Hence an increase in the error of measurements of probability density of duration of excursions above a given level of a broadcast signal envelope for high levels compared to low level.

The experiment confirmed that long excursions are distributed exponentially and that the most probable duration has little, if any, correlation with the threshold level.

7.3 Adaptable Analyzer of Interval Length Distribution for Intervals During Which a Random Signal Remains Within or Goes Beyond Given Boundaries

The problem of finding the distribution of durations for random signal excursions above or below a given threshold level can be extended to that of finding interval length distribution for intervals during which a random signal remains within or goes beyond given boundaries. Those boundaries do not have to be fixed; they may also be expressed by functions of time. This kind of a problem arises in many fields of technology such as radar location, telecommunication, medicine, meteorology, and flaw detection.

Consider a stationary random process $X(t)$, which crosses fixed levels x_1 and x_2 ($x_2 > x_1$) at random time moments. Let $\lambda_1(x_1)$ and $\lambda_2(x_2)$ be the intensities of the process excursions above levels x_1 and x_2, respectively. The total average number N_{av} of intervals during which the signal remains within given boundaries x_1 and x_2 over time T can be obtained as the total intensity of excursions:

$$N_{av} = [\lambda_1(x_1) + \lambda_2(x_2)] * T$$

The mean time $T_{av\ inside}$ during which the signal remains within given boundaries over time T is:

$$T_{av\ inside} = P(x_2 \geq x \geq x_1) * T$$

where $P(x_2 \geq x \geq x_1) = \int_{x_1}^{x_2} W(x)dx$ is the probability that instant values of process $X(t)$ will occur between the two boundaries, and $W(x)$ is the density of instant values distribution.

Similarly, the average time during which the process remains outside the given boundaries over time T is determined as follows:

$$T_{av\ outside} = T - T_{av\ inside} = [1 - P(x_2 \geq x \geq x_1)] * T$$

From this formula we can derive the expressions for the average length of intervals during which the process remains inside and outside the given boundaries:

$$\tau_{av\ inside} = \frac{T_{av\ inside}}{N_{av}} = \frac{P(x_2 \geq x \geq x_1)}{\lambda_1(x_1) + \lambda_1(x_2)} \tag{7.10}$$

$$\tau_{av\ outside} = \frac{T_{av\ outside}}{N_{av}} = \frac{1 - P(x_2 \geq x \geq x_1)}{\lambda_1(x_1) + \lambda_1(x_2)} \tag{7.11}$$

As we can see from Eqs. (7.10) and (7.11), $\tau_{av\ inside}$ and $\tau_{av\ outside}$ depend on the distribution of instant values, the standardized threshold levels x_1/σ andx_2/σ, the width $(x_2 - x_1)/\sigma$ of the area where measurements are taken, and the intensity of excursions above given levels (the width of the process power spectrum). In the general case, $\tau_{av\ inside}$ and $\tau_{av\ outside}$ can differ from each other; therefore, we can expect the distribution densities $W_{inside}(\tau, x_1, x_2)$ and $W_{outside}(\tau, x_1, x_2)$ to be different as well. As a result, we must choose different time sampling interval lengths for experimental measurements of the two densities. By analogy with the results obtained in Section 7.2, the optimal time sampling interval duration $(\Delta\tau)_{opt}$ which is established after the adaptation of the analyzer, must be proportional to the average length of the measurement interval. In other words, we should use the following formulae:

$$(\Delta\tau)_{inside} = k * \tau_{av\ inside}; \quad (\Delta\tau)_{outside} = k * \tau_{av\ outside}$$

We, therefore, recommend that the analyzer should use two parallel channels to estimate, during adaptation to a random process, the average length of intervals during which the input signal level stays inside the given boundaries (hereinafter called "confinement intervals") as well as the length of intervals during which the signal "spills over" the boundaries ("spillover intervals"). On the basis of those estimates the analyzer should set the optimal time sampling interval length. While taking measurements to determine the distribution functions, two channels will be used to accumulate information: one channel will be used to record the length of "confinement" intervals, whereas the other one will be used to accumulate the data on the length of "spillover" intervals. The measurements in the two channels should stop simultaneously the moment when the total count of "confinement" or "spillover" intervals corresponding to the given boundaries reaches a given number.

The kind of adaptable analyzer being discussed in this section is, in fact, a combination of two independent devices which analyze the distribution of excursion duration values as it was described in Section 7.2. The only simplification of the design of the two devices (channels) will consist in the fact that they will share some units such as the counting pulse generator, units which set the length of adaptation and measurement periods, and some other.[63]

To obtain a fair estimate of the length of adaptation and measurement periods, the analyzer's frequency range and speed requirements, let us consider the cases where Rayleigh and Gaussian stationary processes were analyzed, both with a Gaussian power spectrum.

For a Rayleigh process, according to Ref. 8, we have:

$$\lambda_1\left(\frac{x}{\sigma}\right) = 2 * (\Delta F_{\text{eff.}}) * \left(\frac{x}{\sigma}\right) * \exp\left(-\frac{x^2}{2\sigma^2}\right) \tag{7.12}$$

where σ is the standard deviation of the process, and $(\Delta F_{\text{eff.}})$ is the effective width of the power spectrum.

Hence, the formula for the length of the measurement period, that is, the time during which the total number of intervals characterized by the process remaining within given boundaries:

$$T = \frac{E}{2 * \Delta F_{\text{eff.}} * [(x_1/\sigma) * \exp(-(x_1^2/2\sigma^2)) + (x_2/\sigma) * \exp(-(x_2^2/2\sigma^2))]} \tag{7.13}$$

For a Rayleigh process, the probability of the process remaining within given standardized boundaries $(x_1/\sigma, x_2/\sigma)$ equals:

$$P(x_2 \geq x \geq x_1) = \exp\left(-\frac{x_1^2}{2\sigma^2}\right) - \exp\left(-\frac{x_2^2}{2\sigma^2}\right) \tag{7.14}$$

Considering Eqs. (7.12) and (7.14), formulae (7.10) and (7.11) can be rewritten as follows:

$$\tau_{\text{av inside}} = \frac{\exp(-x_1^2/2\sigma^2) - \exp(-x_2^2/2\sigma^2)}{2 * \Delta F_{\text{eff.}} * [(x_1/\sigma)\exp(-x_1^2/2\sigma^2) + (x_2/\sigma)\exp(-x_2^2/2\sigma^2)]} \qquad (7.15)$$

$$\tau_{\text{av outside}} = \frac{1 - \exp(-x_1^2/2\sigma^2) + \exp(-x_2^2/2\sigma^2)}{2 * \Delta F_{\text{eff.}} * [(x_1/\sigma)\exp(-x_1^2/2\sigma^2) + (x_2/\sigma)\exp(-x_2^2/2\sigma^2)]} \qquad (7.16)$$

In the case where $x_2/\sigma \to \infty$ formulae (7.15) and (7.16) express the average duration of excursions above level x_1/σ (interval length $\tau_{\text{av inside}}$) and below this level (interval length $\tau_{\text{av outside}}$), respectively.

At the level $x_1/\sigma = \sqrt{2\ln 2} \approx 1.177$ excursions and pauses have the same average duration. At lower levels an average excursion lasts longer than an average pause, whereas at higher levels the situation is the opposite.

It follows from formula (7.15) that the average interval length $\tau_{\text{av inside}}$ (for $\Delta F_{\text{eff.}} = \text{const}$) is usually determined by the width $(x_2 - x_1)/\sigma$ of the target area (i. e., the area under analysis), and has little or no relation to its location. Widening the target area causes an increase in the time during which the signal stays within the area boundaries: this time increases almost in direct proportion to the area width. Shifting the target area upward causes a slight decrease in $\tau_{\text{av inside}}$.

The relation between the time during which the signal "spills over" the given boundaries, on the one hand, and the width and location of the target area, on the other, appears to be more complicated. It follows from formula (2.16) that $\tau_{\text{av outside}}$ increases when we try to narrow the target area or shift it upward or downward in relation to the zone where $x_1/\sigma = \sqrt{2\ln 2}$ (i.e., the zone characterized by the highest intensity of excursions and the least value of $\tau_{\text{av outside}}$).

The results of calculations made using formulae (7.13)–(7.16) for two widths of the target area are shown in Table 7.2.

Similar calculations were made for a normal random process with a Gaussian spectrum.

Table 7.2 The Rayleigh Process, Two Widths of the Target Area with a Gaussian Spectrum

$\frac{x_2 - x_1}{\sigma} = 0.5$	$\left(\frac{x_1}{\sigma}, \frac{x_2}{\sigma}\right)$	(0;0.5)	(0.5;1)	(1;1.5)	(1.5;2)	(2;2.5)	(2,5.3)	(3;3.5)	(3,5.4)
$E = 10^4$	$(\Delta F_{\text{eff.}}) * T * 10^{-4}$	1.11	0.48	0.46	0.66	1.31	3.49	12.2	55.6
	$(\Delta F_{\text{eff.}}) * \tau_{\text{av inside}}$	0.13	0.13	0.13	0.12	0.12	0.11	0.11	0.10
	$(\Delta F_{\text{eff.}}) * \tau_{\text{av outside}}$	0.98	0.35	0.33	0.54	1.19	3.38	12.1	55.5
$\frac{x_2 - x_1}{\sigma} = 1$	$\left(\frac{x_1}{\sigma}, \frac{x_2}{\sigma}\right)$	(0;1)		(1;2)		(2;3)		(3;4)	
$E = 10^4$	$(\Delta F_{\text{eff.}}) * T * 10^{-4}$	0.824		0.57		1.645		14.423	
	$(\Delta F_{\text{eff.}}) * \tau_{\text{av inside}}$	0.324		0.2686		0.205		0.1554	
	$(\Delta F_{\text{eff.}}) * \tau_{\text{av outside}}$	0.5		0.3014		1.44		14.268	

In this case $\lambda_1(x) = \lambda_1(0) * \exp(-x^2/2\sigma^2)$, $\lambda_1(0) = \sqrt{2/\pi * \Delta F_{\text{eff.}}}$ and from Eq. (7.13) we obtain:

$$T = \frac{E}{\sqrt{2/\pi} * \Delta F_{\text{eff.}} * [\exp(-x_1^2/2\sigma^2) + \exp(-x_2^2/2\sigma^2)]} \tag{7.17}$$

By substituting into Eqs. (7.10) and (7.11) the values for $\lambda_1(x)$ and the probability $P(x_2 \geq x \geq x_1) = F(x_2/\sigma) - F(x_1/\sigma)$ taken from Gaussian function probability integral tables, we obtained the results shown in Table 7.3.

If we compare the data in Tables 7.2 and 7.3, we will notice that, in spite of the difference in the distribution of their instant values, the two processes have similar patterns according to which the length T of the measurement period and the average length of "confinement" or "spillover" intervals change. The lengths of adaptation and measurement time periods are set in inverse proportion to the effective frequency bandwidth of the process: these periods are the longer, the higher the required accuracy of measurement (i.e., the number of samples E) is, and, for a given width of the target area, the higher this area is located.

At the same time, the closer the target area is located to the zone of the most probable instant values (in relation to the mean), the shorter the measurement period and the average "spillover" interval will be.

The average length of "confinement" intervals is almost in direct proportion to the width of the target area. This conclusion is evident. Less evident is the fact that there is little correlation between the average length of "confinement" intervals and the relative height of the target area. The reasons for that are the same as the reasons for the absence of strong correlation between the average duration of above-threshold excursions and the threshold level, which were discussed in the previous section.

From Tables 7.2 and 7.3 we can draw conclusions about the frequency range and response time that the analyzer must have if we want to be sure that the adaptation and measurement periods fall within acceptable time limits. The minimum duration $(\Delta\tau)_{\min}$ of time sampling intervals, which coincides with the length of period of counting pulses generated with frequency $f_{\max}$, can be derived from the least mean length $(\tau_{\text{av inside}})_{\min}$ of an interval during which the process stays within the boundaries of the relatively highest target area.

Table 7.3 The Gaussian Process, Two Widths of the Target Area with a Gaussian Spectrum

<table>
<tr><td>$\frac{x_2 - x_1}{\sigma} = 0.5$</td><td>$\left(\frac{x_1}{\sigma}, \frac{x_2}{\sigma}\right)$</td><td>(0;0.5)</td><td>(0.5;1)</td><td>(1;1.5)</td><td>(1.5;2)</td><td>(2;2.5)</td><td>(2.5;3)</td><td>(3;3.5)</td><td>(3.5;4)</td></tr>
<tr><td>E = 10^4</td><td>$(\Delta F_{\text{eff.}}) * T * 10^{-4}$</td><td>0.67</td><td>0.85</td><td>1.34</td><td>2.73</td><td>6.99</td><td>22.8</td><td>94.3</td><td>497</td></tr>
<tr><td></td><td>$(\Delta F_{\text{eff.}}) * \tau_{\text{av inside}}$</td><td>0.13</td><td>0.13</td><td>0.12</td><td>0.12</td><td>0.12</td><td>0.11</td><td>0.11</td><td>0.1</td></tr>
<tr><td></td><td>$(\Delta F_{\text{eff.}}) * \tau_{\text{av outside}}$</td><td>0.54</td><td>0.72</td><td>1.22</td><td>2.61</td><td>6.87</td><td>22.7</td><td>94.2</td><td>497</td></tr>
<tr><td>$\frac{x_2 - x_1}{\sigma} = 1$</td><td>$\left(\frac{x_1}{\sigma}, \frac{x_2}{\sigma}\right)$</td><td colspan="2">(0;1)</td><td colspan="2">(1;2)</td><td colspan="2">(2;3)</td><td colspan="2">(3;4)</td></tr>
<tr><td>E = 10^4</td><td>$(\Delta F_{\text{eff.}}) * T * 10^{-4}$</td><td colspan="2">0.7801</td><td colspan="2">1.6813</td><td colspan="2">8.5581</td><td colspan="2">109.515</td></tr>
<tr><td></td><td>$(\Delta F_{\text{eff.}}) * \tau_{\text{av inside}}$</td><td colspan="2">0.2663</td><td colspan="2">0.2256</td><td colspan="2">0.1831</td><td colspan="2">0.165</td></tr>
<tr><td></td><td>$(\Delta F_{\text{eff.}}) * \tau_{\text{av outside}}$</td><td colspan="2">0.5138</td><td colspan="2">1.4557</td><td colspan="2">8.375</td><td colspan="2">109.35</td></tr>
</table>

It follows from Tables 7.2 and 7.3, that $(\tau_{\text{av inside}})_{\min} \approx (0.1/(\Delta F_{\text{eff.}})_{\max})$. Considering the results discussed earlier, from this we obtain:

$$(\Delta\tau)_{\min} = k * (\tau_{\text{av inside}})_{\min} = \frac{1}{6} * (\tau_{\text{av inside}})_{\min} \approx \frac{0.1}{6 * (\Delta F_{\text{eff.}})_{\max}}$$

Hence we can derive the speed that the analyzer is required to have, which does not correlate with the distribution of the instant values of the random process:

$$f_{\max} \approx 60 * (\Delta F_{\text{eff.}})_{\max}$$

If, for example, $(\Delta F_{\text{eff.}})_{\max} = 100$ kHz, then $f_{\max} \approx 6$ MHz, that is, at this point the speed does not have to be high.

The lower part of the frequency range $(\Delta F_{\text{eff.}})_{\min}$ is limited by the maximum duration allowed for the measurement period. Thus, according to Table 7.2, in the case of Rayleigh distribution with σ ranging from 3 to 3.5 for $(\Delta F_{\text{eff.}})_{\min} = 100$ Hz and $E = 10^4$ the measurement period lasts about 20.3 min. Downshifting the range of σ to (2.5; 3) for the same conditions results in a decrease in the length of the measurement period to the value of $T = 5.82$ min. Expanding the range of σ by setting its lower boundary at 2 and upper one at 3 will give us $T = 2.75$ min.

From Table 7.3 it follows that in the case of a Gaussian process for $(\Delta F_{\text{eff.}})_{\min} = 100$ Hz and $E = 10^4$ the probability that instant values will occur in the area where σ ranges from 2.5 to 3 is quite small; therefore, in this case, the length of the measurement period reaches its maximum $T_{\max} \approx 38$ min. Downshifting the range of σ to (2; 2.5) for the same conditions results in a decrease in the length of the measurement period to the value of about 11.65 min. Expanding the range of σ by setting its lower boundary at 2 and upper one at 3 also decreases the length of the measurement period compared to its maximum value $T_{\max}$: for σ ranging from 2 to 3 we have $T \approx 14.26$ min.

If, in the experimenters' opinion, the measurement period lasts too long, it can be shortened either at the cost of lowering measurement accuracy (the number E of samples), or at the cost of narrowing the frequency range (by increasing the lower frequency band limit $(\Delta F_{\text{eff.}})_{\min}$ for wide-band processes under investigation). As far as the analyzer adaptation time T_a is concerned, it should be chosen in advance to make a certain percentage of the length T of the measurement period (T_a is chosen in the range from $0.01T$ to $0.1T$, and follows the same pattern as T).

Thus, the analyzer can be designed to cover the frequency range confined between 100 Hz and 100 kHz as long as the speed requirements are not too strict. The length of the measurement period is set automatically; it varies from a few hundredths of a second for processes with the widest frequency range for most probable threshold levels to dozens of minutes for narrowband processes and high standardized threshold levels.

For $E = 10^4$ and the number of analyzer channels $N = 16$, neither statistical, nor system errors of measurement will exceed 5%. The error can be reduced to any given value by increasing E and N; however, the length of the measurement period will also increase.

7.4 The Adaptable Random Signal Amplitude Analyzer

The concept of excursion analyzers described in previous sections suggests knowing in advance the range of random signal amplitude instant values. In those analyzers the optimal values of threshold level ratios x_0/σ and target area boundaries $(x_1/\sigma, x_2/\sigma)$ are derived from the mean and variance of instant values. In real processes, however, this kind of information is either unavailable or incomplete. As a result, choosing amplitude ranges for analysis takes too much extra time. We cannot put up with this situation, considering that real processes only remain stationary during limited time periods and do not last long.

Therefore, we have to ensure that, at the stage of "training," the measuring instrument adapts to both amplitude and frequency parameters of the random process. In a situation where we are not certain about some key parameters of the process which we would like to know in advance, this kind of adaptability allows for effective automatic selection of time and amplitude sampling intervals. In addition to that, we can select the duration of "training" and measurement stages so as to ensure a given measurement error rate.

It is quite obvious that whatever parameter of a random process is being measured, we must know both its amplitude and frequency range. Just like we need to know the signal amplitude range to be able to analyze above-threshold excursion durations, we need to know the frequency properties of the signal to be able to analyze its amplitude distribution: knowing frequency properties is essential for the proper selection of time sampling frequency.

Some amplitude analyzers[52,53] operate at a preset sampling frequency f_s. In our opinion, this solution is rather ineffective. In the case of narrowband processes the absence of frequency adaptability results in close correlation between the neighboring samples of a signal; therefore, the samples do not contain any new information about the process. Having to analyze too many sample sets only makes the equipment structure too complicated and does not give any advantage in terms of measurement accuracy. On the other hand, in the case of broadband processes the value chosen for sampling frequency f_s may prove too low. When samples are not taken frequently enough, essential information on the process behavior in intervals between samples may be lost. Such loss of information can only be prevented by increasing the number of samples and the total time of measurement beyond optimal values.

Some authors (e.g., Refs. 59–61) support the use of noncorrelated (or weakly correlated) samples for measuring the distribution of instant values of a random process. The absence of correlation between neighboring samples can be ensured by the use of a sampling period whose length can be expressed by the formula $T_s = 1/f_s \approx \tau_k$, where τ_k is the length of the time correlation interval of the process under investigation. Since $\tau_k = 1/4\Delta F_{\text{eff.}}$ is in inverse proportion to effective width of the process power spectrum, the optimal frequency f_s must be directly proportional to $\Delta F_{\text{eff.}}$.

As it was shown earlier, the bandwidth $\Delta F_{\text{eff.}}$ can be expressed through the intensity $\lambda_1(x_0/\sigma)$ of signal excursions above the standardized threshold level (x_0/σ). The relation between $\Delta F_{\text{eff.}}$ and $\lambda_1(x_0/\sigma)$, however, is mainly determined by the distribution of instant values, which is *a priori* unknown and therefore should be measured. Apart from the distribution law for instant values we have to find out whether or not the process under investigation has a constant component in it, and if it does, we have to determine the value of such component and the effective value of the signal.

Therefore, we believe that the information on the number of excursions above a certain level (zero or any other) which occur during the adaptation time is, in the general case, not relevant to researchers who investigate frequency properties of a process.

During adaptation time it would be appropriate to measure the average parameters of the random process extrema, namely the average length of intervals between neighboring extrema (i.e., zeros of the first derivative of the process), the average semirange of the signal, that is, the average half-difference between its maxima and minima, and also the mean value (constant component) of the process which is obtained as the average half-sum of the maximum and the minimum values among the signal samples. The first derivative of any process is symmetric about the zero level because the operation of differentiation turns the constant component of any process into zero. Therefore, at the adaptation stage the intensity of the process extrema can be measured as the intensity of zeros of the process first derivative.[49,72–74] The intensity of extrema in a random process is known to be in direct proportion to the effective width of the process power spectrum[8] and, as a result, to optimal time sampling frequency of the process.

The average semirange of the signal can serve as an estimate of its mean square value, whereas the average half-sum of extreme samples can be considered as an estimate of its constant component. In other words, by measuring the average values of the process extrema at the stage of adaptation, it is possible to estimate the ranges for both amplitude and frequency. It means that, even though we do not know in advance the amplitude and frequency characteristics of a random signal, we can choose, on the basis of adaptation results, the best values for the analyzer circuit parameters to be used at the stage of measuring the distribution densities of amplitudes, excursions durations, "confinement" and "spillover" intervals, areas of excursions above a given level, etc.

According to Ref. 8, in a centered Gaussian random process the average number $(\lambda_{\max})_{\text{av}}$ of maximums per unit of time is $\beta = (\sqrt{R^{(4)}(0)/\omega_1^2})$ times higher than the average intensity $\lambda_1(0)$ of excursions above the zero level.

Here $R^{(4)}(0)$ is the fourth derivative of the process autocorrelation function for $\tau = 0$, $\omega_1^2 = -R^{(2)}(0)$ is the mean square of power spectrum frequency of the process, and $R^{(2)}(0)$ is the second derivative of the process autocorrelation function for $\tau = 0$.

For instance, for a Gaussian process with a Gaussian power spectrum, $\beta = \sqrt{3}$, hence

$$(\lambda_{\max})_{\mathrm{av}} = \lambda_1(0) * \sqrt{3} = \sqrt{\frac{6}{\pi}} * \Delta F_{\mathrm{eff.}} \approx 1.382 * \Delta F_{\mathrm{eff.}}$$

For a Gaussian process which has a uniform power spectrum within the frequency range $(0; \Delta F_{\mathrm{eff.}})$, we have $\beta = 3/\sqrt{5}$, the relation between $\lambda_1(0)$ and $\Delta F_{\mathrm{eff.}}$ is given by expression $\lambda_1(0) = 1/\sqrt{3} * \Delta F_{\mathrm{eff.}}$, hence

$$(\lambda_{\max})_{\mathrm{av}} = \sqrt{\frac{3}{5}} * \Delta F_{\mathrm{eff.}} \approx 0.775 \Delta F_{\mathrm{eff.}}$$

The average length τ_{av} of interval between two neighboring extrema (maximum and minimum) can be expressed as follows:

$$\tau_{\mathrm{av}} = \frac{1}{2 * (\lambda_{\max})_{\mathrm{av}}} = \frac{1}{2 * \beta * \lambda_1(0)} = \frac{1}{2 * \beta\gamma * \Delta F_{\mathrm{eff.}}} = \frac{2 * \tau_k}{\beta\gamma} = \alpha * \tau_k.$$

Thus, it is proportional to the time correlation interval τ_k of a random process.

Proportionality factor α depends on the power spectrum shape and the distribution of instant values. Usually it takes a value between 1 and 2. Thus, for a normal process with a Gaussian spectrum, we have $\alpha_1 \approx 1.45$, whereas for a normal process with a uniform spectrum we have $\alpha_2 \approx 2.58$. The respective autocorrelation coefficient values are $R(\tau_{1\mathrm{av}}) \approx 0.193$ and $R(\tau_{2\mathrm{av}}) \approx -0.195$.

In any case, considering the absence of any significant correlation between the values of any two samples of a random signal taken at interval τ_{av}, it is reasonable to set, at the end of the adaption procedure, the time sampling interval length T_s to equal the average length τ_{av} of interval between neighboring extrema of the process $(T_s = \tau_{\mathrm{av}})$.[83]

The analyzer adaptation time is proportional to the length of a time period during which the random process accumulates a sufficiently high number of extrema (zero derivative points), $L \gg 1$:

$$T_a = L * \tau_{\mathrm{av}} = L * \alpha\tau_k = \frac{L * \alpha}{4 * \Delta F_{\mathrm{eff.}}}$$

therefore, it is automatically set in inverse proportion to the effective width of the power spectrum.

The same is true about the length of the period during which the analyzer measures the distribution density of instant values in the process (the measurements stop automatically as soon as the number of extrema reaches a given value $E \gg L$):

$$T = E * \tau_{\mathrm{av}} = (E * \alpha/4 * \Delta F_{\mathrm{eff.}})$$

In the process of adaptation we determine the code $n_{opt} = \tau_{av}/T_{min}$ which sets the optimal input-to-output frequency ratio for a frequency-divider-code-controlled reference signal generator ($f_{max} = 1/T_{min}$).

The frequency f_{max} is selected depending on the analyzer speed requirements: $(\tau_{av})_{min} \geq T_{min}$, where $(\tau_{av})_{min}$ is the length of the shortest mean interval between two extrema in a random process with the widest possible power spectrum (spectrum of effective width $(\Delta F_{eff.})_{max}$). The value $(\Delta F_{eff.})_{max}$ depends on the number of channels N, in which the analyzer can simultaneously operate, and on the speed of the digital analogous converter (DAC) which makes an important part of the analyzer. Calculations show that in the case where DAC makes between 5 and 10 million conversions per second and $N = 16$, $(\Delta F_{eff.})_{max}$ amounts to hundreds of kilohertz.

The minimum width of spectrum $(\Delta F_{eff.})_{min}$ is limited by the maximum acceptable duration T_{max} of the measurement period which is determined by a given statistical error $\xi \leq \sqrt{(N/E)}$. Thus, if we choose $E = 10^5$ and $(\Delta F_{eff.})_{min} = 100$ Hz, then, for a Gaussian process, we will have $T_{max} \approx 6$ min in the case of a Gaussian spectrum, and $T_{max} \approx 10.8$ min in the case of a uniform spectrum. The frequency bandwidth in this kind of analyzer may vary from a few tenths of a Hertz to a few hundred kilohertz.

During the adaptation period, along with the frequency characteristics of a random process its amplitude characteristics are also measured. The latter include the average semirange $SR = ((x_{max})_{av} - (x_{min})_{av})/2$ and constant component $(x)_{av} = ((x_{max})_{av} + (x_{min})_{av})/2$ of the signal, where $(x_{max})_{av}$ and $(x_{min})_{av}$ are, respectively, the average maximum and the average minimum values of the process $X(t)$. Knowing the average semirange of the signal makes it possible to pick the best level sampling interval (quantization step) $\Delta x = \mu * \pi/N$ after the adaptation. Given that semirange SR is comparable with the mean square σ (for a Gaussian process $SR = \sqrt{\pi/2} * \sigma \approx 1.25 * \sigma$), it seems reasonable that factor μ should be chosen in the range (2; 3).

Increasing μ to a value greater than 3 will cause an increase in the system error of the analyzer since widening the sampling interval beyond reasonable limits shifts the estimate of amplitude distribution density. Reducing μ to a value lower than 2 will either cause an instrumentation error resulting from the truncation of "tails" in the distribution densities measured, or require an increase in the number of channels N, in which the analyzer can operate, which, in its turn will increase the statistical error for a given measurement period duration.

The process can be centered by subtracting the mean $(x)_{av}$ from its current values. For centered random processes symmetric about the mean, a higher accuracy of instant values distribution density measurement can be achieved by using the technique described in Ref. 53. By "straightening" the centered process (i.e., by reversing the negative parts of the signal) we can reduce the signal range almost by one half and narrow down the sampling interval without increasing the number of channels in the analyzer.

To extend the widest acceptable dynamic range of amplitudes in processes under investigation, we must ensure the adaptability of the analyzer transmitting voltage response. The use of a programmable amplifier with autoranging over several decades at the analyzer input allows for the selection of the optimal gain (attenuation factor) at the end of the adaptation stage. This technique is widely used in modern measuring devices designated for the analysis of deterministic signals.

The search of the number of decades required in the programmable amplifier, in which the options are examined starting from the highest possible number, extends the adaptation period significantly and, in some cases, nearly doubles the adaptation time. At the beginning of the search, when the amplifier is likely to get overloaded, the signal often goes beyond the acceptable limits $\pm U_{max}$, and ranging over $(i-1)$ decades will not take long. The search will not slow down until the amplifier switches to the last (ith) decade. In the worst-case scenario, the amplified signal will not go beyond the set thresholds until the very end of the adaptation period, just before the number of extrema reaches L. If that happens, a decision is made to exclude the last (ith) decade; then the adaptation starts all over again for the last time, which results in nearly doubling the total time of "training."

By reducing the dynamic range of the random output signal to a value close to 10 in a programmable amplifier with a large dynamic range of the input signal (the range of the order of 10^n, where $n \geq 3$) we can make the technical requirements to the analyzer less strict, and the implementation of many of its key parts more simple. Examples of such parts include the signal centering circuit, the modulus shaper, the unit for measuring the average signal range, and the differentiator.

In spite of an increase in the total adaptation time $T_a \leq 2 * (L/E) * T$, it is still relatively short in comparison with the total measurement time, because estimates made at the adaptation stage usually do not have to be very accurate (the ratio L/E can be selected from the range (0.01; 0.1)).

By the time the measurements start, the analyzer will have automatically set the optimal transmitted voltage response $(K_i)_{opt} = 10^{n-i}$ and the optimal time sampling interval length $T_s = \tau_{av}$, the random process mean value will have been found, the process will have been centered; the optimal level sampling interval Δx, which is directly proportional to the average semirange of the amplified signal, will also have been selected automatically. In addition to that, arrangements will have been made to ensure that the measurements end after time T as soon as the number of extrema in the random process reaches a given value E.

All these prearrangements eliminate the need for the use of a number of switches in the instant values distribution analyzer, such as the ones that set the gain, measurement time and passband; they also make the measurements more accurate and more effective in a situation where some important data about the signal are missing. The proposed algorithm for adaptation and measurement can be programmed on a PC, which allows for real-time processing of random signals.

7.5 An Adaptable Analyzer of Areas Under Above-Threshold Excursions of Random Processes

The need to calculate the areas under excursions arises in solving the problem of overloading which occurs in a number of fields such as broadcasting, multichannel telecommunication, construction, and medicine. Because of its complexity, the problem of finding the laws of distribution of areas under above-threshold excursions has not been studied sufficiently even for the case of stationary Gaussian processes. For high threshold-to-variance ratios, areas under excursion curves can be approximated as areas of triangles, if a joint distribution of excursion amplitudes and durations is used.[53]

In the general case, however, estimation of areas under excursions can be more complicated due to a more sophisticated relation between amplitude and duration of excursions. When trying to determine the law of distribution of areas under excursions above a given level we have to take into account the correlation between instant values and excursion durations.

Considering the importance of the problem for practical applications, it is worth trying to find the sought characteristics of random processes experimentally. However, the analyzers of areas under excursions[53,64] are difficult to implement because of their extremely wide dynamical range which is approximately equal to the product of two dynamical ranges of random processes being investigated: that of amplitudes and that of frequencies.

In the situation where the amplitude and time-related parameters of the random process are not known in advance, we can expect that the dynamic range for the area S of a shape formed by an excursion curve and a threshold line will be $D = 20\,lg(S_{max}/S_{min}) = 150$ decibels or higher. This means that we have to either increase the number of channels and the number of switchable ranges of signal level and frequency in the analyzer, which will complicate its implementation, or deal with lower accuracy of measurement caused by irrational choice of sampling interval duration. In addition to that, the more complicated the analyzer design gets, the longer it will take to find the optimal amplitude and frequency ranges. In the cases where the processes under investigation are nonrepetitive and limited in time, the very possibility of taking real-time measurements is threatened. The only possible compromise between the complexity of the analyzer design, on the one hand, and its speed and accuracy of measurement, on the other, is a flexible self-adapting device.

The proposed adaptable analyzer of areas under excursions[64] has two important features: in the process of self-adaptation it searches for the best transmitted voltage response value and also estimates the optimal sampling interval $(\Delta S)_{opt}$ by measuring the average total area confined under L above-threshold excursions which took place during the adaptation time. In this case, just like in the case of analyzers described in previous sections of this chapter, we determine the adaptation-to-measurement time ratio by dividing the number of excursions which took place during the adaptation period into the number of excursions that took place during the measurement period.

If we need to measure the area S under the excursion curve, a voltage-to-frequency converter will ideally suit for that purpose.[53] The linear conversion of voltage $U(t)$ to frequency $f(t)$ is described by the expression $f(t) = \kappa * U(t)$, where proportionality constant κ is chosen on the basis of the analyzer speed requirement.

The area S_i under an excursion of duration τ_i proves to be in direct proportion to the number n_i of pulses that came from the voltage-to-frequency converter:

$$S_i = \int_0^{\tau_i} U(t)\mathrm{d}t = \frac{1}{\kappa} * \int_0^{\tau_i} f(t)\mathrm{d}t = \frac{n_i}{\kappa}$$

The total count of pulses over the adaptation period is proportional to the total area under L above-threshold excursions:

$$\sum_{i=1}^{L} S_i = \frac{1}{\kappa} * \sum_{i=1}^{L} n_i = \frac{1}{\kappa} * L * n_{\mathrm{av}} = L * S_{\mathrm{av}}$$

where $S_{\mathrm{av}} = n_{\mathrm{av}}/\kappa$ is the estimate of the average area under one excursion.

By choosing the optimal sampling interval proportional to S_{av}, that is, $(\Delta S)_{\mathrm{opt}} = \alpha * S_{\mathrm{av}}$, and the number of channels in the analyzer $i = S_i/(\Delta S)_{\mathrm{opt}}$, we obtain:

$$i = \frac{S_i}{\alpha * S_{\mathrm{av}}} = \frac{n_i}{\alpha * n_{\mathrm{av}}} = \frac{n_i}{K_{\mathrm{opt}}}, \quad N \geq i \geq 1$$

Here N is the number of channels in the analyzer, and K_{opt} is the optimal division ratio of the programmable frequency divider which is connected to the voltage-to-frequency converter output. The division ratio K_{opt} is determined by a code which is proportional to the average area under one excursion curve. Given that the dynamic range of the signal excursion amplitude is stabilized in the analyzer by a programmable amplifier, the maximum division ratio $(K_{\mathrm{opt}})_{\max}$ is usually derived from the dynamic range of excursion durations.

Let us see how the range of the average areas under excursion curves is estimated in the case where $(K_{\mathrm{opt}})_{\max} = (S_{av})_{\max}/(S_{av})_{\min}$ for $(K_{\mathrm{opt}})_{\min} = 1$, using a Gaussian random process as an example. Let us assume that the signal level stabilized after the adaptation and the effective value σ is the same for all the processes under investigation. The average area under one excursion curve is expected to reach its maximum value $(S_{\mathrm{av}})_{\max}$ at the zero level for the random signals of the lowest frequency with power spectrum of minimum width $(\Delta F_{\mathrm{eff.}})_{\min}$.

The average area under one excursion curve can be approximated as the product of the average excursion duration τ_{av} and he average modulus $|x|_{\mathrm{av}}$ of a Gaussian process:

$$(S_{\mathrm{av}})_{\max} \approx \tau_{\mathrm{av}} * |x|_{\mathrm{av}} = \frac{1}{2\lambda_1(0)} * \sqrt{\frac{2}{\pi}} * \sigma = \frac{1}{2\gamma * (\Delta F_{\mathrm{eff.}})_{\min}} * \sqrt{\frac{2}{\pi}} * \sigma \approx \frac{0{,}4\sigma}{\gamma * (\Delta F_{\mathrm{eff.}})_{\min}}$$

The minimum value $(S_{\mathrm{av}})_{\min}$ of the average area under one excursion curve is likely to be reached at the highest levels in processes with the widest power spectrum $(\Delta F_{\mathrm{eff.}})_{\max}$. Considering that at high levels excursions have triangular shape, from the approximate formula (7.9) which relates the average excursion duration $\tau_{av}(x_0)$ to threshold x_0/σ, we obtain for $x_0/\sigma = 3$:

$$(S_{av})_{min} \approx \frac{|x|_{av}}{16\gamma * (\Delta F_{eff.})_{max}}$$

Calculations show that for $x_0/\sigma = 3$ the average level of excursions equals $|x|_{av} \approx 0.2\sigma$, hence:

$$(S_{av})_{min} \approx \frac{\sigma}{80\gamma * (\Delta F_{eff.})_{max}}$$

Thus, we can estimate the dynamic range of the average areas under excursion curves as follows:

$$(K_{opt})_{max} = \frac{(S_{av})_{max}}{(S_{av})_{min}} \approx 32 * \frac{(\Delta F_{eff.})_{max}}{(\Delta F_{eff.})_{min}}$$

If, for instance, $(\Delta F_{eff.})_{min} = 100\text{ Hz}$ and $(\Delta F_{eff.})_{max} = 100\text{ kHz}$, then $(K_{opt})_{max} \approx 3.2 \cdot 10^4$, that is, the dynamic range of the average areas under excursion curves equals approximately 90 decibels. To implement a programmable divider with a variable division ratio, a binary counter with no less than 15 digits will be required.

Let us note that, without preliminary stabilization of excursion amplitudes, the dynamic range of the average areas under excursion curves would have a further increase by 3 or 4 orders of magnitude and would constitute, in the general case, no less than 150 decibels.

From the maximum effective width $(\Delta F_{eff.})_{max}$ of a random process power spectrum we can derive the dynamic working range of the voltage-to-frequency converter. The constant $\kappa = f(t)/U(t)$ is chosen so as to fit $(1/\alpha)$ sampling intervals in the minimum area under an average excursion curve $(\Delta S)_{min} = \alpha * (S_{av})_{min}$. Assuming, as we did for the adaptable analyzers described in the previous sections, that $\alpha = 1/6$, we will have $(\Delta S)_{min} = (S_{av})_{min}/6 = 1/\kappa$. Hence,

$$\kappa = \frac{480\gamma * (\Delta F_{eff.})_{max}}{\sigma}$$

Assuming further that, after the stabilization of excursion amplitudes, the highest voltage level acceptable is $U_{max} = 10\text{ V}$, for a Gaussian process with a Gaussian spectrum where $(\Delta F_{eff.})_{max} = 100\text{ KHz}$, $\gamma \approx 0.8$ and $\sigma \approx 3.3\text{ V}$ we obtain $\kappa \approx 1.2 * 10^7 (1/B * c)$.

Thus, for excursion amplitudes $U(t)$ ranging from 0.1 to 10 V, the voltage-to-frequency converter will produce frequency $f(t) = \kappa * U(t)$ ranging from 1.2 to 120 MHz.

In random processes under analysis, the highest achievable value of $f(t)$ is the upper limit $(\Delta F_{\text{eff.}})_{\max}$ of power spectrum width.

The lower boundary $(\Delta F_{\text{eff.}})_{\min}$ of the frequency range is determined by the longest measurement period duration with which a given level of accuracy can be ensured. In the case of a Gaussian process with a Gaussian spectrum, choosing $(\Delta F_{\text{eff.}})_{\min} = 100$ Hz and $E = 10^4$ gives the following maximum values of measurement period duration: $T_{\max} \approx 2.1$ min for $x_0/\sigma = 0$; $T_{\max} \approx 3.43$ min for $x_0/\sigma = 1$; $T_{\max} \approx 15.3$ min for $x_0/\sigma = 2$, and $T_{\max} \approx 188$ min for $x_0/\sigma = 3$.

At high threshold levels ($x_0/\sigma \geq 2$) we can reduce the measurement time by restricting ourselves to measuring the average areas under excursion curves during the adaptation period only. We have this opportunity due to the fact that, at high levels, the numbers of level crossings tend to be distributed in accordance with the Poisson law, whereas excursion durations tend to be distributed according to the exponential law. In the case of a high threshold, the triangular shape of above-threshold excursions and stabilization of excursion amplitude around the average value explain why the areas under excursion curves are also distributed according to exponential law. Given that the exponential characteristics are entirely determined by the mean values of distribution functions, we can restrict ourselves to the estimation of the average areas under excursion curves by reducing the measurement period duration by one or two orders of magnitude.

7.6 One Way to Measure the Variance in a Broadband Centered Gaussian Random Process

The need to measure the variance arises quite often when we try to estimate the power of a Gaussian noise at the output of low-frequency linear radio engineering links. It is known that analogous devices designed for measuring the power (effective value) of a signal of an arbitrary waveform contain a squarer and a meaner (integrator). In the situation where the amplitude and frequency ranges of a random signal are not known in advance, the implementation of high-accuracy squarers and integrators can be difficult. A squarer must have the range of input levels twice as wide as the maximum range of the analyzer input levels, and an integrator must keep linearity over wide amplitude and frequency ranges. When the time correlation interval or the frequency bandwidth of a low-frequency signal under analysis is unknown, the measurement (integration) period duration remains uncertain. Although it reduces the statistical error of averaging, the measurement period prolongation also increases the instrumental error of the integrator. Shortening the measurement period too much should also be avoided, as it causes an increase in pulsation rate (the statistical error of averaging).

In Ref. 53, a device measuring the power of Gaussian signals is described, which uses the direct proportion between the average modulus and effective value of the process. Knowing the signal waveform (e.g., a sinusoid) eliminates the need to use a squarer. The latter can be replaced with a device which forms the modulus of the process instant values. We cannot, however, do without the integrator, which means that we have to face all the above-mentioned difficulties of finding the optimal design for the variance analyzer in the absence of information on the random process frequency characteristics.

The variance analyzer design should provide for an automatic adjustment of the measurement period duration to ensure that it remains in inverse proportion to the effective width of the signal frequency band or in direct proportion to the time correlation interval. The number of independent samples of the process taken during the measurement period remains the same. As a result, the statistical error of the variance analyzer will be known in advance and will remain the same, irrespective of the frequency characteristics of the broadband centered stationary Gaussian process at the analyzer input.[65] The adaptation of the analyzer to frequency properties of the random process should, therefore, be done by compromising between the frequency bandwidth and the measurement period duration.

The proposed method of variance measurement uses the relation between the number $\lambda(x_0)$ of points where the centered Gaussian process crosses a given threshold level x_0, and variance σ^2 of the process:

$$\lambda(x_0) = \lambda(0) * \exp\left(-\frac{x_0^2}{2\sigma^2}\right)$$

The average number of points where level x_0 is crossed by the Gaussian process over a time period of length T is expressed as follows:

$$N_{av}(x_0) = \lambda(x_0) * T = E * \exp\left(-\frac{x_0^2}{2\sigma^2}\right)$$

where $E = \lambda(0) * T$ is the average number of zero crossings which occur during time T. Hence we can easily obtain the relation for variance:

$$\sigma^2 = \frac{x_0^2}{2\ln(E/N)} \tag{7.18}$$

While counting the number of zero crossings until it reaches the counter capacity E (a fixed constant), we have to count at the same time, using a second counter, the number N of points where the signal crosses the threshold level x_0. When the measurements come to an end (i.e., at the moment $T = E/\lambda(0)$) and the total count of zero crossings reaches the counter capacity, the second counter reading, which is the number N of points where the signal crosses the threshold level x_0, can be substituted into formula (2.18) to calculate the variance σ^2 of the Gaussian process.

As we can see from formula (7.18) the relation between normalized variance (σ^2/x_0^2) and normalized count of level-crossing points (N/E), that is, the number of points where the signal crosses the threshold level x_0 over the measurement period of length T, is described by a logarithmic function.

For measurement purposes, the most suitable range of values (N/E) is from 0.1 to 0.8. The corresponding range of values for normalized variance (σ^2/x_0^2) will be from 0.22 to 2.2 (the upper limit is 10 times higher than the lower one). Within this range the logarithmic scale of the measuring instrument is not much different from the linear one. As (N/E) increases beyond the upper boundary of the range $((N/E) > 0.8)$, the nonlinear character of the relation becomes more and more obvious. Going beyond the lower limit $((N/E) < 0.1)$ should be avoided because of the reduction in the number N when E is fixed. A reduction in N is known to cause an increase in statistical error of variance analyzer $(\delta \leq 1/\sqrt{N})$ and make less uniform the distribution of error along (N/E) scale in the case where maximum-to-minimum variance ratio remains the same, $\sigma^2_{\max}/\sigma^2_{\min} = 10$.

As (N/E) tends to 1, statistical error decreases and its values become evenly distributed over the entire range of variance σ^2, but at the same time we see a increase in instrumental error caused by the nonlinearity of relation $\sigma^2 = f(N)$.

The range of variances measured can be extended beyond 10 decibels by introducing automatic control of the random input signal amplification/attenuation for a given threshold x_0. Similarly, it is possible to control the threshold level x_0 when the input signal level is fixed. In any case, it is desirable to have a variance analyzer which adapts not only to the frequency range but also to the amplitude range of the random process. When part of adaptation is automatic selection of transmission efficiency of the programmable amplifier/attenuator at the input stage of the analyzer, we can be sure that, when variance is measured for a given level x_0, the normalized variance (σ^2/x_0^2) will remain within a given range.

Automatic selection (for instance, by means of a DAC) of threshold level x_0 which is set by code N_a of the number of points where certain initial threshold $(x_0)_{\min}$ is crossed during the adaptation period, allows for the optimization of the range (N/E) of the measuring device scale.

The measurement period duration $T = E/\lambda(0) = E/\gamma * (\Delta F_{\text{eff.}})$ is established automatically in inverse proportion to the effective width $(\Delta F_{\text{eff.}})$ of the process power spectrum; it has very little relation with the power spectrum density shape factor γ. In the worst case where $(N/E) = 0.1$, choosing $E = 10^4$ will keep statistical error of variance measurement under 3% $(\delta \leq 1/\sqrt{N})$.

Once E is preset, the measurement time (measurement period duration) is mainly determined by the width of the process power spectrum. If, for example, the pulse-counter capacity $E = 10^4$ and the random signal fed to the analyzer input is a wide-band Gaussian process with rectangular spectrum $(\gamma = 1/\sqrt{3})$ of effective width $(\Delta F_{\text{eff.}}) = 100$ Hz, then the time required for the pulse count to reach the counter capacity will be $T \approx 173$ s. Under the same conditions, the measurements of variance of a Gaussian process with a Gaussian spectrum $(\gamma = \sqrt{2/\pi})$ will take much less time ($T \approx 125$ s) because of a higher intensity of zero crossings.

The search for the optimal voltage transfer ratio for the variance analyzer or the optimal threshold level x_0 will require the time T_a which varies approximately between 0.01 T and 0.1 T. In the above examples it will not exceed 13 s in the case of a Gaussian spectrum and 18 s in the case of a rectangular spectrum, respectively. Under the same conditions, reducing the upper limit for acceptable statistical error of the analyzer to 1% will require a 10-fold increase in the adaptation period duration; the pulse-counter capacity E in this case should amount to 10^5.

The proposed method of variance measurement suggests that the maximum width of the Gaussian process spectrum, $(\Delta F_{\text{eff.}})_{\max}$, is mainly determined by the response time of the comparator of threshold level x_0. In the worst-case scenario where the relative threshold reaches its highest value ($\sigma^2/x_0^2 \approx 0.22$, $x_0/\sigma \approx 2.2$), the approximate average duration of excursions above the threshold will equal

$$\tau_{\text{av}}(2.2) \approx \frac{\tau_{\text{av}}(0)}{1 + (x_0/\sigma)} = \frac{1}{3.2 * \gamma * (\Delta F_{\text{eff.}})_{\max}}$$

Allowing for the possibility of omitting short excursions, that is, excursions of duration $\tau \leq t_{\text{comp}} = 0.01 * \tau_{\text{av}}(2.2)$ where t_{comp} is the threshold comparator response time, we obtain:

$$(\Delta F_{\text{eff.}})_{\max} = \frac{1}{320\gamma * t_{\text{comp}}}$$

If, for example, we use a CMOS-based voltage comparator with an average response time $t_{\text{comp}} = 100$ ns, then for $\gamma = 0.8$ in a process with a Gaussian spectrum we will obtain $(\Delta F_{\text{eff.}})_{\max}$ under 40 kHz.

Thus, the frequency range of a digital variance analyzer based on microchips of an average response time extends approximately from a value of the order of 100 Hz to a value of the order of 40 kHz.

The amplitude range of the measuring device, considering that it adapts to the level, can vary from 60 to 80 dB, which means that the rms value of σ will range from a few tenths of a millivolt (or even a few millivolts) to a few volts.

Appendix 1: PC Simulations of Gaussian and Rayleigh Random Processes

Tables A1.1 and A1.2 present the values calculated on a PC for the first approximations of the distribution of time intervals (τ) between successive upward or downward crossings of zero level by centered Gaussian random processes with an arbitrarily chosen $q_1(\tau, 0)$. The direction of crossings (e.g., upward) is determined by the sign of the first derivative denoted as $q_{1+}(\tau, 0)$.

Approximation functions were calculated for broadband signals with two kinds of power spectrum distribution densities (PSDs): a Gaussian one and a rectangular one.

The said tables also provide the data calculated using formulae (3.16) and (3.17), which represent the improved estimates of distribution densities for time intervals between successive zeros with an arbitrary $W(\tau, 0)$ and a given sign $W_+(\tau, 0)$ of the first derivative. Along with the values calculated for $W(\tau, 0)$, tables also provide distribution densities $W_M(\tau, 0)$ of time intervals between zeros, calculated using McFadden formula (3.19).

In addition to the above-mentioned data, tables show the results of statistical PC simulation of distribution densities $W_{\exp}(\tau, 0)$ and $W_{+\exp}(\tau, 0)$, which allow to judge about the accuracy of approximations $q_1(\tau, 0)$ and $q_{1+}(\tau, 0)$, as well as $W(\tau, 0)$, $W_M(\tau, 0)$, and $W_+(\tau, 0)$.

For convenience, all distribution density values in Tables A1.1 and A1.2 as well as other tables were increased 100-fold.

Tables A1.3 and A1.4 show the values obtained experimentally (i.e., by means of statistical simulation on a PC) for the distribution density $W_{\exp}(\tau, x_0)$ of durations in the excursions of a Gaussian process above the standardized thresholds $x_0/\sigma = 0.5, 1.0, 1.5, 2, 2.5, 3$. The same tables show the values obtained experimentally for the densities $W_{+\exp}(\tau, x_0)$ of distribution of time intervals between successive upward crossings of the given standardized thresholds.

Approximation functions were calculated for broadband signals with two kinds of PSDs: a Gaussian one and a rectangular one.

The functions $W_3(\tau, 0)$ and $W_4(\tau, 0)$ whose values are shown in Tables A1.5 and A1.6 are PC-simulated distribution densities of the total length of, respectively, three and four intervals between successive zeros of a centered Gaussian process.

Tables A1.7 and A1.8 present the values obtained by statistical simulation for the distribution densities $W_{\exp}(\tau, x_0)$ of durations in the excursions of a Rayleigh process above the standardized thresholds $x_0/\sigma = 0.5, 1.0, 1.5, 2.0, 3.0$. The same

Table A1.1 Gaussian PSD

τ/τ_0	0.2	0.4	0.6	0.8	1.0	1.2	1.4	1.6	1.8
$q_1(\tau,0)$	9.936	19.48	28.10	35.03	39.31	40.29	38.26	34.47	30.44
$W(\tau,0)$	9.936	19.48	28.10	35.02	39.26	40.17	37.96	33.85	29.29
$W_M(\tau,0)$	9.936	19.48	28.10	35.02	39.27	40.19	37.99	33.90	29.36
$W_{\exp}(\tau,0)$	9.942	19.49	28.10	35.00	39.23	40.20	38.07	34.08	29.62
$q_{1+}(\tau,0)$	0.005	0.073	0.363	1.112	2.575	4.921	8.088	11.71	15.22
$W_+(\tau,0)$	0.005	0.069	0.363	1.112	2.575	4.920	8.086	11.70	15.20
$W_{+\exp}(\tau,0)$	0.005	0.069	0.363	1.099	2.562	4.831	8.04	11.60	15.08

τ/τ_0	2.0	2.2	2.4	2.6	2.8	3.0	3.2	3.4	3.6
$q_1(\tau,0)$	27.17	24.98	23.69	23.02	22.71	22.58	22.3	22.52	22.51
$W(\tau,0)$	25.27	22.07	19.56	17.5	15.71	14.08	12.58	11.21	9.971
$W_M(\tau,0)$	25.35	22.16	19.66	17.61	15.82	14.19	12.69	11.31	10.07
$W_{\exp}(\tau,0)$	25.66	22.39	19.76	17.54	15.63	13.95	12.37	11.00	9.769
$q_{1+}(\tau,0)$	18.10	20.12	21.35	22.00	22.30	22.43	22.48	22.50	22.51
$W_+(\tau,0)$	18.05	20.01	21.13	21.62	21.67	21.44	21.02	20.45	19.77
$W_{+\exp}(\tau,0)$	18.04	19.9	21.16	21.71	21.70	21.56	21.14	20.57	19.84

τ/τ_0	3.8	4.0	4.2	4.4	4.6	4.8	5.0	5.2	5.4
$q_1(\tau,0)$	22.51	22.51	22.51	22.51	22.51	22.51	22.51	22.51	22.51
$W(\tau,0)$	8.862	7.877	7.003	6.225	5.534	4.917	4.367	3.877	3.440
$W_M(\tau,0)$	8.959	7.970	7.091	6.309	5.613	4.991	4.437	3.942	3.501
$W_{\exp}(\tau,0)$	8.681	7.702	6.842	6.080	5.400	4.790	4.242	3.787	3.364
$q_{1+}(\tau,0)$	22.51	22.51	22.51	22.51	22.51	22.51	22.51	22.51	22.51
$W_+(\tau,0)$	18.99	18.14	17.24	16.32	15.39	14.46	13.55	12.66	11.8
$W_{+\exp}(\tau,0)$	19.06	18.12	17.21	16.31	15.18	14.34	13.51	12.48	11.97

τ/τ_0	5.6	5.8	6.0	6.2	6.4	6.6	6.8	7.0	7.2
$q_1(\tau,0)$	22.51	22.51	22.51	22.51	22.51	22.51	22.51	22.51	22.51
$W(\tau,0)$	3.052	2.706	2.398	2.125	1.882	1.666	1.475	1.305	1.154
$W_M(\tau,0)$	3.108	2.759	2.447	2.170	1.924	1.705	1.510	1.337	1.184
$W_{\exp}(\tau,0)$	2.968	2.650	2.348	2.084	1.853	1.649	1.458	1.304	1.154
$q_{1+}(\tau,0)$	22.51	22.51	22.51	22.51	22.51	22.51	22.51	22.51	22.51
$W_+(\tau,0)$	10.98	10.19	9.442	8.732	8.063	7.433	6.843	6.291	5.776
$W_{+\exp}(\tau,0)$	10.95	10.13	9.424	8.61	8.077	7.452	6.886	6.238	5.782

τ/τ_0	7.6	8.0	8.4	9.0	10.0	11.0	12.0	13.0	14.0
$q_1(\tau,0)$	22.51	22.51	22.51	22.51	22.51	22.51	22.51	22.51	22.51
$W(\tau,0)$	0.901	0.702	0.547	0.374	0.197	0.102	0.052	0.026	0.013
$W_M(\tau,0)$	0.926	0.724	0.564	0.387	0.205	0.107	0.055	0.028	0.014
$W_{\exp}(\tau,0)$	0.912	0.717	0.567	0.395	0.218	0.122	0.068	0.037	0.021
$q_{1+}(\tau,0)$	22.51	22.51	22.51	22.51	22.51	22.51	22.51	22.51	22.51
$W_+(\tau,0)$	4.853	4.06	3.383	2.557	1.58	0.961	0.577	0.342	0.200
$W_{+\exp}(\tau,0)$	4.75	4.034	3.49	2.577	1.586	0.969	0.617	0.363	0.207

tables show the values of distribution densities $W_{\_exp}(\tau, x_0)$ of time intervals between the excursions above the same standardized thresholds. All the data were obtained for broadband signals with two kinds of PSD: a Gaussian one and a rectangular one.

Tables A1.9 and A1.10 show the values obtained by means of statistical simulation for distribution densities $W_{max}(\tau, 0)$ of time intervals between successive extrema (maxima and minima) of a Rayleigh process.

To obtain those data, we experimented with the Gaussian random process representing the first derivative of the original Rayleigh process. Since the extrema of the Rayleigh function correspond to zeros of its first derivative, the distribution density $W_{max}(\tau, 0)$ is the distribution of intervals between successive zeros in the Gaussian process with the same type of power spectrum.

Table A1.2 Rectangular PSD

τ/τ_0	0.2	0.4	0.6	0.8	1.0	1.2	1.4	1.6	1.8
$q_1(\tau, 0)$	4.069	8.564	13.95	20.74	29.30	39.39	49.22	54.96	52.94
$W(\tau, 0)$	4.069	8.564	13.95	20.74	29.30	39.39	49.21	54.93	52.87
$W_M(\tau, 0)$	4.069	8.564	13.95	20.74	29.30	39.39	49.22	54.94	52.88
$W_{exp}(\tau, 0)$	4.066	8.563	13.83	20.73	29.29	39.35	49.11	54.77	52.72
$q_{1+}(\tau, 0)$	–	0.004	0.022	0.075	0.204	0.482	1.045	2.122	4.087
$W_+(\tau, 0)$	–	0.004	0.022	0.075	0.204	0.482	1.045	2.122	4.087
$W_{+exp}(\tau, 0)$	–	0.004	0.022	0.074	0.204	0.482	1.045	2.116	4.100

τ/τ_0	2.0	2.2	2.4	2.6	2.8	3.0	3.2	3.4	3.6
$q_1(\tau, 0)$	43.64	31.68	21.50	14.79	11.27	10.19	11.02	13.56	17.67
$W(\tau, 0)$	43.49	31.39	20.93	13.73	9.436	7.176	6.356	6.752	8.259
$W_M(\tau, 0)$	43.50	31.41	20.96	13.78	9.499	7.25	6.434	6.823	8.318
$W_{exp}(\tau, 0)$	43.59	31.67	21.44	14.58	10.70	8.85	8.229	8.286	8.584
$q_{1+}(\tau, 0)$	7.451	12.69	19.81	27.85	34.82	38.46	37.52	32.78	26.43
$W_+(\tau, 0)$	7.450	12.68	19.80	27.83	34.80	38.43	37.46	32.66	26.20
$W_{+exp}(\tau, 0)$	7.425	12.64	19.76	27.79	34.72	38.34	37.42	32.77	26.42

τ/τ_0	3.8	4.0	4.2	4.4	4.6	4.8	5.0	5.2	5.4
$q_1(\tau, 0)$	22.89	28.18	32.08	33.35	31.67	27.89	23.42	19.50	16.81
$W(\tau, 0)$	10.54	12.79	13.85	12.84	9.74	5.55	1.604	–	–
$W_M(\tau, 0)$	10.59	12.82	13.88	12.86	9.763	5.57	1.631	–	–
$W_{exp}(\tau, 0)$	8.807	8.654	8.010	6.914	5.628	4.40	3.395	2.700	2.244
$q_{1+}(\tau, 0)$	20.58	16.39	14.10	13.10	14.68	17.18	20.78	24.86	28.44
$W_+(\tau, 0)$	20.17	15.67	12.90	11.66	11.70	12.79	14.61	16.60	17.93
$W_{+exp}(\tau, 0)$	20.56	16.27	13.66	12.63	12.56	13.24	14.07	14.68	14.63

τ/τ_0	5.6	6.0	7.0	8.0	9.0	10.0	11.0	12.0	13.0
$q_1(\tau, 0)$	15.56	17.18	29.09	18.53	23.07	24.77	18.90	25.95	20.91
$W_{exp}(\tau, 0)$	1.963	1.756	1.134	0.455	0.316	0.157	–	–	–
$q_{1+}(\tau, 0)$	30.48	28.25	16.70	26.64	21.68	20.20	26.39	19.26	24.05
$W_+(\tau, 0)$	17.79	12.40	1.408	–	–	–	–	–	–
$W_{+exp}(\tau, 0)$	13.73	10.29	4.806	4.143	1.924	1.292	0.792	0.385	0.276

Table A1.3 Gaussian PSD

$x_0/\sigma = 0.5$									
τ/τ_0	0.1	0.2	0.3	0.4	0.5	0.6	0.7	0.8	0.9
$W_{exp}(\tau, x_0)$	8.569	16.81	24.81	32.26	39.09	44.78	49.62	52.86	54.80
τ/τ_0	1.0	1.1	1.2	1.3	1.4	1.5	1.6	1.70	1.8
$W_{exp}(\tau, x_0)$	55.15	54.00	51.65	48.37	44.39	40.37	36.36	32.45	29.01
τ/τ_0	2.0	2.2	2.4	2.6	2.8	3.0	3.2	3.4	3.6
$W_{exp}(\tau, x_0)$	23.28	19.02	15.76	13.22	10.97	9.079	7.598	6.325	5.193
τ/τ_0	3.8	4.0	4.2	4.4	4.6	4.8	5.0	5.4	6.0
$W_{exp}(\tau, x_0)$	4.264	3.549	2.911	2.454	2.058	1.679	1.399	0.924	0.537
τ/τ_0	0.4	0.8	1.2	1.4	1.6	1.8	2.0	2.2	2.4
$W_{+exp}(\tau, x_0)$	0.051	0.963	4.346	7.118	10.25	13.29	15.77	17.51	18.59
τ/τ_0	2.6	2.8	3.0	3.4	4.0	5.0	6.0	7.0	8.0
$W_{+exp}(\tau, x_0)$	19.04	19.15	19.00	18.25	16.36	12.81	9.528	6.948	5.010
τ/τ_0	9.0	10.0	11.0	12.0	13.0	14.0	15.0	16.0	17.0
$W_{+exp}(\tau, x_0)$	3.566	2.547	1.809	1.278	0.922	0.656	0.454	0.321	0.230
$x_0/\sigma = 1.0$									
τ/τ_0	0.1	0.2	0.3	0.4	0.5	0.6	0.7	0.8	0.9
$W_{exp}(\tau, x_0)$	13.41	26.60	38.63	49.20	58.39	65.36	69.32	70.87	70.15
τ/τ_0	1.0	1.1	1.2	1.3	1.4	1.5	1.6	1.70	1.8
$W_{exp}(\tau, x_0)$	66.64	61.72	55.66	48.93	42.32	36.39	30.93	26.39	22.47
τ/τ_0	2.0	2.2	2.4	2.6	3.0	3.4	3.8	4.2	4.6
$W_{exp}(\tau, x_0)$	16.48	12.37	9.411	7.164	4.157	2.412	1.407	0.806	0.471
τ/τ_0	0.4	0.8	1.2	1.6	2.0	2.4	2.8	3.2	3.6
$W_{+exp}(\tau, x_0)$	0.043	0.770	3.280	7.350	10.94	12.81	13.21	12.98	12.52
τ/τ_0	4.0	5.0	6.0	7.0	8.0	9.0	10.0	11.0	12.0
$W_{+exp}(\tau, x_0)$	11.85	10.15	8.531	7.155	5.975	5.006	4.193	3.507	2.917
τ/τ_0	13.0	14.0	15.0	16.0	17.0	18.0	19.0	20.0	21.0
$W_{+exp}(\tau, x_0)$	2.422	2.046	1.716	1.425	1.199	0.987	0.823	0.698	0.581
τ/τ_0	22.0	23.0	24.0	25.0	26.0	27.0	28.0	29.0	30.0
$W_{+exp}(\tau, x_0)$	0.483	0.407	0.338	0.287	0.232	0.194	0.165	0.135	0.116
$x_0/\sigma = 1.5$									
τ/τ_0	0.1	0.2	0.3	0.4	0.5	0.6	0.7	0.8	0.9
$W_{exp}(\tau, x_0)$	20.46	39.52	56.65	70.52	80.72	85.94	87.35	84.55	78.29
τ/τ_0	1.0	1.1	1.2	1.3	1.4	1.5	1.6	1.70	1.8
$W_{exp}(\tau, x_0)$	70.10	60.37	50.55	41.18	33.26	26.58	21.28	16.84	13.63
τ/τ_0	1.9	2.0	2.1	2.2	2.4	2.6	2.8	3.0	3.2
$W_{exp}(\tau, x_0)$	10.98	9.023	7.434	6.033	4.173	2.838	1.901	1.329	0.876
τ/τ_0	0.4	0.8	1.2	1.6	2.0	2.4	2.8	3.2	3.6
$W_{+exp}(\tau, x_0)$	0.031	0.544	1.209	4.444	6.255	7.160	7.381	7.313	7.230
τ/τ_0	4.0	5.0	6.0	7.0	8.0	9.0	10.0	11.0	12.0
$W_{+exp}(\tau, x_0)$	6.986	6.479	5.976	5.478	5.075	4.668	4.306	3.976	3.668
τ/τ_0	13.0	14.0	15.0	16.0	17.0	18.0	19.0	20.0	22.0
$W_{+exp}(\tau, x_0)$	3.368	3.105	2.867	2.639	2.438	2.228	2.072	1.902	1.613
τ/τ_0	24.0	26.0	28.0	30.0	32.0	34.0	36.0	38.0	40.0
$W_{+exp}(\tau, x_0)$	1.376	1.165	1.000	0.836	0.719	0.605	0.516	0.439	0.375

(*Continued*)

Table A1.3 (Continued)

$x_0/\sigma = 2.0$									
τ/τ_0	0.05	0.1	0.15	0.2	0.25	0.3	0.35	0.4	0.5
$W_{exp}(\tau, x_0)$	14.68	29.14	43.38	55.59	68.14	77.58	87.17	94.53	102.8
τ/τ_0	0.6	0.7	0.8	0.9	1.0	1.1	1.2	1.3	1.4
$W_{exp}(\tau, x_0)$	104.4	100.3	89.86	76.73	63.01	50.07	38.29	28.74	21.27
τ/τ_0	1.5	1.6	1.7	1.8	1.9	2.0	2.2	2.4	2.6
$W_{exp}(\tau, x_0)$	15.80	11.87	8.917	6.510	4.956	3.809	2.322	1.384	0.830
τ/τ_0	0.8	1.2	1.6	2.0	2.4	2.8	3.2	3.6	4.0
$W_{+exp}(\tau, x_0)$	0.271	1.058	1.987	2.643	2.974	3.079	3.081	3.045	3.018
τ/τ_0	6.0	8.0	10.0	12.0	14.0	16.0	18.0	20.0	22.0
$W_{+exp}(\tau, x_0)$	2.859	2.689	2.540	2.404	2.265	2.130	1.999	1.901	1.790
τ/τ_0	24.0	26.0	28.0	30.0	32.0	34.0	36.0	40.0	44.0
$W_{+exp}(\tau, x_0)$	1.693	1.593	1.496	1.410	1.327	1.249	1.186	1.050	0.931
τ/τ_0	48.0	52.0	56.0	60.0	64.0	68.0	72.0	76.0	80.0
$W_{+exp}(\tau, x_0)$	0.828	0.733	0.661	0.584	0.517	0.463	0.406	0.361	0.326
$x_0/\sigma = 2.5$									
τ/τ_0	0.05	0.1	0.15	0.2	0.25	0.3	0.35	0.4	0.45
$W_{exp}(\tau, x_0)$	20.46	40.09	58.16	74.81	90.04	101.8	110.8	118.0	121.6
τ/τ_0	0.50	0.55	0.60	0.7	0.8	0.9	1.0	1.1	1.2
$W_{exp}(\tau, x_0)$	122.4	120.2	116.6	103.7	85.97	67.47	50.52	35.85	24.91
τ/τ_0	1.3	1.4	1.5	1.6	1.7	1.8	1.9	2.0	2.1
$W_{exp}(\tau, x_0)$	17.32	11.83	8.031	5.579	3.780	2.716	1.947	1.385	0.963
τ/τ_0	0.8	1.2	1.6	2.0	2.4	2.8	3.2	3.6	4.0
$W_{+exp}(\tau, x_0)$	0.229	0.808	1.283	1.567	1.683	1.700	1.719	1.738	1.724
τ/τ_0	6.0	8.0	10.0	12.0	14.0	16.0	18.0	20.0	22.0
$W_{+exp}(\tau, x_0)$	1.702	1.672	1.640	1.608	1.588	1.530	1.510	1.494	1.481
τ/τ_0	24.0	26.0	28.0	30.0	32.0	34.0	36.0	40.0	44.0
$W_{+exp}(\tau, x_0)$	1.444	1.414	1.388	1.361	1.336	1.323	1.307	1.268	1.197
τ/τ_0	48.0	52.0	56.0	60.0	64.0	68.0	72.0	76.0	80.0
$W_{+exp}(\tau, x_0)$	1.156	1.136	1.088	1.049	0.993	0.958	0.932	0.911	0.894
$x_0/\sigma = 3.0$									
τ/τ_0	0.05	0.10	0.15	0.20	0.25	0.30	0.35	0.40	0.45
$W_{exp}(\tau, x_0)$	26.30	51.71	74.48	94.80	112.1	124.1	133.1	137.0	137.7
τ/τ_0	0.50	0.55	0.60	0.65	0.70	0.75	0.80	0.9	1.0
$W_{exp}(\tau, x_0)$	133.1	127.2	119.1	108.9	97.74	86.27	74.11	53.62	36.71
τ/τ_0	1.1	1.2	1.3	1.4	1.5	1.6	1.7	1.8	1.9
$W_{exp}(\tau, x_0)$	24.75	16.36	10.74	7.000	4.714	3.149	2.232	1.493	0.987
τ/τ_0	0.8	1.2	1.6	3.2	4.8	20.0	40.0	60.0	80.0
$W_{+exp}(\tau, x_0)$	0.310	0.901	1.202	1.374	1.380	1.309	1.289	1.202	1.143
τ/τ_0	100.0	120.0	140.0	160.0	180.0	200.0	240.0	280.0	320.0
$W_{+exp}(\tau, x_0)$	1.088	1.035	0.984	0.936	0.891	0.847	0.767	0.694	0.628
τ/τ_0	360.0	400.0	480.0	560.0	640.0	720.0	800.0	880.0	960.0
$W_{+exp}(\tau, x_0)$	0.568	0.514	0.421	0.344	0.282	0.231	0.189	0.155	0.127

Table A1.4 Rectangular PSD

$x_0/\sigma = 0.5$

τ/τ_0	0.1	0.2	0.3	0.4	0.5	0.6	0.7	0.8	0.9
$W_{exp}(\tau, x_0)$	4.545	9.148	13.75	18.80	23.96	29.50	35.40	41.41	47.47
τ/τ_0	1.0	1.1	1.2	1.3	1.4	1.5	1.6	1.70	1.8
$W_{exp}(\tau, x_0)$	53.33	58.53	62.73	65.39	65.72	63.19	58.97	52.67	45.36
τ/τ_0	1.9	2.0	2.2	2.4	2.6	2.8	3.0	3.2	3.4
$W_{exp}(\tau, x_0)$	37.59	30.62	19.27	12.15	8.263	6.227	5.424	5.001	4.763
τ/τ_0	3.6	3.8	4.0	4.2	4.4	4.6	4.8	5.0	5.2
$W_{exp}(\tau, x_0)$	4.431	3.972	3.322	2.599	1.959	1.398	1.014	0.724	0.597
τ/τ_0	0.8	1.2	1.6	2.0	2.2	2.4	2.6	2.8	3.0
$W_{+exp}(\tau, x_0)$	0.061	0.403	1.753	6.134	10.60	16.90	24.28	30.99	34.75
τ/τ_0	3.1	3.2	3.4	3.6	3.8	4.0	4.2	4.4	4.6
$W_{+exp}(\tau, x_0)$	34.87	34.03	29.51	23.58	18.1	14.22	11.89	11.01	11.15
τ/τ_0	4.8	5.0	5.4	6.0	7.0	7.8	9.0	10.0	11.0
$W_{+exp}(\tau, x_0)$	11.99	13.17	14.58	10.93	5.354	5.726	2.930	2.255	1.720
τ/τ_0	12.0	13.0	14.0	15.0	16.0	17.0	18.0	19.0	20.0
$W_{+exp}(\tau, x_0)$	0.960	0.856	0.507	0.365	0.283	0.167	0.137	0.086	0.062

$x_0/\sigma = 1.0$

τ/τ_0	0.1	0.2	0.3	0.4	0.5	0.6	0.7	0.8	0.9
$W_{exp}(\tau, x_0)$	8.819	17.53	26.37	35.30	43.95	52.18	60.01	66.95	72.64
τ/τ_0	1.0	1.1	1.2	1.3	1.4	1.5	1.6	1.70	1.8
$W_{exp}(\tau, x_0)$	75.85	76.80	74.79	69.98	62.85	53.68	43.92	34.75	26.56
τ/τ_0	1.9	2.0	2.1	2.2	2.3	2.4	2.5	2.6	2.7
$W_{exp}(\tau, x_0)$	19.67	14.69	10.76	8.103	6.207	4.842	3.966	3.336	2.854
τ/τ_0	2.8	2.9	3.0	3.1	3.2	3.4	3.6	3.8	4.0
$W_{exp}(\tau, x_0)$	2.532	2.261	2.123	1.939	1.794	1.512	1.201	0.884	0.627
τ/τ_0	0.8	1.2	1.6	2.0	2.2	2.4	2.6	2.8	3.0
$W_{+exp}(\tau, x_0)$	0.037	0.234	0.988	3.473	6.218	10.54	16.30	22.04	25.42
τ/τ_0	3.1	3.2	3.4	3.6	3.8	4.0	4.2	4.4	4.6
$W_{+exp}(\tau, x_0)$	25.48	25.17	21.59	16.73	12.38	9.411	7.759	7.216	7.549
τ/τ_0	4.8	5.0	5.4	6.0	6.4	7.0	8.0	9.6	10.6
$W_{+exp}(\tau, x_0)$	8.564	10.07	12.84	13.04	7.278	5.340	7.395	3.628	4.396
τ/τ_0	12.2	13.0	15.0	15.6	17.4	18.0	20.0	24.0	28.0
$W_{+exp}(\tau, x_0)$	2.357	2.663	1.524	1.643	0.963	1.010	0.612	0.323	0.155

$x_0/\sigma = 1.5$

τ/τ_0	0.05	0.10	0.15	0.20	0.25	0.30	0.35	0.40	0.45
$W_{exp}(\tau, x_0)$	7.640	15.16	22.55	30.22	37.38	44.32	51.07	57.38	63.21
τ/τ_0	0.50	0.55	0.60	0.65	0.70	0.75	0.80	0.85	0.90
$W_{exp}(\tau, x_0)$	69.70	74.73	78.66	82.69	86.02	88.44	89.81	90.43	90.60
τ/τ_0	1.0	1.1	1.2	1.3	1.4	1.5	1.6	1.70	1.8
$W_{exp}(\tau, x_0)$	86.25	78.70	67.38	55.23	42.93	31.76	22.24	15.13	10.32
τ/τ_0	1.9	2.0	2.1	2.2	2.3	2.4	2.5	2.6	2.7
$W_{exp}(\tau, x_0)$	6.981	4.829	3.319	2.330	1.770	1.434	1.174	0.967	0.766
τ/τ_0	1.0	1.5	2.0	2.2	2.4	2.6	2.8	3.0	3.1
$W_{+exp}(\tau, x_0)$	0.047	0.292	1.362	2.565	4.847	8.528	12.89	15.92	16.34

(Continued)

Table A1.4 (Continued)

τ/τ_0	3.2	3.4	3.6	3.8	4.0	4.4	5.0	5.6	6.2
$W_{+\exp}(\tau, x_0)$	15.99	13.19	9.673	6.711	4.829	3.543	5.830	9.430	6.494
τ/τ_0	7.0	8.2	9.5	10.7	12.0	13.2	14.6	15.8	17.3
$W_{+\exp}(\tau, x_0)$	3.616	6.638	3.245	5.004	2.793	3.872	2.342	3.034	1.956
τ/τ_0	18.6	19.8	20.9	22.3	23.4	25.0	25.8	27.5	28.5
$W_{+\exp}(\tau, x_0)$	2.333	1.613	1.916	1.329	1.526	1.084	1.226	0.880	0.978
$x_0/\sigma = 2.0$									
τ/τ_0	0.05	0.10	0.15	0.20	0.25	0.30	0.35	0.40	0.45
$W_{\exp}(\tau, x_0)$	11.78	23.83	35.39	46.96	57.43	67.13	76.10	84.63	90.75
τ/τ_0	0.50	0.55	0.60	0.65	0.70	0.75	0.80	0.85	0.90
$W_{\exp}(\tau, x_0)$	97.18	101.2	104.0	105.2	106.5	104.0	100.8	97.53	91.50
τ/τ_0	0.95	1.0	1.05	1.1	1.15	1.2	1.25	1.3	1.35
$W_{\exp}(\tau, x_0)$	86.04	78.65	70.27	62.48	54.22	46.13	39.59	32.55	26.57
τ/τ_0	1.4	1.45	1.5	1.6	1.7	1.8	1.9	2.0	2.1
$W_{\exp}(\tau, x_0)$	21.49	17.06	13.33	8.053	4.878	2.940	1.780	1.182	0.774
τ/τ_0	1.0	1.5	2.0	2.2	2.4	2.6	2.8	3.0	3.1
$W_{+\exp}(\tau, x_0)$	0.026	0.103	0.450	0.922	2.042	4.221	7.337	9.899	10.34
τ/τ_0	3.2	3.4	3.6	3.8	4.0	4.4	5.0	5.7	6.2
$W_{+\exp}(\tau, x_0)$	10.05	8.037	5.465	3.455	2.298	1.580	2.977	6.175	4.149
τ/τ_0	7.0	8.2	9.6	10.8	12.2	13.3	14.7	15.9	17.2
$W_{+\exp}(\tau, x_0)$	1.993	4.783	2.120	4.041	2.133	3.500	2.051	3.101	1.978
τ/τ_0	18.5	19.8	21.2	22.4	23.7	25.0	26.2	27.6	28.6
$W_{+\exp}(\tau, x_0)$	2.793	1.868	2.504	1.768	2.277	1.642	2.073	1.520	1.887
$x_0/\sigma = 2.5$									
τ/τ_0	0.05	0.10	0.15	0.20	0.25	0.30	0.35	0.40	0.45
$W_{\exp}(\tau, x_0)$	17.49	34.62	51.10	66.20	80.08	92.53	103.4	111.2	117.8
τ/τ_0	0.50	0.55	0.60	0.65	0.70	0.75	0.80	0.85	0.90
$W_{\exp}(\tau, x_0)$	122.0	122.6	121.9	118.4	113.8	106.6	99.02	88.72	79.77
τ/τ_0	0.95	1.0	1.05	1.1	1.15	1.2	1.25	1.3	1.35
$W_{\exp}(\tau, x_0)$	68.76	58.61	48.74	39.62	31.67	25.54	19.46	14.98	11.14
τ/τ_0	1.4	1.45	1.5	1.55	1.6	1.65	1.7	1.75	1.8
$W_{\exp}(\tau, x_0)$	8.549	6.058	4.598	3.381	2.584	1.891	1.457	1.198	0.902
τ/τ_0	1.0	1.5	2.0	2.2	2.4	2.6	2.8	3.0	3.1
$W_{+\exp}(\tau, x_0)$	0.027	0.061	0.188	0.404	1.009	2.629	5.471	8.122	8.678
τ/τ_0	3.2	3.4	3.6	3.8	4.0	4.3	5.0	5.7	6.2
$W_{+\exp}(\tau, x_0)$	8.530	6.408	3.969	2.246	1.345	0.846	1.811	4.999	3.076
τ/τ_0	7.0	8.3	9.6	10.9	12.2	13.4	14.7	15.9	17.3
$W_{+\exp}(\tau, x_0)$	1.228	3.940	1.451	3.435	1.564	3.087	1.634	2.828	1.651
τ/τ_0	18.5	19.8	21.1	22.4	23.7	25.0	26.3	27.5	28.8
$W_{+\exp}(\tau, x_0)$	2.699	1.676	2.575	1.680	2.460	1.644	2.360	1.637	2.252
$x_0/\sigma = 3.0$									
τ/τ_0	0.05	0.10	0.15	0.20	0.25	0.30	0.35	0.40	0.45
$W_{\exp}(\tau, x_0)$	23.62	46.08	67.29	86.57	103.6	117.4	127.9	134.1	137.8

(Continued)

Table A1.4 (Continued)

τ/τ_0	0.50	0.55	0.60	0.65	0.70	0.75	0.80	0.85	0.90
$W_{exp}(\tau, x_0)$	138.1	134.2	127.0	118.8	108.3	96.03	83.73	71.11	59.20
τ/τ_0	0.95	1.0	1.05	1.1	1.15	1.2	1.25	1.3	1.35
$W_{exp}(\tau, x_0)$	48.37	38.49	30.14	23.34	17.95	13.32	10.41	7.864	6.027
τ/τ_0	1.4	1.45	1.5	1.55	1.6	1.65	1.7	1.75	1.8
$W_{exp}(\tau, x_0)$	4.625	3.718	2.989	2.336	2.002	1.548	1.323	1.056	0.794
τ/τ_0	1.0	1.5	2.0	2.2	2.4	2.6	2.8	3.0	3.1
$W_{+exp}(\tau, x_0)$	0.083	0.106	0.139	0.234	0.678	2.013	5.207	8.592	9.730
τ/τ_0	3.2	3.4	3.6	3.8	4.0	4.5	5.0	5.7	6.3
$W_{+exp}(\tau, x_0)$	9.454	6.579	3.724	1.870	1.065	0.577	1.372	5.129	2.329
τ/τ_0	6.9	8.2	9.6	10.9	12.2	13.5	14.9	16.1	17.3
$W_{+exp}(\tau, x_0)$	0.886	3.799	1.109	3.255	1.176	3.014	1.377	2.879	1.452
τ/τ_0	18.5	19.8	21.0	22.3	23.7	25.1	26.3	27.6	28.9
$W_{+exp}(\tau, x_0)$	2.771	1.456	2.636	1.566	2.546	1.564	2.391	1.646	2.415

Table A1.5 Gaussian PSD

τ/τ_0	1.2	1.5	1.8	2.1	2.4	2.7	3.0	3.3	3.6
$W_3(\tau, 0)$	0.163	0.531	1.258	2.439	4.055	5.978	8.022	9.992	11.74
τ/τ_0	3.9	4.2	4.5	4.8	5.1	5.4	5.7	6.0	6.6
$W_3(\tau, 0)$	13.18	14.27	15.01	15.45	15.57	15.48	15.20	14.74	13.47
τ/τ_0	7.8	9.0	10.2	11.4	12.6	13.8	15.0	16.5	18.0
$W_3(\tau, 0)$	10.36	7.361	4.947	3.192	2.001	1.223	0.732	0.375	0.190
τ/τ_0	2.0	2.4	2.8	3.2	3.6	4.0	4.4	4.8	5.2
$W_4(\tau, 0)$	0.065	0.246	0.663	1.409	2.507	3.906	5.488	7.104	8.631
τ/τ_0	5.6	6.0	6.4	6.8	7.2	7.6	8.0	8.8	9.6
$W_4(\tau, 0)$	9.981	11.10	11.91	12.45	12.73	12.76	12.59	11.74	10.45
τ/τ_0	10.4	11.2	12.0	13.6	15.2	16.8	18.4	20.0	21.6
$W_4(\tau, 0)$	8.992	7.503	6.121	3.831	2.263	1.273	0.692	0.364	0.187

Table A1.6 Rectangular PSD

τ/τ_0	1.6	1.8	2.0	2.2	2.4	2.8	3.2	3.6	4.0
$W_3(\tau, 0)$	0.111	0.213	0.407	0.726	1.202	2.935	5.957	10.11	14.62
τ/τ_0	4.4	4.8	5.0	5.2	5.6	6.0	6.4	7.0	8.0
$W_3(\tau, 0)$	18.20	19.81	19.91	19.57	18.14	16.69	15.43	13.74	10.09
τ/τ_0	9.0	10.0	11.0	12.0	13.0	14.0	15.0	16.0	17.0
$W_3(\tau, 0)$	7.047	4.821	3.128	2.042	1.289	0.797	0.488	0.304	0.183
τ/τ_0	3.0	3.4	3.8	4.2	4.6	5.0	5.4	5.8	6.2
$W_4(\tau, 0)$	0.189	0.517	1.226	2.463	4.433	6.997	9.678	12.19	14.07
τ/τ_0	6.6	7.0	7.4	8.0	9.0	10.0	11.0	12.0	13.0
$W_4(\tau, 0)$	14.97	15.27	15.20	14.57	12.70	10.12	7.695	5.592	3.956
τ/τ_0	14.0	15.0	16.0	17.0	18.0	19.0	20.0	21.0	22.0
$W_4(\tau, 0)$	2.712	1.826	1.221	0.786	0.505	0.323	0.205	0.124	0.072

Table A1.7 Gaussian PSD

$x_0/\sigma = 0.5$

τ/τ_0	0.05	0.10	0.15	0.20	0.25	0.30	0.35	0.40	0.45
$W_{exp}(\tau, x_0)$	1.826	3.676	5.413	7.174	8.844	10.44	12.01	13.45	14.96
τ/τ_0	0.50	0.60	0.70	0.80	0.90	1.0	1.10	1.20	1.40
$W_{exp}(\tau, x_0)$	16.31	18.90	21.11	22.86	24.34	25.28	25.74	25.88	24.99
τ/τ_0	1.50	1.60	1.70	1.80	2.0	3.0	4.0	5.0	6.0
$W_{exp}(\tau, x_0)$	24.26	23.30	22.37	21.73	20.03	14.54	10.46	7.522	5.443
τ/τ_0	7,0	8.0	9.0	10.0	11.0	12.0	13.0	14.0	15.0
$W_{exp}(\tau, x_0)$	3.925	2.867	2.045	1.458	1.045	0.774	0.558	0.400	0.294
τ/τ_0	0.02	0.04	0.06	0.08	0.10	0.12	0.14	0.16	0.18
$W_{\_exp}(\tau, x_0)$	11.57	23.48	35.54	47.56	60.75	75.47	91.96	110.7	133.6
τ/τ_0	0.20	0.22	0.26	0.30	0.34	0.38	0.42	0.46	0.50
$W_{\_exp}(\tau, x_0)$	159.5	185.4	223.4	231.9	217.2	190.5	160.0	132.5	107.8
τ/τ_0	0.54	0.58	0.62	0.66	0.70	0.74	0.78	0.82	0.90
$W_{\_exp}(\tau, x_0)$	87.50	72.41	59.83	50.87	43.60	38.19	34.15	31.15	26.07
τ/τ_0	1.00	1.10	1.20	1.30	1.40	1.50	1.60	1.70	1.80
$W_{\_exp}(\tau, x_0)$	21.00	16.57	12.53	9.582	7.327	5.570	4.314	3.325	2.545

$x_0/\sigma = 1.0$

τ/τ_0	0.10	0.20	0.30	0.40	0.50	0.60	0.70	0.80	0.90
$W_{exp}(\tau, x_0)$	7.914	15.55	22.72	29.43	35.63	40.75	44.86	47.60	48.94
τ/τ_0	0.95	1.0	1.2	1.4	1.6	1.8	2.0	2.2	2.4
$W_{exp}(\tau, x_0)$	49.12	48.94	45.75	39.78	33.49	28.08	23.75	20.47	17.57
τ/τ_0	2.6	2.8	3.0	3.4	3.8	4.2	4.6	5.0	5.4
$W_{exp}(\tau, x_0)$	15.05	13.00	11.10	8.224	6.134	4.478	3.324	2.442	1.797
τ/τ_0	5.8	6.2	6.6	7.0	7.4	7.8	8.2	8.6	9.0
$W_{exp}(\tau, x_0)$	1.325	0.996	0.737	0.529	0.397	0.284	0.209	0.166	0.120
τ/τ_0	0.05	0.10	0.15	0.20	0.25	0.30	0.35	0.40	0.45
$W_{\_exp}(\tau, x_0)$	6.094	12.40	18.86	25.92	33.84	43.19	55.22	69.58	83.63
τ/τ_0	0.50	0.55	0.60	0.65	0.70	0.75	0.80	0.85	0.90
$W_{\_exp}(\tau, x_0)$	93.97	98.79	98.22	93.27	86.07	77.47	69.34	62.07	55.58
τ/τ_0	1.0	1.1	1.2	1.4	1.6	1.8	2.0	2.2	2.4
$W_{\_exp}(\tau, x_0)$	46.14	40.02	35.92	30.00	24.48	19.50	15.55	12.54	10.06
τ/τ_0	2.6	2.8	3.0	3.2	3.4	3.6	3.8	4.0	4.6
$W_{\_exp}(\tau, x_0)$	8.023	6.483	5.231	4.165	3.351	2.695	2.165	1.739	0.892

$x_0/\sigma = 1.5$

τ/τ_0	0.1	0.2	0.3	0.4	0.5	0.6	0.7	0.8	0.9
$W_{exp}(\tau, x_0)$	13.86	27.30	39.62	50.52	59.26	65.74	69.46	70.24	68.52
τ/τ_0	1.0	1.1	1.2	1.3	1.4	1.5	1.6	1.70	1.8
$W_{exp}(\tau, x_0)$	64.80	59.41	52.93	46.31	40.30	34.70	29.86	25.75	22.25
τ/τ_0	1.9	2.0	2.1	2.2	2.3	2.4	2.5	2.6	2.8
$W_{exp}(\tau, x_0)$	19.33	16.92	14.84	13.02	11.46	10.08	8.833	7.785	5.987
τ/τ_0	3.0	3.2	3.4	3.6	3.8	4.0	4.2	4.4	4.6
$W_{exp}(\tau, x_0)$	4.580	3.558	2.698	2.088	1.619	1.235	0.980	0.728	0.579

(Continued)

Table A1.7 (Continued)

τ/τ_0	0.1	0.2	0.3	0.4	0.5	0.6	0.7	0.8	0.9
W_exp(τ, x_0)	3.972	8.073	12.53	17.84	25.07	34.49	42.86	46.33	44.79
τ/τ_0	1.0	1.2	1.4	1.6	1.8	2.0	2.2	2.4	2.8
W_exp(τ, x_0)	40.62	32.76	28.51	26.12	24.04	21.95	19.94	18.08	14.94
τ/τ_0	3.2	3.6	4.0	4.4	4.8	5.2	5.6	6.0	6.4
W_exp(τ, x_0)	12.33	10.22	8.434	6.968	5.758	4.776	3.928	3.237	2.682
τ/τ_0	6.8	7.2	7.6	8.0	8.4	8.8	9.2	9.6	10.0
W_exp(τ, x_0)	2.214	1.838	1.513	1.256	1.034	0.856	0.704	0.580	0.481
$x_0/\sigma = 2.0$									
τ/τ_0	0.05	0.10	0.15	0.20	0.25	0.30	0.35	0.40	0.45
$W_{exp}(\tau, x_0)$	11.20	22.26	32.71	42.99	52.55	61.02	68.82	75.26	80.84
τ/τ_0	0.50	0.55	0.60	0.65	0.70	0.75	0.80	0.85	0.90
$W_{exp}(\tau, x_0)$	85.33	88.35	89.95	90.44	90.13	87.91	85.58	82.20	77.52
τ/τ_0	0.95	1.0	1.1	1.2	1.3	1.4	1.5	1.6	1.7
$W_{exp}(\tau, x_0)$	72.84	67.53	57.10	47.06	37.82	30.29	24.04	19.19	15.43
τ/τ_0	1.8	1.9	2.0	2.2	2.4	2.6	2.8	3.0	3.2
$W_{exp}(\tau, x_0)$	12.58	10.25	8.399	5.618	3.862	2.572	1.740	1.173	0.819
τ/τ_0	0.1	0.2	0.3	0.4	0.5	0.6	0.7	0.8	0.9
W_exp(τ, x_0)	1.385	2.785	4.224	5.698	7.450	9.585	12.46	15.78	18.41
τ/τ_0	1.0	1.1	1.2	1.4	1.6	1.8	2.0	2.4	2.8
W_exp(τ, x_0)	19.65	19.34	18.43	16.50	15.42	14.86	14.38	13.34	12.33
τ/τ_0	3.2	3.6	4.0	5.0	6.0	7.0	8.0	9.0	10.0
W_exp(τ, x_0)	11.43	10.59	9.810	8.100	6.693	5.486	4.562	3.738	3.081
τ/τ_0	12.0	14.0	16.0	18.0	20.0	22.0	24.0	26.0	28.0
W_exp(τ, x_0)	2.104	1.440	0.973	0.655	0.437	0.309	0.210	0.140	0.095
$x_0/\sigma = 3.0$									
τ/τ_0	0.05	0.10	0.15	0.20	0.25	0.30	0.35	0.40	0.45
$W_{exp}(\tau, x_0)$	22.95	45.69	66.68	84.56	100.0	113.2	121.7	128.0	130.6
τ/τ_0	0.50	0.55	0.60	0.65	0.70	0.75	0.80	0.85	0.90
$W_{exp}(\tau, x_0)$	131.0	126.1	119.6	111.9	103.2	91.55	81.51	70.86	60.69
τ/τ_0	0.95	1.0	1.05	1.1	1.2	1.3	1.4	1.5	1.6
$W_{exp}(\tau, x_0)$	51.72	42.80	35.74	29.41	19.44	12.71	8.430	5.563	3.692
τ/τ_0	1.7	1.8	1.9	2.0	2.1	2.2	2.3	2.4	2.5
$W_{exp}(\tau, x_0)$	2.453	1.672	1.120	0.797	0.538	0.358	0.277	0.241	0.143
τ/τ_0	0.2	0.4	0.6	0.8	1.0	1.2	1.5	2.0	5.0
W_exp(τ, x_0)	0.307	0.590	0.822	1.027	1.405	1.789	1.985	1.848	1.771
τ/τ_0	10.0	15.0	20.0	25.0	30.0	35.0	40.0	45.0	50.0
W_exp(τ, x_0)	1.628	1.472	1.335	1.210	1.105	1.006	0.911	0.835	0.758
τ/τ_0	55.0	60.0	65.0	70.0	75.0	80.0	85.0	90.0	95.0
W_exp(τ, x_0)	0.681	0.623	0.566	0.515	0.461	0.420	0.380	0.335	0.312
τ/τ_0	100.0	105.0	110.0	115.0	120.0	125.0	130.0	135.0	140.0
W_exp(τ, x_0)	0.285	0.250	0.229	0.210	0.189	0.167	0.156	0.142	0.125

Table A1.8 Rectangular PSD

$x_0/\sigma = 0.5$

τ/τ_0	0.10	0.20	0.30	0.40	0.5	0.60	0.70	0.80	0.90
$W_{exp}(\tau, x_0)$	2.035	4.015	6.112	8.343	10.57	13.13	15.77	18.73	22.02
τ/τ_0	1.00	1.10	1.20	1.30	1.40	1.50	1.60	1.80	2.00
$W_{exp}(\tau, x_0)$	25.17	28.48	31.37	34.04	35.37	35.83	34.81	29.86	23.63
τ/τ_0	2.2	2.4	2.8	3.2	3.6	4.0	4.4	4.8	5.2
$W_{exp}(\tau, x_0)$	19.16	16.75	15.71	13.73	11.22	10.03	8.730	7.336	6.327
τ/τ_0	5,6	6.0	7.0	8.0	9.0	10.0	11.0	12.0	13.0
$W_{exp}(\tau, x_0)$	5.619	4.822	3.438	2.389	1.659	1.185	0.837	0.598	0.415
τ/τ_0	0.05	0.10	0.15	0.20	0.25	0.30	0.35	0.40	0.45
$W\_{exp}(\tau, x_0)$	24.32	51.35	87.01	141.2	197.4	221.2	210.5	181.8	148.8
τ/τ_0	0.50	0.55	0.60	0.65	0.70	0.75	0.80	0.85	0.90
$W\_{exp}(\tau, x_0)$	118.4	92.92	72.51	56.66	44.74	35.75	29.16	24.43	20.96
τ/τ_0	0.95	1.00	1.10	1.20	1.30	1.40	1.50	1.60	1.70
$W\_{exp}(\tau, x_0)$	18.69	16.92	14.70	13.09	11.42	9.761	8.025	6.480	5.190
τ/τ_0	1.80	1.90	2.00	2.20	2.40	2.60	2.80	3.00	3.20
$W\_{exp}(\tau, x_0)$	4.156	3.375	2.740	2.059	1.572	1.240	0.934	0.664	0.520

$x_0/\sigma = 1.0$

τ/τ_0	0.1	0.2	0.3	0.4	0.5	0.6	0.7	0.8	0.9
$W_{exp}(\tau, x_0)$	4.848	9.718	14.68	19.89	25.21	30.81	36.60	42.19	47.72
τ/τ_0	1.0	1.1	1.2	1.25	1.3	1.4	1.6	1.8	2.0
$W_{exp}(\tau, x_0)$	52.42	56.08	58.10	58.22	57.83	55.45	45.34	33.08	23.88
τ/τ_0	2.2	2.4	2.6	2.8	3.0	3.2	3.4	3.6	3.8
$W_{exp}(\tau, x_0)$	18.67	16.06	14.51	12.82	10.81	8.932	7.182	6.073	5.202
τ/τ_0	4.0	4.2	4.4	4.6	4.8	5.2	5.6	6.0	6.4
$W_{exp}(\tau, x_0)$	4.474	3.791	3.150	2.642	2.194	1.604	1.132	0.799	0.568
τ/τ_0	0.05	0.10	0.15	0.20	0.25	0.30	0.35	0.40	0.45
$W\_{exp}(\tau, x_0)$	4.515	9.174	14.37	20.09	27.13	36.36	49.16	65.69	83.19
τ/τ_0	0.50	0.55	0.60	0.65	0.70	0.75	0.80	0.85	0.90
$W\_{exp}(\tau, x_0)$	97.96	107.0	109.6	107.1	100.1	91.05	80.80	70.50	60.63
τ/τ_0	1.0	1.1	1.2	1.3	1.4	1.6	1.8	2.0	2.2
$W\_{exp}(\tau, x_0)$	44.42	33.31	26.81	23.31	21.95	21.18	19.48	15.62	11.82
τ/τ_0	2.4	2.6	2.8	3.0	3.4	3.8	4.2	4.6	5.0
$W\_{exp}(\tau, x_0)$	8.992	7.444	6.748	5.832	3.972	2.723	2.067	1.458	1.051

$x_0/\sigma = 1.5$

τ/τ_0	0.05	0.10	0.15	0.20	0.25	0.30	0.35	0.40	0.45
$W_{exp}(\tau, x_0)$	4.900	9.767	14.72	19.60	24.42	29.42	34.11	38.97	43.59
τ/τ_0	0.50	0.60	0.70	0.80	0.90	1.0	1.1	1.2	1.3
$W_{exp}(\tau, x_0)$	48.30	56.82	64.54	70.87	75.20	76.52	74.89	70.20	62.72
τ/τ_0	1.4	1.5	1.6	1.7	1.8	1.9	2.0	2.2	2.4
$W_{exp}(\tau, x_0)$	53.52	43.86	34.94	27.07	21.20	16.76	13.69	10.00	7.951
τ/τ_0	2.6	2.8	3.0	3.2	3.4	3.6	3.8	4.0	4.2
$W_{exp}(\tau, x_0)$	6.359	4.834	3.461	2.500	1.784	1.322	1.016	0.729	0.558

(Continued)

Table A1.8 (Continued)

τ/τ_0	0.10	0.20	0.30	0.40	0.50	0.60	0.70	0.80	0.85
$W\_{exp}(\tau, x_0)$	2.361	5.003	8.320	13.09	21.67	36.22	52.28	61.93	63.03
τ/τ_0	0.90	1.00	1.10	1.20	1.30	1.40	1.50	1.60	1.8
$W\_{exp}(\tau, x_0)$	61.98	54.71	44.21	33.86	26.23	21.22	18.87	18.32	20.00
τ/τ_0	2.05	2.2	2.4	2.6	2.8	3.0	3.4	3.8	4.2
$W\_{exp}(\tau, x_0)$	21.65	20.34	16.98	13.60	11.93	11.46	10.85	8.192	6.872
τ/τ_0	4.6	5.0	5.8	6.6	7.4	8.2	9.0	9.8	10.6
$W\_{exp}(\tau, x_0)$	6.259	5.075	3.723	2.568	1.936	1.418	1.001	0.761	0.565
$x_0/\sigma = 2.0$									
τ/τ_0	0.1	0.2	0.3	0.4	0.5	0.6	0.7	0.8	0.9
$W_{exp}(\tau, x_0)$	7.628	15.32	23.01	30.87	38.72	46.31	53.45	59.99	65.25
τ/τ_0	1.0	1.1	1.2	1.3	1.4	1.5	1.6	1.7	1.8
$W_{exp}(\tau, x_0)$	68.55	69.68	68.01	63.68	57.15	49.04	40.72	33.10	26.69
τ/τ_0	1.9	2.0	2.1	2.2	2.4	2.6	2.8	3.0	3.2
$W_{exp}(\tau, x_0)$	21.63	18.01	15.38	13.46	11.20	9.365	7.573	5.911	4.427
τ/τ_0	3.4	3.6	3.8	4.0	4.2	4.4	4.6	4.8	5.0
$W_{exp}(\tau, x_0)$	3.380	2.613	2.095	1.671	1.291	1.009	0.771	0.620	0.470
τ/τ_0	0.1	0.2	0.3	0.4	0.5	0.6	0.7	0.8	0.9
$W\_{exp}(\tau, x_0)$	3.904	8.441	14.26	23.69	40.13	60.60	74.36	76.39	68.16
τ/τ_0	1.0	1.1	1.2	1.3	1.4	1.5	1.6	1.7	1.8
$W\_{exp}(\tau, x_0)$	55.20	42.39	32.32	26.05	22.20	20.88	20.62	21.39	21.99
τ/τ_0	1.9	2.0	2.2	2.4	2.8	3.2	3.6	4.0	4.4
$W\_{exp}(\tau, x_0)$	22.32	21.81	18.70	14.81	10.70	9.667	7.279	5.481	4.717
τ/τ_0	4.8	5.2	5.6	6.0	6.4	6.8	7.2	7.6	8.0
$W\_{exp}(\tau, x_0)$	3.754	2.918	2.406	1.974	1.536	1.287	1.055	0.840	0.696
$x_0/\sigma = 3.0$									
τ/τ_0	0.05	0.10	0.15	0.20	0.25	0.30	0.35	0.40	0.45
$W_{exp}(\tau, x_0)$	8.400	16.69	24.94	32.94	40.97	48.46	55.83	62.76	69.08
τ/τ_0	0.50	0.55	0.60	0.65	0.70	0.75	0.80	0.85	0.90
$W_{exp}(\tau, x_0)$	74.86	80.19	84.46	87.98	90.55	92.18	92.78	91.91	90.17
τ/τ_0	0.95	1.0	1.05	1.1	1.2	1.3	1.4	1.5	1.6
$W_{exp}(\tau, x_0)$	87.35	83.75	78.96	73.29	60.99	48.03	35.96	25.94	18.38
τ/τ_0	1.7	1.8	1.9	2.0	2.2	2.4	2.6	2.8	3.0
$W_{exp}(\tau, x_0)$	12.94	9.372	7.057	5.484	3.624	2.550	1.730	1.097	0.671
τ/τ_0	0.1	0.2	0.3	0.4	0.5	0.6	0.7	0.8	0.9
$W\_{exp}(\tau, x_0)$	0.726	1.480	2.385	3.654	5.385	8.638	14.80	23.50	31.70
τ/τ_0	1.0	1.1	1.2	1.3	1.4	1.5	1.6	1.7	2.0
$W\_{exp}(\tau, x_0)$	36.25	35.29	30.24	23.60	18.07	13.86	11.67	11.06	13.79
τ/τ_0	2.3	2.7	3.1	3.5	4.0	5.0	6.0	7.0	8.0
$W\_{exp}(\tau, x_0)$	16.40	12.16	9.917	11.25	8.995	7.873	6.261	5.019	4.091
τ/τ_0	9.0	10.0	11.0	12.0	13.0	14.0	15.0	16.0	17.0
$W\_{exp}(\tau, x_0)$	3.422	2.972	2.366	1.994	1.673	1.396	1.154	0.970	0.801

Table A1.9 presents simulation results for a Rayleigh process with a Gaussian PSD. The same table shows the values of the distribution density $W_{+\max}(\tau, 0)$ of time intervals between successive maxima (or minima) of the Rayleigh process. Each of those intervals is a combination of two adjacent intervals between successive zeros of the first derivative.

Table A1.10 shows similar data obtained for a Rayleigh process with a Rectangular PSD.

Table A1.9 Gaussian PSD: Distribution Densities of Time Intervals Between Successive Extrema (Maxima and Minima) of a Rayleigh Process

τ/τ_0	0.05	0.10	0.15	0.20	0.25	0.30	0.35	0.40	0.45
$W_{\max}(\tau, 0)$	2.540	5.046	7.669	10.28	13.16	16.09	19.26	22.68	26.32
τ/τ_0	0.50	0.55	0.60	0.65	0.70	0.75	0.80	0.85	0.90
$W_{\max}(\tau, 0)$	30.38	34.73	39.60	44.90	50.48	56.39	62.72	68.76	74.56
τ/τ_0	0.95	1.0	1.05	1.10	1.15	1.20	1.30	1.40	1.50
$W_{\max}(\tau, 0)$	79.51	83.33	85.44	85.63	83.91	80.42	69.93	57.69	47.03
τ/τ_0	1.60	1.80	2.00	2.40	2.80	3.20	3.60	4.00	4.4
$W_{\max}(\tau, 0)$	38.67	28.21	22.03	12.17	5.560	2.489	1.051	0.401	0.156
τ/τ_0	0.40	0.60	0.80	1.00	1.20	1.40	1.60	1.80	2.00
$W_{+\max}(\tau, 0)$	0.164	0.883	3.013	7.695	16.14	27.21	37.52	43.96	45.90
τ/τ_0	2.20	2.40	2.60	2.80	3.00	3.20	3.40	3.60	3.80
$W_{+\max}(\tau, 0)$	44.66	41.52	37.97	34.26	30.00	25.83	21.68	17.99	14.56
τ/τ_0	4.00	4.20	4.40	4.60	4.80	5.00	5.20	5.40	5.60
$W_{+\max}(\tau, 0)$	11.64	9.210	7.184	5.527	4.161	3.177	2.364	1.737	1.267
τ/τ_0	5.80	6.00	6.20	6.40	6.60	6.80	7.00	7.20	7.40
$W_{+\max}(\tau, 0)$	0.913	0.651	0.447	0.310	0.225	0.150	0.099	0.067	0.045

Table A1.10 Rectangular PSD: Distribution Densities of Time Intervals Between Successive Extrema (Maxima and Minima) of a Rayleigh Process

τ/τ_0	0.10	0.20	0.30	0.40	0.50	0.60	0.70	0.80	0.90
$W_{\max}(\tau, 0)$	0.901	1.782	2.805	4.051	5.463	7.358	9.809	13.23	18.13
τ/τ_0	1.00	1.10	1.20	1.30	1.40	1.50	1.55	1.60	1.70
$W_{\max}(\tau, 0)$	25.16	35.70	51.45	73.29	99.32	120.3	124.6	123.1	104.3
τ/τ_0	1.80	1.90	2.00	2.10	2.20	2.30	2.40	2.60	2.80
$W_{\max}(\tau, 0)$	77.14	52.39	34.91	23.33	16.02	11.64	8.852	6.196	5.423
τ/τ_0	3.00	3.20	3.40	3.60	3.80	4.00	4.20	4.40	4.60
$W_{\max}(\tau, 0)$	5.388	5.488	5.178	3.999	2.490	1.351	0.733	0.431	0.287
τ/τ_0	1.00	1.20	1.40	1.60	1.80	1.90	2.00	2.10	2.20
$W_{+\max}(\tau, 0)$	0.191	0.475	1.109	2.452	5.167	7.306	10.21	13.79	18.30
τ/τ_0	2.30	2.40	2.50	2.60	2.70	2.80	2.90	3.00	3.10
$W_{+\max}(\tau, 0)$	23.88	30.48	38.37	47.19	56.22	64.57	70.20	71.82	68.67
τ/τ_0	3.20	3.30	3.40	3.60	3.80	4.00	4.20	4.40	4.60
$W_{+\max}(\tau, 0)$	61.70	52.56	43.15	27.74	18.36	13.44	11.29	10.68	10.82
τ/τ_0	4.70	5.00	5.20	5.40	5.60	6.00	6.40	7.00	8.00
$W_{+\max}(\tau, 0)$	10.91	10.14	8.340	6.052	4.036	1.806	1.040	0.706	0.137

Tables A1.11 and A1.12 present the data obtained experimentally (i.e., by means of statistical simulation) for the distribution density of time intervals during which the momentary values of a Rayleigh process are confined to the corridor formed by two standardized thresholds $(x_1/\sigma, x_2/\sigma)$ (in the case of Table A1.11) or remain out of this corridor (in the case of Table A1.12). The corridor boundaries were chosen within the interval (0.5, 3) with the distance between the lower and upper boundaries taking one of two values $(x_1 - x_2)/\sigma = 0.5, 1.0$. The data were obtained for broadband processes with two kinds of PSDs: a Gaussian one and a rectangular one.

Table A1.11 Gaussian PSD

$(x_1/\sigma)/(x_2/\sigma) = 1.0/0.5$									
τ/τ_0	0.02	0.04	0.06	0.08	0.10	0.12	0.14	0.16	0.18
$W_{inside}(\tau,0)$	1.748	3.422	5.283	9.923	36.00	102.2	188.4	258.1	295.3
τ/τ_0	0.20	0.22	0.24	0.26	0.28	0.30	0.32	0.34	0.38
$W_{inside}(\tau,0)$	304.7	294.1	272.9	246.9	220.7	195.0	173.4	153.0	122.2
τ/τ_0	0.42	0.46	0.50	0.60	0.70	0.80	0.90	1.00	1.10
$W_{inside}(\tau,0)$	99.86	83.72	72.48	53.27	42.10	33.40	26.64	21.42	17.07
τ/τ_0	1.20	1.30	1.40	1.50	1.60	1.80	2.00	2.20	2.40
$W_{inside}(\tau,0)$	13.55	10.70	8.681	7.022	5.588	3.499	2.278	1.482	0.949
τ/τ_0	0.05	0.10	0.15	0.20	0.25	0.30	0.35	0.40	0.45
$W_{outside}(\tau,0)$	14.73	30.61	49.86	76.44	101.1	110.0	104.0	91.09	77.87
τ/τ_0	0.50	0.55	0.60	0.70	0.80	0.90	1.00	1.20	1.40
$W_{outside}(\tau,0)$	66.39	57.56	51.42	44.50	41.32	39.31	37.16	31.87	26.09
τ/τ_0	1.60	1.80	2.00	2.20	2.40	2.60	2.80	3.00	3.40
$W_{outside}(\tau,0)$	21.21	17.34	14.40	12.25	10.41	8.862	7.618	6.475	4.781
τ/τ_0	3.80	4.20	4.60	5.00	5.40	5.80	6.20	6.60	7.00
$W_{outside}(\tau,0)$	3.560	2.597	1.926	1.415	1.041	0.768	0.577	0.427	0.307
$(x_1/\sigma)/(x_2/\sigma) = 1.5/1.0$									
τ/τ_0	0.02	0.04	0.06	0.08	0.10	0.12	0.14	0.16	0.18
$W_{inside}(\tau,0)$	1.259	2.420	3.714	7.846	34.12	101.4	189.4	262.1	305.0
τ/τ_0	0.20	0.22	0.24	0.26	0.28	0.30	0.32	0.34	0.38
$W_{inside}(\tau,0)$	316.6	308.8	286.6	261.9	233.3	204.7	179.4	156.3	119.0
τ/τ_0	0.42	0.46	0.50	0.60	0.70	0.80	0.90	1.00	1.10
$W_{inside}(\tau,0)$	92.36	75.11	63.53	49.85	42.91	35.18	27.40	20.90	16.19
τ/τ_0	1.20	1.30	1.40	1.50	1.60	1.80	2.00	2.20	2.40
$W_{inside}(\tau,0)$	13.01	10.23	8.110	6.568	5.147	3.336	2.140	1.336	0.876
τ/τ_0	0.05	0.10	0.15	0.20	0.25	0.30	0.35	0.40	0.45
$W_{outside}(\tau,0)$	6.515	13.05	19.69	26.53	33.82	41.60	50.79	61.09	70.99
τ/τ_0	0.50	0.55	0.60	0.70	0.80	0.90	1.00	1.10	1.20
$W_{outside}(\tau,0)$	78.52	82.79	83.76	78.67	69.74	61.34	54.45	48.66	43.49
τ/τ_0	1.40	1.60	1.80	2.00	2.20	2.40	2.60	2.80	3.00
$W_{outside}(\tau,0)$	34.59	26.87	20.73	16.16	12.75	10.07	7.917	6.262	4.941
τ/τ_0	3.20	3.40	3.60	4.00	4.40	4.80	5.20	5.60	6.00
$W_{outside}(\tau,0)$	3.895	3.060	2.425	1.514	0.944	0.596	0.374	0.239	0.148

(Continued)

Table A1.11 (Continued)

$(x_1/\sigma)/(x_2/\sigma) = 2.0/1.5$

τ/τ_0	0.02	0.04	0.06	0.08	0.10	0.12	0.14	0.16	0.18
$W_{inside}(\tau, 0)$	1.885	3.841	5.671	10.67	36.55	101.3	187.6	257.0	296.9
τ/τ_0	0.20	0.22	0.24	0.26	0.28	0.30	0.32	0.34	0.38
$W_{inside}(\tau, 0)$	309.8	302.5	282.7	255.8	228.8	202.6	176.4	153.5	117.7
τ/τ_0	0.42	0.46	0.50	0.60	0.70	0.80	0.90	1.00	1.10
$W_{inside}(\tau, 0)$	93.60	78.28	68.34	58.39	51.07	39.94	28.36	19.73	14.32
τ/τ_0	1.20	1.30	1.40	1.50	1.60	1.80	2.00	2.20	2.40
$W_{inside}(\tau, 0)$	10.73	8.506	6.465	5.052	3.971	2.476	1.421	0.855	0.528
τ/τ_0	0.05	0.10	0.15	0.20	0.25	0.30	0.35	0.40	0.45
$W_{outside}(\tau, 0)$	5.312	10.57	15.63	20.61	25.45	30.04	34.41	38.56	42.52
τ/τ_0	0.50	0.55	0.60	0.65	0.70	0.75	0.80	0.90	1.00
$W_{outside}(\tau, 0)$	46.78	50.89	54.61	57.88	60.30	61.18	60.93	57.04	50.56
τ/τ_0	1.10	1.20	1.30	1.40	1.60	1.80	2.00	2.40	2.80
$W_{outside}(\tau, 0)$	43.91	38.13	33.07	29.35	23.78	20.05	17.24	13.12	10.28
τ/τ_0	3.20	3.60	4.00	5.00	6.00	7.00	8.00	9.00	10.00
$W_{outside}(\tau, 0)$	8.276	6.741	5.506	3.394	2.081	1.294	0.806	0.503	0.312

$(x_1/\sigma)/(x_2/\sigma) = 2.5/2.0$

τ/τ_0	0.02	0.04	0.06	0.08	0.10	0.12	0.14	0.16	0.18
$W_{inside}(\tau, 0)$	3.279	6.394	9.614	15.86	41.27	104.5	183.1	251.6	288.7
τ/τ_0	0.20	0.22	0.24	0.26	0.28	0.30	0.32	0.34	0.36
$W_{inside}(\tau, 0)$	297.8	291.2	270.2	247.5	219.0	194.2	169.8	148.8	132.0
τ/τ_0	0.38	0.40	0.44	0.48	0.52	0.56	0.60	0.70	0.80
$W_{inside}(\tau, 0)$	118.8	106.7	92.57	82.72	78.06	75.67	72.01	59.30	41.50
τ/τ_0	0.90	1.00	1.10	1.20	1.30	1.40	1.50	1.60	1.80
$W_{inside}(\tau, 0)$	26.34	17.09	11.57	8.033	5.978	4.554	3.283	2.506	1.513
τ/τ_0	0.10	0.20	0.30	0.40	0.50	0.60	0.65	0.70	0.80
$W_{outside}(\tau, 0)$	10.42	19.92	27.85	33.61	37.01	38.31	38.38	38.19	36.91
τ/τ_0	1.00	1.20	1.40	1.60	1.80	2.00	2.40	2.80	3.20
$W_{outside}(\tau, 0)$	30.91	22.81	16.95	13.77	12.11	11.11	9.813	8.914	8.168
τ/τ_0	3.60	4.00	4.40	5.00	6.00	7.00	8.00	9.00	10.00
$W_{outside}(\tau, 0)$	7.564	6.986	6.442	5.758	4.731	3.900	3.224	2.641	2.180
τ/τ_0	11.00	12.00	13.00	14.00	15.00	16.00	17.00	18.00	19.00
$W_{outside}(\tau, 0)$	1.818	1.498	1.225	1.007	0.824	0.688	0.563	0.463	0.386

$(x_1/\sigma)/(x_2/\sigma) = 3.0/2.5$

τ/τ_0	0.02	0.04	0.06	0.08	0.10	0.12	0.14	0.16	0.18
$W_{inside}(\tau, 0)$	5.339	10.50	15.62	22.78	48.55	108.1	184.0	246.6	279.6
τ/τ_0	0.20	0.22	0.24	0.26	0.28	0.30	0.32	0.34	0.36
$W_{inside}(\tau, 0)$	286.1	279.5	260.8	239.0	212.6	188.1	167.0	147.1	134.1
τ/τ_0	0.38	0.40	0.44	0.48	0.52	0.56	0.60	0.64	0.70
$W_{inside}(\tau, 0)$	121.3	114.5	102.6	97.39	93.81	90.98	85.51	78.93	63.45
τ/τ_0	0.80	0.90	1.00	1.10	1.20	1.30	1.40	1.50	1.60
$W_{inside}(\tau, 0)$	38.43	21.41	12.60	7.682	5.041	3.526	2.520	1.838	1.260
τ/τ_0	0.05	0.10	0.15	0.20	0.25	0.30	0.35	0.40	0.45
$W_{outside}(\tau, 0)$	5.689	11.33	16.53	21.00	24.88	28.35	30.77	32.13	32.92

(Continued)

Table A1.11 (Continued)

τ/τ_0	0.50	0.55	0.60	0.70	0.80	0.90	1.00	1.20	1.40
$W_{\text{outside}}(\tau, 0)$	33.17	32.17	30.80	27.35	22.75	18.59	15.02	10.01	7.352
τ/τ_0	1.60	1.80	2.00	4.00	6.00	8.00	10.00	12.00	14.00
$W_{\text{outside}}(\tau, 0)$	5.912	5.271	4.972	4.197	3.664	3.200	2.804	2.456	2.110
τ/τ_0	16.00	18.00	20.00	22.00	24.00	26.00	28.00	30.00	32.00
$W_{\text{outside}}(\tau, 0)$	1.838	1.625	1.420	1.252	1.090	0.960	0.840	0.723	0.622

$(x_1/\sigma)/(x_2/\sigma) = 2.0/1.0$

τ/τ_0	0.05	0.10	0.15	0.20	0.25	0.30	0.35	0.40	0.45
$W_{\text{inside}}(\tau, 0)$	2.940	5.901	9.647	25.02	64.84	109.3	134.5	136.6	125.1
τ/τ_0	0.50	0.55	0.60	0.65	0.70	0.80	0.90	1.00	1.10
$W_{\text{inside}}(\tau, 0)$	108.1	91.67	78.57	68.87	62.27	55.03	51.07	47.04	41.82
τ/τ_0	1.20	1.30	1.40	1.60	1.80	2.0	2.20	2.40	2.60
$W_{\text{inside}}(\tau, 0)$	36.07	30.78	26.17	19.72	15.26	11.76	9.216	7.050	5.431
τ/τ_0	2.80	3.00	3.20	3.40	3.60	3.80	4.20	4.60	5.00
$W_{\text{inside}}(\tau, 0)$	4.176	3.212	2.468	1.915	1.477	1.153	0.666	0.410	0.237
τ/τ_0	0.05	0.10	0.15	0.20	0.25	0.30	0.35	0.40	0.45
$W_{\text{outside}}(\tau, 0)$	7.669	15.44	23.13	31.18	39.61	48.69	59.42	71.33	82.77
τ/τ_0	0.50	0.55	0.60	0.65	0.70	0.80	0.90	1.00	1.10
$W_{\text{outside}}(\tau, 0)$	91.31	95.57	95.67	92.40	87.32	74.35	62.34	52.74	45.29
τ/τ_0	1.20	1.30	1.40	1.60	1.80	2.00	2.20	2.40	2.60
$W_{\text{outside}}(\tau, 0)$	39.35	34.37	30.10	22.85	17.37	13.35	10.40	8.147	6.342
τ/τ_0	2.80	3.00	3.20	3.60	4.00	4.40	4.80	5.20	5.60
$W_{\text{outside}}(\tau, 0)$	5.020	3.980	3.134	1.980	1.252	0.794	0.520	0.334	0.216

$(x_1/\sigma)/(x_2/\sigma) = 3.0/2.0$

τ/τ_0	0.05	0.10	0.15	0.20	0.25	0.30	0.35	0.40	0.45
$W_{\text{inside}}(\tau, 0)$	9.984	19.84	29.76	47.08	78.96	111.4	128.8	130.4	122.9
τ/τ_0	0.50	0.55	0.60	0.70	0.80	0.90	1.00	1.10	1.20
$W_{\text{inside}}(\tau, 0)$	113.1	104.5	98.01	89.13	81.15	70.62	56.96	43.16	31.60
τ/τ_0	1.30	1.40	1.50	1.60	1.70	1.80	1.90	2.00	2.10
$W_{\text{inside}}(\tau, 0)$	23.04	17.18	13.17	10.34	8.260	6.687	5.469	4.431	3.529
τ/τ_0	2.20	2.30	2.40	2.50	2.60	2.70	2.80	2.90	3.00
$W_{\text{inside}}(\tau, 0)$	2.787	2.290	1.816	1.423	1.118	0.892	0.707	0.560	0.452
τ/τ_0	0.10	0.20	0.30	0.40	0.50	0.60	0.70	0.80	0.90
$W_{\text{outside}}(\tau, 0)$	6.245	11.82	16.22	19.19	21.06	21.71	22.37	23.02	23.17
τ/τ_0	1.00	1.20	1.40	1.60	1.80	2.00	3.00	4.00	5.00
$W_{\text{outside}}(\tau, 0)$	22.22	18.71	15.70	14.24	13.43	12.99	10.73	8.756	7.276
τ/τ_0	6.00	7.00	8.00	9.00	10.00	11.00	12.00	13.00	14.00
$W_{\text{outside}}(\tau, 0)$	5.984	4.883	4.037	3.313	2.762	2.253	1.859	1.546	1.249
τ/τ_0	15.00	16.00	17.00	18.00	19.00	20.00	21.00	22.00	23.00
$W_{\text{outside}}(\tau, 0)$	1.036	0.863	0.707	0.584	0.479	0.397	0.327	0.266	0.220

Table A1.12 Rectangular PSD

$(x_1/\sigma)/(x_2/\sigma) = 1.0/0.5$

τ/τ_0	0.05	0.10	0.15	0.20	0.25	0.30	0.35	0.40	0.45
$W_{inside}(\tau,0)$	3.501	50.63	217.4	302.1	267.3	205.3	154.1	117.4	91.51
τ/τ_0	0.50	0.55	0.60	0.65	0.70	0.75	0.80	0.90	1.00
$W_{inside}(\tau,0)$	72.87	59.31	49.19	41.65	35.99	31.68	28.30	23.14	19.02
τ/τ_0	1.10	1.20	1.30	1.40	1.50	1.60	1.70	1.80	1.90
$W_{inside}(\tau,0)$	15.65	12.93	10.60	8.621	6.954	5.599	4.539	3.690	3.042
τ/τ_0	2.00	2.20	2.40	2.60	2.80	3.00	3.20	3.40	3.60
$W_{inside}(\tau,0)$	2.505	1.767	1.241	0.871	0.600	0.422	0.290	0.208	0.146
τ/τ_0	0.05	0.10	0.15	0.20	0.25	0.30	0.35	0.40	0.45
$W_{outside}(\tau,0)$	11.99	25.16	42.08	67.14	93.07	104.9	101.6	90.60	77.62
τ/τ_0	0.50	0.55	0.60	0.70	0.80	0.90	1.00	1.20	1.40
$W_{outside}(\tau,0)$	65.91	56.34	49.02	40.15	36.50	36.04	36.92	38.44	35.50
τ/τ_0	1.60	1.80	2.00	2.20	2.40	2.60	2.80	3.00	3.40
$W_{outside}(\tau,0)$	28.37	20.45	14.65	11.42	9.731	8.716	7.631	6.378	4.227
τ/τ_0	3.80	4.20	4.60	5.00	5.40	5.80	6.20	6.60	7.00
$W_{outside}(\tau,0)$	3.051	2.209	1.540	1.073	0.783	0.541	0.398	0.284	0.199

$(x_1/\sigma)/(x_2/\sigma) = 1.5/1.0$

τ/τ_0	0.05	0.10	0.15	0.20	0.25	0.30	0.35	0.40	0.45
$W_{inside}(\tau,0)$	2.389	49.12	220.0	312.5	280.9	214.5	155.1	110.5	80.05
τ/τ_0	0.50	0.55	0.60	0.65	0.70	0.75	0.80	0.90	1.00
$W_{inside}(\tau,0)$	60.69	49.10	42.43	38.62	36.55	34.98	33.40	28.47	22.27
τ/τ_0	1.10	1.20	1.30	1.40	1.50	1.60	1.70	1.80	1.90
$W_{inside}(\tau,0)$	16.85	12.88	9.967	7.802	6.126	4.858	3.929	3.170	2.646
τ/τ_0	2.00	2.10	2.20	2.30	2.40	2.50	2.60	2.70	2.80
$W_{inside}(\tau,0)$	2.223	1.844	1.532	1.252	1.031	0.838	0.680	0.563	0.459
τ/τ_0	0.05	0.10	0.15	0.20	0.25	0.30	0.35	0.40	0.45
$W_{outside}(\tau,0)$	4.685	9.435	14.52	19.87	25.94	33.30	42.54	53.93	65.76
τ/τ_0	0.50	0.55	0.60	0.65	0.70	0.80	0.90	1.00	1.20
$W_{outside}(\tau,0)$	76.11	83.13	86.38	86.90	84.45	76.43	67.04	58.55	45.90
τ/τ_0	1.40	1.60	1.80	2.00	2.20	2.40	2.60	2.80	3.00
$W_{outside}(\tau,0)$	35.84	27.24	20.24	14.77	11.02	8.534	6.966	5.905	4.789
τ/τ_0	3.20	3.40	3.60	4.00	4.40	4.80	5.20	5.60	6.00
$W_{outside}(\tau,0)$	3.847	3.009	2.374	1.640	1.160	0.772	0.565	0.427	0.301

$(x_1/\sigma)/(x_2/\sigma) = 2.0/1.5$

τ/τ_0	0.05	0.10	0.15	0.20	0.25	0.30	0.35	0.40	0.45
$W_{inside}(\tau,0)$	3.767	50.14	215.6	304.8	273.3	209.2	149.9	105.5	77.19
τ/τ_0	0.50	0.55	0.60	0.65	0.70	0.75	0.80	0.85	0.90
$W_{inside}(\tau,0)$	61.26	53.44	49.92	49.12	48.50	47.31	44.48	39.87	34.38
τ/τ_0	0.95	1.00	1.10	1.20	1.30	1.40	1.50	1.60	1.70
$W_{inside}(\tau,0)$	28.63	23.42	15.50	10.61	7.555	5.613	4.223	3.274	2.597
τ/τ_0	1.80	1.90	2.00	2.10	2.20	2.30	2.40	2.50	2.60
$W_{inside}(\tau,0)$	2.089	1.734	1.416	1.147	0.881	0.685	0.549	0.421	0.330
τ/τ_0	0.05	0.10	0.15	0.20	0.25	0.30	0.35	0.40	0.45
$W_{outside}(\tau,0)$	3.829	7.632	11.44	15.35	19.23	23.11	27.21	31.41	36.01

(Continued)

Table A1.12 (Continued)

τ/τ_0	0.50	0.55	0.60	0.65	0.70	0.75	0.80	0.85	0.90
$W_{outside}(\tau, 0)$	41.39	47.37	54.03	60.64	66.52	70.85	73.26	73.54	72.16
τ/τ_0	1.00	1.10	1.20	1.40	1.60	1.80	2.00	2.40	2.80
$W_{outside}(\tau, 0)$	64.97	54.00	42.94	25.93	17.94	15.99	15.66	11.75	8.048
τ/τ_0	3.20	3.60	4.00	5.00	6.00	7.00	8.00	9.00	10.00
$W_{outside}(\tau, 0)$	7.473	6.273	4.732	3.275	2.227	1.466	0.986	0.646	0.449

$(x_1/\sigma)/(x_2/\sigma) = 2.5/2.0$

τ/τ_0	0.02	0.04	0.06	0.08	0.10	0.12	0.14	0.16	0.18
$W_{inside}(\tau, 0)$	3.388	5.107	7.401	12.64	37.68	100.6	180.9	250.5	287.8
τ/τ_0	0.20	0.22	0.24	0.26	0.28	0.30	0.32	0.34	0.36
$W_{inside}(\tau, 0)$	300.7	294.0	275.2	251.5	224.8	198.8	173.8	152.2	131.6
τ/τ_0	0.38	0.40	0.44	0.48	0.52	0.56	0.60	0.64	0.70
$W_{inside}(\tau, 0)$	115.0	102.2	83.32	74.04	69.10	67.58	67.49	67.31	66.02
τ/τ_0	0.80	0.90	1.00	1.10	1.20	1.30	1.40	1.60	1.80
$W_{inside}(\tau, 0)$	52.55	33.83	19.25	11.01	6.578	4.348	2.988	1.599	0.869
τ/τ_0	0.10	0.20	0.30	0.40	0.50	0.60	0.70	0.80	0.90
$W_{outside}(\tau, 0)$	8.462	16.45	23.52	29.41	34.14	37.77	41.04	44.26	46.09
τ/τ_0	1.00	1.10	1.20	1.30	1.40	1.60	1.80	2.00	2.30
$W_{outside}(\tau, 0)$	44.70	39.43	31.58	23.32	16.62	9.423	8.115	9.481	11.31
τ/τ_0	2.60	3.00	3.55	4.00	5.00	6.00	7.00	8.00	9.00
$W_{outside}(\tau, 0)$	9.254	6.680	7.581	6.174	5.454	4.428	3.538	2.914	2.533
τ/τ_0	10.00	11.00	12.00	13.00	14.00	15.00	16.00	18.00	20.00
$W_{outside}(\tau, 0)$	2.125	1.778	1.459	1.243	1.051	0.896	0.772	0.536	0.356

$(x_1/\sigma)/(x_2/\sigma) = 3.0/2.5$

τ/τ_0	0.02	0.04	0.06	0.08	0.10	0.12	0.14	0.16	0.18
$W_{inside}(\tau, 0)$	5.324	8.381	12.97	19.22	44.57	102.8	181.0	243.9	277.5
τ/τ_0	0.20	0.22	0.24	0.26	0.28	0.30	0.32	0.34	0.36
$W_{inside}(\tau, 0)$	288.0	280.8	261.9	237.7	214.9	189.1	165.9	145.9	129.7
τ/τ_0	0.38	0.40	0.44	0.48	0.52	0.56	0.60	0.64	0.68
$W_{inside}(\tau, 0)$	114.3	104.2	95.41	89.89	87.96	88.10	88.19	86.82	81.45
τ/τ_0	0.72	0.76	0.80	0.88	0.96	1.04	1.12	1.20	1.30
$W_{inside}(\tau, 0)$	73.96	61.88	48.67	28.12	15.56	8.740	5.185	3.189	1.812
τ/τ_0	0.05	0.10	0.15	0.20	0.25	0.30	0.35	0.40	0.45
$W_{outside}(\tau, 0)$	4.810	9.529	14.20	18.37	22.44	25.39	28.06	30.01	31.37
τ/τ_0	0.50	0.55	0.60	0.70	0.80	0.90	1.00	1.20	1.40
$W_{outside}(\tau, 0)$	32.16	31.77	31.46	28.73	25.58	22.62	20.79	16.81	9.965
τ/τ_0	1.60	1.85	2.45	3.20	3.70	4.45	5.05	5.55	10.00
$W_{outside}(\tau, 0)$	4.814	3.143	6.204	3.512	4.752	3.370	3.990	3.310	2.641
τ/τ_0	14.00	18.00	22.00	26.00	30.00	34.00	38.00	42.00	46.00
$W_{outside}(\tau, 0)$	2.017	1.538	1.176	0.931	0.720	0.564	0.448	0.349	0.281

$(x_1/\sigma)/(x_2/\sigma) = 2.0/1.0$

τ/τ_0	0.05	0.10	0.15	0.20	0.25	0.30	0.35	0.40	0.45
$W_{inside}(\tau, 0)$	1.793	3.574	6.302	21.48	63.00	110.8	138.8	143.7	132.9
τ/τ_0	0.50	0.55	0.60	0.65	0.70	0.80	0.90	1.00	1.10
$W_{inside}(\tau, 0)$	115.4	96.15	79.40	66.37	57.55	48.82	46.72	46.55	45.37

(Continued)

Table A1.12 (Continued)

τ/τ_0	1.20	1.30	1.40	1.60	1.80	2.0	2.20	2.40	2.60
$W_{inside}(\tau, 0)$	41.26	35.09	28.89	19.53	13.77	10.19	8.088	6.573	5.227
τ/τ_0	2.80	3.00	3.20	3.40	3.60	3.80	4.20	4.60	5.00
$W_{inside}(\tau, 0)$	3.998	3.054	2.369	1.811	1.442	1.132	0.677	0.407	0.250
τ/τ_0	0.05	0.10	0.15	0.20	0.25	0.30	0.35	0.40	0.45
$W_{outside}(\tau, 0)$	5.767	11.60	17.80	24.33	31.73	40.47	51.72	65.40	79.56
τ/τ_0	0.50	0.55	0.60	0.65	0.70	0.80	0.90	1.00	1.10
$W_{outside}(\tau, 0)$	91.71	99.48	102.6	101.9	97.78	84.74	69.69	56.24	45.06
τ/τ_0	1.20	1.30	1.40	1.60	1.80	2.00	2.20	2.40	2.60
$W_{outside}(\tau, 0)$	36.65	30.33	25.73	20.00	16.24	12.43	9.264	6.958	5.653
τ/τ_0	2.80	3.00	3.20	3.60	4.00	4.40	4.80	5.20	5.60
$W_{outside}(\tau, 0)$	5.012	4.248	3.536	2.281	1.661	1.236	0.843	0.639	0.499

$(x_1/\sigma)/(x_2/\sigma) = 3.0/2.0$

τ/τ_0	0.05	0.10	0.15	0.20	0.25	0.30	0.35	0.40	0.45
$W_{inside}(\tau, 0)$	7.729	15.35	23.59	39.84	72.36	106.2	125.5	128.8	121.1
τ/τ_0	0.50	0.55	0.60	0.70	0.80	0.90	1.00	1.10	1.20
$W_{inside}(\tau, 0)$	109.9	99.60	93.21	88.41	86.39	81.98	71.99	55.03	37.66
τ/τ_0	1.30	1.40	1.50	1.60	1.70	1.80	1.90	2.00	2.10
$W_{inside}(\tau, 0)$	24.53	15.80	10.39	7.248	5.325	4.066	3.316	2.756	2.256
τ/τ_0	2.20	2.30	2.40	2.50	2.60	2.70	2.80	2.90	3.00
$W_{inside}(\tau, 0)$	1.868	1.509	1.158	0.924	0.691	0.527	0.411	0.318	0.239
τ/τ_0	0.10	0.20	0.30	0.40	0.50	0.60	0.70	0.80	0.90
$W_{outside}(\tau, 0)$	4.856	9.446	13.26	16.23	18.55	20.45	23.67	28.73	33.41
τ/τ_0	1.00	1.10	1.20	1.40	1.60	1.75	2.00	2.30	3.00
$W_{outside}(\tau, 0)$	35.20	33.12	27.72	15.64	9.832	9.097	11.37	13.86	8.289
τ/τ_0	3.55	4.40	6.00	7.00	8.00	9.00	10.00	11.00	12.00
$W_{outside}(\tau, 0)$	9.424	6.795	5.506	4.363	3.661	3.142	2.685	2.235	1.802
τ/τ_0	13.00	14.00	15.00	16.00	17.00	18.00	19.00	20.00	21.00
$W_{outside}(\tau, 0)$	1.564	1.326	1.114	0.955	0.794	0.667	0.552	0.474	0.413

Appendix 2: Simulation of the Distribution of Time to the Next Boundary Crossing in Gaussian and Rayleigh Random Processes

Table A2.1 shows values obtained from the simulation of the cumulative function and distribution density for the time to the next downward crossing of the set boundary in a Gaussian random process with a Gaussian power spectrum. It was assumed that at the arbitrarily chosen starting moment the signal was above the set threshold level; in other words, we are dealing excursions above level. The standardized threshold level was gradually increased from 0.5 to 3.0 in 0.5 incremental steps.

Similarly, Table A2.2 shows values obtained from the simulation of a Gaussian process with a rectangular power spectrum.

Table A2.3 is different from Table A2.1 in that the values it shows were obtained for the negative excursion (i.e., the initial momentary value was below the set threshold and the next threshold crossing was an upward one).

Similarly, Table A2.4 is different from Table A2.2 in that the simulation data it presents were obtained for the negative excursion.

Due to the symmetry of the centered Gaussian process, the data obtained by simulation for the case where the threshold is zero are not affected by the initial signal value or the direction in which the threshold is crossed for the first time. This is the reason why the simulation data for zero threshold presented in Tables A2.1 and A2.3 are missing from Tables A2.2 and A2.4.

Tables A2.5–A2.8 show the time to the next crossing of a boundary in a given direction (either upward or downward) measured in a simulated Rayleigh process. The data in tables are presented for processes with Gaussian and rectangular power spectra, respectively.

Because all the momentary values of a Rayleigh signal are positive, all the tables presenting data for Rayleigh processes show them for the standardized threshold levels varying from 0.5 to 3.0.

Tables A2.5 and A2.6 show the simulation data for the cumulative function and distribution density of time to the next downward crossing of a set threshold.

Tables A2.7 and A2.8 show the same type of data for negative excursions (i.e., distribution density of time to the next upward crossing of a set threshold).

Tables A2.5 and A2.7 present the results obtained by simulation of processes with a Gaussian power spectrum, whereas Tables A2.6 and A2.8 present the data obtained by simulation of processes with a rectangular power spectrum.

Table A2.1 Gaussian Random Process

Gaussian power spectrum excursions above level $x_0/\sigma = 0.0$.

τ/τ_0	0.00	0.10	0.20	0.30	0.40	0.50	0.60	0.70	0.80
$P_0(\tau,(x_0/\sigma)) * 100$	100.0	95.51	91.03	86.61	82.24	77.97	73.81	69.77	65.87
τ/τ_0	0.90	1.00	1.10	1.20	1.30	1.40	1.60	1.80	2.00
$P_0(\tau,(x_0/\sigma)) * 100$	62.13	58.56	55.17	51.95	48.92	46.06	40.86	36.26	32.20
τ/τ_0	2.20	2.40	2.60	3.00	3.40	3.80	4.20	4.60	5.00
$P_0(\tau,(x_0/\sigma)) * 100$	28.60	25.41	22.57	17.81	14.04	11.07	9.004	6.909	5.512
τ/τ_0	5.50	6.00	6.50	7.00	7.50	8.00	8.50	9.00	10.00
$P_0(\tau,(x_0/\sigma)) * 100$	4.038	3.007	2.236	1.663	1.236	0.919	0.684	0.511	0.299
τ/τ_0	0.00	0.10	0.20	0.30	0.40	0.50	0.60	0.70	0.80
$W_{2+}(\tau,x_0)$	44.95	44.72	44.28	43.61	42.73	41.66	40.38	38.97	37.41
τ/τ_0	0.90	1.00	1.10	1.20	1.30	1.40	1.60	1.80	2.00
$W_{2+}(\tau,x_0)$	35.69	33.94	32.14	30.34	28.56	26.85	23.67	20.89	18.25
τ/τ_0	2.20	2.40	2.60	3.00	3.40	3.80	4.20	4.60	5.00
$W_{2+}(\tau,x_0)$	16.42	14.73	13.03	10.26	8.025	6.368	5.014	3.922	3.112
τ/τ_0	5.50	6.00	6.50	7.00	7.50	8.00	8.50	9.00	10.00
$W_{2+}(\tau,x_0)$	2.309	1.723	1.286	0.960	0.716	0.534	0.399	0.297	0.166

Gaussian power spectrum excursions above level $x_0/\sigma = 0.5$.

τ/τ_0	0.00	0.10	0.20	0.30	0.40	0.50	0.60	0.70	0.80
$P_0(\tau,(x_0/\sigma)) * 100$	100.0	93.57	87.20	80.94	74.83	68.94	63.29	57.93	52.90
τ/τ_0	0.90	1.00	1.10	1.20	1.30	1.40	1.60	1.80	2.00
$P_0(\tau,(x_0/\sigma)) * 100$	48.20	43.85	39.86	36.21	32.90	29.89	24.72	20.48	16.99
τ/τ_0	2.20	2.40	2.60	2.80	3.00	3.20	3.40	3.60	3.80
$P_0(\tau,(x_0/\sigma)) * 100$	14.09	11.69	9.693	8.038	6.661	5.533	4.586	3.804	3.151
τ/τ_0	4.00	4.20	4.40	4.60	4.80	5.00	5.20	5.40	6.00
$P_0(\tau,(x_0/\sigma)) * 100$	2.614	2.166	1.796	1.489	1.235	1.024	0.848	0.703	0.400
τ/τ_0	0.00	0.10	0.20	0.30	0.40	0.50	0.60	0.70	0.80
$W_{2+}(\tau,x_0)$	64.26	63.72	62.64	61.04	58.96	56.46	53.57	50.38	47.01
τ/τ_0	0.90	1.00	1.10	1.20	1.30	1.40	1.60	1.80	2.00
$W_{2+}(\tau,x_0)$	43.47	39.91	36.45	33.14	30.04	27.16	22.25	18.44	15.07

τ/τ_0	2.20	2.40	2.60	2.80	3.00	3.20	3.40	3.60	3.80
$W_{2+}(\tau, x_0)$	12.57	10.37	8.678	7.135	5.925	4.940	4.090	3.422	2.809
τ/τ_0	4.00	4.20	4.40	4.60	4.80	5.00	5.20	5.40	6.00
$W_{2+}(\tau, x_0)$	2.369	1.972	1.629	1.345	1.111	0.922	0.758	0.626	0.355

Gaussian power spectrum excursions above level $(x_0/\sigma) = 1.0$.

τ/τ_0	0.00	0.05	0.10	0.15	0.20	0.25	0.30	0.35	0.40
$P_0(\tau, (x_0/\sigma)) * 100$	100.0	95.70	91.42	87.16	82.95	78.79	74.72	70.70	66.81
τ/τ_0	0.45	0.50	0.55	0.60	0.65	0.70	0.75	0.80	0.90
$P_0(\tau, (x_0/\sigma)) * 100$	63.00	59.33	55.75	52.35	49.04	45.93	42.92	40.11	34.89
τ/τ_0	1.00	1.20	1.40	1.60	1.80	2.00	2.20	2.40	2.60
$P_0(\tau, (x_0/\sigma)) * 100$	30.27	22.71	17.05	12.86	9.744	7.393	5.604	4.258	3.229
τ/τ_0	2.80	3.00	3.20	3.40	3.60	3.80	4.00	4.50	5.00
$P_0(\tau, (x_0/\sigma)) * 100$	2.445	1.853	1.399	1.062	0.804	0.608	0.459	0.231	0.116
τ/τ_0	0.00	0.05	0.10	0.15	0.20	0.25	0.30	0.35	0.40
$W_{2+}(\tau, x_0)$	85.81	85.26	84.65	83.70	82.37	81.01	79.05	76.90	74.81
τ/τ_0	0.45	0.50	0.55	0.60	0.65	0.70	0.75	0.80	0.90
$W_{2+}(\tau, x_0)$	72.27	69.81	66.95	64.20	61.14	58.23	55.14	52.22	46.18
τ/τ_0	1.00	1.20	1.40	1.60	1.80	2.00	2.20	2.40	2.60
$W_{2+}(\tau, x_0)$	40.39	30.39	22.49	16.71	12.58	9.504	7.180	5.544	4.210
τ/τ_0	2.80	3.00	3.20	3.40	3.60	3.80	4.00	4.50	5.00
$W_{2+}(\tau, x_0)$	3.167	2.463	1.803	1.393	1.052	0.776	0.567	0.272	0.131

Gaussian power spectrum excursions above level $(x_0/\sigma) = 1.5$.

τ/τ_0	0.00	0.05	0.10	0.15	0.20	0.25	0.30	0.35	0.40
$P_0(\tau, (x_0/\sigma)) * 100$	100.0	94.54	89.11	83.74	78.45	73.26	68.21	63.31	58.59
τ/τ_0	0.45	0.50	0.55	0.60	0.65	0.70	0.75	0.80	0.85
$P_0(\tau, (x_0/\sigma)) * 100$	54.07	49.74	45.65	41.76	38.12	34.74	31.57	28.64	25.94
τ/τ_0	0.90	0.95	1.00	1.10	1.20	1.30	1.40	1.50	1.60
$P_0(\tau, (x_0/\sigma)) * 100$	23.50	21.23	19.17	15.64	12.73	10.37	8.474	6.945	5.703
τ/τ_0	1.70	1.80	1.90	2.00	2.20	2.40	2.60	2.80	3.00
$P_0(\tau, (x_0/\sigma)) * 100$	4.688	3.882	3.181	2.622	1.779	1.207	0.818	0.555	0.376

(Continued)

Table A2.1 (Continued)

τ/τ_0	0.00	0.05	0.10	0.15	0.20	0.25	0.30	0.35	0.40
$W_{2+}(\tau, x_0)$	109.2	108.9	108.0	106.7	104.8	102.4	99.52	96.21	92.39
τ/τ_0	0.45	0.50	0.55	0.60	0.65	0.70	0.75	0.80	0.85
$W_{2+}(\tau, x_0)$	88.46	84.24	79.86	75.22	70.21	65.57	61.01	56.22	51.33
τ/τ_0	0.90	0.95	1.00	1.10	1.20	1.30	1.40	1.50	1.60
$W_{2+}(\tau, x_0)$	47.13	43.30	39.31	32.23	26.01	21.25	16.91	13.77	11.21
τ/τ_0	1.70	1.80	1.90	2.00	2.20	2.40	2.60	2.80	3.00
$W_{2+}(\tau, x_0)$	9.213	7.586	6.246	5.143	3.487	2.364	1.603	1.087	0.737

Gaussian power spectrum excursions above level $(x_0/\sigma) = 2.0$.

τ/τ_0	0.00	0.04	0.08	0.12	0.16	0.20	0.24	0.28	0.32
$P_0(\tau, (x_0/\sigma)) * 100$	100.0	94.69	89.40	84.15	79.00	73.94	68.99	64.22	59.54
τ/τ_0	0.36	0.40	0.44	0.48	0.52	0.56	0.60	0.64	0.68
$P_0(\tau, (x_0/\sigma)) * 100$	55.06	50.78	46.68	42.72	39.12	35.69	32.45	29.45	26.62
τ/τ_0	0.72	0.76	0.80	0.88	0.96	1.04	1.12	1.20	1.28
$P_0(\tau, (x_0/\sigma)) * 100$	24.08	21.78	19.56	15.79	12.69	10.20	8.179	6.563	5.286
τ/τ_0	1.36	1.44	1.52	1.60	1.76	1.92	2.08	2.24	2.40
$P_0(\tau, (x_0/\sigma)) * 100$	4.246	3.438	2.784	2.254	1.498	0.996	0.662	0.440	0.289
τ/τ_0	0.00	0.04	0.08	0.12	0.16	0.20	0.24	0.28	0.32
$W_{2+}(\tau, x_0)$	132.9	132.5	131.5	130.0	127.6	125.2	121.5	118.2	113.6
τ/τ_0	0.36	0.40	0.44	0.48	0.52	0.56	0.60	0.64	0.68
$W_{2+}(\tau, x_0)$	108.8	104.3	99.59	94.53	89.90	83.58	77.31	71.11	65.02
τ/τ_0	0.72	0.76	0.80	0.88	0.96	1.04	1.12	1.20	1.28
$W_{2+}(\tau, x_0)$	59.25	53.67	49.16	40.10	31.32	24.76	19.58	15.48	12. 24
τ/τ_0	1.36	1.44	1.52	1.60	1.76	1.92	2.08	2.24	2.40
$W_{2+}(\tau, x_0)$	9.678	7.592	5.956	4.672	3.665	2.875	2.256	1.775	1.388

Gaussian power spectrum excursions above level $(x_0/\sigma) = 2.5$.

τ/τ_0	0.00	0.04	0.08	0.12	0.16	0.20	0.24	0.28	0.32
$P_0(\tau, (x_0/\sigma)) * 100$	100.0	93.79	87.63	81.56	75.57	69.75	64.11	58.72	53.50

τ/τ_0	0.36	0.40	0.44	0.48	0.52	0.56	0.60	0.64	0.68
$P_0(\tau,(x_0/\sigma)) * 100$	48.58	44.00	39.59	35.52	31.74	28.26	25.11	22.21	19.66
τ/τ_0	0.72	0.76	0.80	0.88	0.96	1.04	1.12	1.20	1.28
$P_0(\tau,(x_0/\sigma)) * 100$	17.27	15.08	13.27	10.18	7.769	5.923	4.497	3.443	2.644
τ/τ_0	1.36	1.44	1.52	1.60	1.68	1.76	1.84	1.92	2.00
$P_0(\tau,(x_0/\sigma)) * 100$	2.030	1.557	1.206	0.934	0.723	0.560	0.441	0.337	0.258
(τ/τ_0)	0.00	0.04	0.08	0.12	0.16	0.20	0.24	0.28	0.32
$W_{2+}(\tau,x_0)$	155.3	154.7	153.5	150.8	147.2	143.2	138.2	132.7	126.2
τ/τ_0	0.36	0.40	0.44	0.48	0.52	0.56	0.60	0.64	0.68
$W_{2+}(\tau,x_0)$	119.4	112.4	105.2	98.08	90.49	82.90	75.66	68.20	61.69
τ/τ_0	0.72	0.76	0.80	0.88	0.96	1.04	1.12	1.20	1.28
$W_{2+}(\tau,x_0)$	55.85	50.01	44.30	34.43	26.23	19.96	15.45	11.34	8.704
τ/τ_0	1.36	1.44	1.52	1.60	1.68	1.76	1.84	1.92	2.00
$W_{2+}(\tau,x_0)$	6.554	5.039	3.876	2.981	2.293	1.752	1.339	1.023	0.806

Gaussian power spectrum excursions above level $(x_0/\sigma) = 3.0$.

τ/τ_0	0.00	0.04	0.08	0.12	0.16	0.20	0.24	0.28	0.32
$P_0(\tau,(x_0/\sigma)) * 100$	100.0	93.34	86.73	80.23	73.87	67.71	61.78	56.08	50.78
τ/τ_0	0.36	0.40	0.44	0.48	0.52	0.56	0.60	0.64	0.68
$P_0(\tau,(x_0/\sigma)) * 100$	45.76	41.07	36.70	32.72	29.04	25.73	22.71	19.99	17.59
τ/τ_0	0.72	0.76	0.80	0.88	0.96	1.04	1.12	1.20	1.28
$P_0(\tau,(x_0/\sigma)) * 100$	15.44	13.54	11.87	9.093	6.949	5.339	4.057	3.105	2.373
τ/τ_0	1.36	1.44	1.52	1.60	1.68	1.76	1.84	1.92	2.00
$P_0(\tau,(x_0/\sigma)) * 100$	1.814	1.374	1.030	0.772	0.579	0.434	0.326	0.242	0.178
τ/τ_0	0.00	0.04	0.08	0.12	0.16	0.20	0.24	0.28	0.32
$W_{2+}(\tau,x_0)$	166.7	165.9	164.3	160.8	156.1	151.1	144.5	137.4	129.5
τ/τ_0	0.36	0.40	0.44	0.48	0.52	0.56	0.60	0.64	0.68
$W_{2+}(\tau,x_0)$	121.3	113.0	104.3	95.82	87.49	79.15	71.74	63.88	56.87
τ/τ_0	0.72	0.76	0.80	0.88	0.96	1.04	1.12	1.20	1.28
$W_{2+}(\tau,x_0)$	50.58	44.69	39.67	30.25	23.48	17.56	13.18	10.40	8.210
τ/τ_0	1.36	1.44	1.52	1.60	1.68	1.76	1.84	1.92	2.00
$W_{2+}(\tau,x_0)$	6.200	4.831	3.764	2.885	2.111	1.614	1.277	0.991	0.776

Table A2.2 Gaussian Random Process

Rectangular power spectrum excursions above level $(x_0/\sigma) = 0.0$.

τ/τ_0	0.00	0.20	0.40	0.60	0.80	1.00	1.20	1.40	1.60
$P_0(\tau,(x_0/\sigma)) * 100$	100.0	91.01	82.10	73.35	64.84	56.71	49.12	42.23	36.22
τ/τ_0	1.80	2.00	2.20	2.40	2.60	2.80	3.00	3.20	3.40
$P_0(\tau,(x_0/\sigma)) * 100$	31.18	27.09	23.77	21.03	18.69	16.60	14.71	12.98	11.39
τ/τ_0	3.60	3.80	4.00	4.20	4.40	4.60	4.80	5.00	5.40
$P_0(\tau,(x_0/\sigma)) * 100$	9.967	8.683	7.572	6.614	5.778	5.093	4.500	3.982	3.127
τ/τ_0	5.80	6.20	6.60	7.00	7.40	7.80	8.20	8.60	9.00
$P_0(\tau,(x_0/\sigma)) * 100$	2.429	1.875	1.430	1.100	0.865	0.675	0.528	0.408	0.315
τ/τ_0	0.00	0.20	0.40	0.60	0.80	1.00	1.20	1.40	1.60
$W_{2+}(\tau,x_0)$	44.97	44.74	44.17	43.15	41.58	39.32	36.22	32.25	27.62
τ/τ_0	1.80	2.00	2.20	2.40	2.60	2.80	3.00	3.20	3.40
$W_{2+}(\tau,x_0)$	22.81	18.52	15.16	12.71	11.07	9.930	9.050	8.299	7.542
τ/τ_0	3.60	3.80	4.00	4.20	4.40	4.60	4.80	5.00	5.40
$W_{2+}(\tau,x_0)$	6.777	5.986	5.236	4.487	3.841	3.195	2.779	2.431	1.935
τ/τ_0	5.80	6.20	6.60	7.00	7.40	7.80	8.20	8.60	9.00
$W_{2+}(\tau,x_0)$	1.558	1.236	0.942	0.714	0.542	0.415	0.318	0.243	0.186

Rectangular power spectrum excursions above level $(x_0/\sigma) = 0.5$.

τ/τ_0	0.00	0.10	0.20	0.30	0.40	0.50	0.60	0.70	0.80
$P_0(\tau,(x_0/\sigma)) * 100$	100.0	93.56	87.16	80.81	74.55	68.42	62.44	56.65	51.08
τ/τ_0	0.90	1.00	1.10	1.20	1.30	1.40	1.60	1.80	2.00
$P_0(\tau,(x_0/\sigma)) * 100$	45.79	40.80	36.15	31.89	28.02	24.57	18.93	14.81	11.84
τ/τ_0	2.20	2.40	2.60	2.80	3.00	3.20	3.40	3.60	3.80
$P_0(\tau,(x_0/\sigma)) * 100$	9.681	7.986	6.627	5.488	4.495	3.661	2.960	2.358	1.877
τ/τ_0	4.00	4.20	4.40	4.60	4.80	5.00	5.20	5.40	5.60
$P_0(\tau,(x_0/\sigma)) * 100$	1.489	1.212	0.987	0.803	0.654	0.537	0.441	0.362	0.298
τ/τ_0	0.00	0.10	0.20	0.30	0.40	0.50	0.60	0.70	0.80
$W_{2+}(\tau,x_0)$	64.39	64.21	63.77	63.02	61.97	60.58	58.85	56.77	54.28
τ/τ_0	0.90	1.00	1.10	1.20	1.30	1.40	1.60	1.80	2.00
$W_{2+}(\tau,x_0)$	51.43	48.21	44.53	40.63	36.59	32.36	24.23	17.43	12.83

τ/τ_0	2.20	2.40	2.60	2.80	3.00	3.20	3.40	3.60	3.80
$W_{2+}(\tau, x_0)$	9.630	7.564	6.207	5.287	4.506	3.869	3.364	2.589	2.157
τ/τ_0	4.00	4.20	4.40	4.60	4.80	5.00	5.20	5.40	5.60
$W_{2+}(\tau, x_0)$	1.676	1.313	1.029	0.823	0.632	0.485	0.372	0.286	0.220
Rectangular power spectrum excursions above level $(x_0/\sigma) = 1.0$.									
τ/τ_0	0.00	0.10	0.20	0.30	0.40	0.50	0.60	0.70	0.80
$P_0(\tau, (x_0/\sigma)) * 100$	100.0	91.41	82.89	74.53	66.39	58.55	51.09	44.09	37.60
τ/τ_0	0.90	1.00	1.10	1.20	1.30	1.40	1.50	1.60	1.70
$P_0(\tau, (x_0/\sigma)) * 100$	31.68	26.39	21.75	17.77	14.43	11.70	9.505	7.782	6.420
τ/τ_0	1.80	1.90	2.00	2.10	2.20	2.30	2.40	2.50	2.60
$P_0(\tau, (x_0/\sigma)) * 100$	5.367	4.527	3.862	3.336	2.881	2.500	2.179	1.898	1.652
τ/τ_0	2.70	2.80	2.90	3.00	3.20	3.40	3.60	3.80	4.00
$P_0(\tau, (x_0/\sigma)) * 100$	1.430	1.236	1.066	0.914	0.659	0.470	0.327	0.228	0.166
τ/τ_0	0.00	0.10	0.20	0.30	0.40	0.50	0.60	0.70	0.80
$W_{2+}(\tau, x_0)$	86.01	85.54	84.40	82.51	79.86	76.48	72.32	67.49	62.04
τ/τ_0	0.90	1.00	1.10	1.20	1.30	1.40	1.50	1.60	1.70
$W_{2+}(\tau, x_0)$	56.03	49.67	43.10	36.58	30.34	24.63	19.59	15.42	12.07
τ/τ_0	1.80	1.90	2.00	2.10	2.20	2.30	2.40	2.50	2.60
$W_{2+}(\tau, x_0)$	9.465	7.522	6.074	4.905	4.150	3.512	3.011	2.634	2.339
τ/τ_0	2.70	2.80	2.90	3.00	3.20	3.40	3.60	3.80	4.00
$W_{2+}(\tau, x_0)$	2.080	1.823	1.608	1.438	1.090	0.830	0.593	0.414	0.282
Rectangular power spectrum excursions above level $(x_0/\sigma) = 1.5$.									
τ/τ_0	0.00	0.04	0.08	0.12	0.16	0.20	0.24	0.28	0.32
$P_0(\tau, (x_0/\sigma)) * 100$	100.0	95.63	91.28	86.95	82.64	78.37	74.17	70.02	65.94
τ/τ_0	0.36	0.40	0.44	0.48	0.52	0.56	0.60	0.64	0.68
$P_0(\tau, (x_0/\sigma)) * 100$	61.95	58.05	54.25	50.55	46.98	43.53	40.21	37.03	33.99
τ/τ_0	0.72	0.76	0.80	0.84	0.88	0.96	1.04	1.12	1.20
$P_0(\tau, (x_0/\sigma)) * 100$	31.12	28.42	25.79	23.36	21.10	17.04	13.60	10.73	8.388

(Continued)

Table A2.2 (Continued)

τ/τ_0	1.28	1.36	1.44	1.60	1.76	1.92	2.08	2.24	2.40
$P_0(\tau,(x_0/\sigma))*100$	6.527	5.069	3.973	2.466	1.612	1.103	0.784	0.572	0.426
τ/τ_0	0.00	0.04	0.08	0.12	0.16	0.20	0.24	0.28	0.32
$W_{2+}(\tau,x_0)$	109.2	109.1	108.7	108.0	107.1	105.9	104.5	102.8	100.8
τ/τ_0	0.36	0.40	0.44	0.48	0.52	0.56	0.60	0.64	0.68
$W_{2+}(\tau,x_0)$	98.63	96.28	93.75	90.87	87.74	84.55	81.30	77.76	73.91
τ/τ_0	0.72	0.76	0.80	0.84	0.88	0.96	1.04	1.12	1.20
$W_{2+}(\tau,x_0)$	70.31	66.62	62.62	58.63	54.73	46.93	39.57	32.50	25. 83
τ/τ_0	1.28	1.36	1.44	1.60	1.76	1.92	2.08	2.24	2.40
$W_{2+}(\tau,x_0)$	20.66	16.02	12.19	7.023	4.084	2.461	1.629	1.076	0.774
Rectangular power spectrum excursions above level $(x_0/\sigma) = 2.0$.									
τ/τ_0	0.00	0.04	0.08	0.12	0.16	0.20	0.24	0.28	0.32
$P_0(\tau,(x_0/\sigma))*100$	100.0	94.70	89.42	84.16	79.00	73.90	68.89	64.00	59.25
τ/τ_0	0.36	0.40	0.44	0.48	0.52	0.56	0.60	0.64	0.68
$P_0(\tau,(x_0/\sigma))*100$	54.64	50.18	45.92	41.80	37.97	34.29	30.84	27.60	24.57
τ/τ_0	0.72	0.76	0.80	0.88	0.96	1.04	1.12	1.20	1.28
$P_0(\tau,(x_0/\sigma))*100$	21.79	19.29	16.87	12.81	9.542	6.988	5.055	3.604	2.571
τ/τ_0	1.36	1.44	1.52	1.60	1.68	1.76	1.84	1.92	2.00
$P_0(\tau,(x_0/\sigma))*100$	1.841	1.321	0.979	0.726	0.542	0.417	0.330	0.255	0.201
τ/τ_0	0.00	0.04	0.08	0.12	0.16	0.20	0.24	0.28	0.32
$W_{2+}(\tau,x_0)$	132.6	132.3	131.5	130.3	128.5	126.3	123.6	120.6	117.1
τ/τ_0	0.36	0.40	0.44	0.48	0.52	0.56	0.60	0.64	0.68
$W_{2+}(\tau,x_0)$	113.4	108.9	104.8	99.42	93.93	89.13	83.72	78.33	72.60
τ/τ_0	0.72	0.76	0.80	0.88	0.96	1.04	1.12	1.20	1.28
$W_{2+}(\tau, x_0)$	67.04	61.52	56.34	45.65	36.31	27.89	20.88	15.30	11.00
τ/τ_0	1.36	1.44	1.52	1.60	1.68	1.76	1.84	1.92	2.00
$W_{2+}(\tau,x_0)$	7.695	5.284	3.703	2.608	1.932	1.410	1.037	0.762	0.561
Rectangular power spectrum excursions above level $(x_0/\sigma) = 2.5$.									
τ/τ_0	0.00	0.04	0.08	0.12	0.16	0.20	0.24	0.28	0.32
$P_0(\tau,(x_0/\sigma))*100$	100.0	93.81	87.65	81.55	75.57	69.71	64.00	58.50	53.19

τ/τ_0	0.36	0.40	0.44	0.48	0.52	0.56	0.60	0.64	0.68
$P_0(\tau,(x_0/\sigma)) * 100$	48.13	43.23	38.77	34.57	30.58	26.91	23.55	20.50	17.72
τ/τ_0	0.72	0.76	0.80	0.84	0.88	0.92	0.96	1.00	1.04
$P_0(\tau,(x_0/\sigma)) * 100$	15.24	12.94	11.07	9.368	7.889	6.619	5.516	4.603	3.818
τ/τ_0	1.08	1.12	1.16	1.20	1.28	1.36	1.44	1.52	1.60
$P_0(\tau,(x_0/\sigma)) * 100$	3.176	2.632	2.183	1.810	1.256	0.883	0.634	0.460	0.337
τ/τ_0	0.00	0.04	0.08	0.12	0.16	0.20	0.24	0.28	0.32
$W_{2+}(\tau,x_0)$	155.0	154.4	153.0	151.0	148.0	144.5	140.3	135.2	129.3
τ/τ_0	0.36	0.40	0.44	0.48	0.52	0.56	0.60	0.64	0.68
$W_{2+}(\tau,x_0)$	123.2	116.9	110.0	102.4	95.24	87.91	80.14	72.86	65.79
τ/τ_0	0.72	0.76	0.80	0.84	0.88	0.92	0.96	1.00	1.04
$W_{2+}(\tau,x_0)$	58.79	52.09	45.74	39.77	34.36	29.65	25.19	21.23	17.84
τ/τ_0	1.08	1.12	1.16	1.20	1.28	1.36	1.44	1.52	1.60
$W_{2+}(\tau,x_0)$	14.83	12.42	10.27	8.385	5.806	3.839	2.558	1.873	1.371

Rectangular power spectrum excursions above level $(x_0/\sigma) = 3.0$.

τ/τ_0	0.00	0.04	0.08	0.12	0.16	0.20	0.24	0.28	0.32
$P_0(\tau,(x_0/\sigma)) * 100$	100.0	93.35	86.75	80.24	73.86	67.69	61.72	55.96	50.49
τ/τ_0	0.36	0.40	0.44	0.48	0.52	0.56	0.60	0.64	0.68
$P_0(\tau,(x_0/\sigma)) * 100$	45.33	40.51	36.02	31.85	28.05	24.59	21.50	18.70	16.20
τ/τ_0	0.72	0.76	0.80	0.84	0.88	0.92	0.96	1.00	1.04
$P_0(\tau,(x_0/\sigma)) * 100$	13.99	12.10	10.38	8.894	7.673	6.606	5.650	4.855	4.182
τ/τ_0	1.08	1.12	1.16	1.20	1.28	1.36	1.44	1.52	1.60
$P_0(\tau,(x_0/\sigma)) * 100$	3.551	3.089	2.627	2.280	1.690	1.238	0.900	0.652	0.462
τ/τ_0	0.00	0.04	0.08	0.12	0.16	0.20	0.24	0.28	0.32
$W_{2+}(\tau,x_0)$	166.4	165.7	163.9	161.1	156.9	152.4	146.5	139.9	133.1
τ/τ_0	0.36	0.40	0.44	0.48	0.52	0.56	0.60	0.64	0.68
$W_{2+}(\tau,x_0)$	124.8	116.1	107.3	98.61	90.21	81.96	73.80	65.81	58.02
τ/τ_0	0.72	0.76	0.80	0.84	0.88	0.92	0.96	1.00	1.04
$W_{2+}(\tau,x_0)$	51.91	45.07	39.86	33.82	28.60	24.74	21.55	18.43	16.13
τ/τ_0	1.08	1.12	1.16	1.20	1.28	1.36	1.44	1.52	1.60
$W_{2+}(\tau,x_0)$	13.96	12.07	10.44	9.056	6.786	5.086	3.812	2.857	2.141

Table A2.3 Gaussian Random Process

Gaussian power spectrum excursions below level $(x_0/\sigma) = 0.5$.									
τ/τ_0	0.00	0.20	0.40	0.60	0.80	1.00	1.20	1.40	1.60
$P_0(\tau,(x_0/\sigma)) * 100$	100.0	94.30	88.66	83.14	77.81	72.71	67.88	63.34	59.11
τ/τ_0	1.80	2.00	2.40	2.80	3.20	3.60	4.00	4.40	4.80
$P_0(\tau,(x_0/\sigma)) * 100$	55.18	51.51	44.90	39.14	34.12	29.74	25.93	22.60	19.70
τ/τ_0	5.20	5.60	6.00	6.40	6.80	7.20	7.60	8.00	8.40
$P_0(\tau,(x_0/\sigma)) * 100$	17.17	14.97	13.05	11.38	9.918	8.647	7.537	6.566	5.725
τ/τ_0	8.80	9.20	9.60	10.00	12.00	14.00	16.00	18.00	20.00
$P_0(\tau,(x_0/\sigma)) * 100$	4.986	4.352	3.791	3.306	1.660	0.832	0.420	0.211	0.107
τ/τ_0	0.00	0.20	0.40	0.60	0.80	1.00	1.20	1.40	1.60
$W_{2-}(\tau,x_0)$	28.73	28.20	27.58	26.67	25.51	24.15	22.67	21.14	19.69
τ/τ_0	1.80	2.00	2.40	2.80	3.20	3.60	4.00	4.40	4.80
$W_{2-}(\tau,x_0)$	18.35	17.09	14.90	13.00	11.32	9.860	8.621	7.486	6.547
τ/τ_0	5.20	5.60	6.00	6.40	6.80	7.20	7.60	8.00	8.40
$W_{2-}(\tau,x_0)$	5.677	4.980	4.327	3.765	3.294	2.882	2.523	2.174	1.913
τ/τ_0	8.80	9.20	9.60	10.00	12.00	14.00	16.00	18.00	20.00
$W_{2-}(\tau,x_0)$	1.626	1.457	1.258	1.088	0.549	0.275	0.138	0.069	0.034

Gaussian power spectrum excursions below level $(x_0/\sigma) = 1.0$.									
τ/τ_0	0.00	0.20	0.40	0.60	0.80	1.00	1.20	1.40	1.60
$P_0(\tau,(x_0/\sigma)) * 100$	100.0	96.79	93.59	90.43	87.33	84.29	81.33	78.46	75.69
τ/τ_0	2.00	2.40	2.80	3.20	3.60	4.00	4.40	4.80	5.20
$P_0(\tau,(x_0/\sigma)) * 100$	70.44	65.57	61.03	56.81	52.88	49.22	45.81	42.64	39.69
τ/τ_0	5.60	6.00	6.40	6.80	7.20	7.60	8.00	10.00	12.00
$P_0(\tau,(x_0/\sigma)) * 100$	36.95	34.39	32.01	29.79	27.73	25.82	24.03	16.79	11.73
τ/τ_0	14.00	16.00	18.00	20.00	22.00	24.00	26.00	28.00	30.00
$P_0(\tau,(x_0/\sigma)) * 100$	8.192	5.723	4.004	2.793	1.954	1.363	0.954	0.665	0.463
τ/τ_0	0.00	0.20	0.40	0.60	0.80	1.00	1.20	1.40	1.60
$W_{2-}(\tau,x_0)$	16.23	16.03	15.80	15.54	15.20	14.80	14.34	13.86	13.36

τ/τ_0	2.00	2.40	2.80	3.20	3.60	4.00	4.40	4.80	5.20
$W_{2-}(\tau,x_0)$	12.40	11.54	10.74	10.00	9.319	8.646	8.064	7.509	7.000
τ/τ_0	5.60	6.00	6.40	6.80	7.20	7.60	8.00	10.00	12.00
$W_{2-}(\tau,x_0)$	6.501	6.059	5.638	5.230	4.898	4.540	4.214	2.961	2.081
τ/τ_0	14.00	16.00	18.00	20.00	22.00	24.00	26.00	28.00	30.00
$W_{2-}(\tau,x_0)$	1.423	0.990	0.704	0.481	0.356	0.251	0.172	0.115	0.082
Gaussian power spectrum excursions below level $(x_0/\sigma) = 1.5$.									
τ/τ_0	0.00	0.40	0.80	1.20	1.60	2.00	2.40	2.80	3.20
$P_0(\tau,(x_0/\sigma)) * 100$	100.0	96.92	93.87	90.88	87.97	85.15	82.43	79.79	77.23
τ/τ_0	3.60	4.00	4.40	4.80	5.20	5.60	6.00	6.40	6.80
$P_0(\tau,(x_0/\sigma)) * 100$	74.76	72.36	70.04	67.80	65.63	63.53	61.49	59.53	57.62
τ/τ_0	7.20	7.60	8.00	8.40	8.80	9.20	9.60	10.00	14.00
$P_0(\tau,(x_0/\sigma)) * 100$	55.77	53.98	52.26	50.58	48.96	47.40	45.88	44.40	32.07
τ/τ_0	18.00	22.00	26.00	30.00	34.00	38.00	42.00	46.00	50.00
$P_0(\tau,(x_0/\sigma)) * 100$	23.17	16.73	12.08	8.725	6.299	4.551	3.284	2.372	1.714
τ/τ_0	0.00	0.40	0.80	1.20	1.60	2.00	2.40	2.80	3.20
$W_{2-}(\tau,x_0)$	7.723	7.663	7.534	7.347	7.123	6.891	6.667	6.453	6.256
τ/τ_0	3.60	4.00	4.40	4.80	5.20	5.60	6.00	6.40	6.80
$W_{2-}(\tau,x_0)$	6.047	5.859	5.651	5.473	5.304	5.160	4.968	4.824	4.650
τ/τ_0	7.20	7.60	8.00	8.40	8.80	9.20	9.60	10.00	14.00
$W_{2-}(\tau,x_0)$	4.532	4.371	4.216	4.091	3.957	3.847	3.704	3.586	2.598
τ/τ_0	18.00	22.00	26.00	30.00	34.00	38.00	42.00	46.00	50.00
$W_{2-}(\tau,x_0)$	1.896	1.341	0.952	0.681	0.487	0.348	0.249	0.178	0.127
Gaussian power spectrum excursions below level $(x_0/\sigma) = 2.0$.									
τ/τ_0	0.00	1.00	2.00	3.00	4.00	5.00	6.00	7.00	8.00
$P_0(\tau,(x_0/\sigma)) * 100$	100.0	96.96	93.99	91.10	88.30	85.59	82.96	80.42	77.95
τ/τ_0	9.00	10.00	11.00	12.00	13.00	14.00	15.00	16.00	17.00
$P_0(\tau,(x_0/\sigma)) * 100$	75.55	73.23	70.99	68.81	66.69	64.65	62.66	60.74	58.87
τ/τ_0	18.00	20.00	24.00	28.00	32.00	36.00	40.00	44.00	48.00
$P_0(\tau,(x_0/\sigma)) * 100$	57.07	53.61	47.33	41.77	36.87	32.55	28.72	25.28	22.31

(Continued)

Table A2.3 (Continued)

τ/τ_0	52.00	56.00	60.00	70.00	80.00	90.00	100.0	110.0	120.0
$P_0(\tau,(x_0/\sigma)) * 100$	19.70	17.38	15.35	11.22	8.214	6.013	4.402	3.222	2.358
τ/τ_0	0.00	1.00	2.00	3.00	4.00	5.00	6.00	7.00	8.00
$W_{2-}(\tau,x_0)$	3.058	3.015	2.930	2.840	2.750	2.670	2.558	2.504	2.427
τ/τ_0	9.00	10.00	11.00	12.00	13.00	14.00	15.00	16.00	17.00
$W_{2-}(\tau,x_0)$	2.358	2.287	2.194	2.146	2.075	2.044	1.948	1.885	1.855
τ/τ_0	18.00	20.00	24.00	28.00	32.00	36.00	40.00	44.00	48.00
$W_{2-}(\tau,x_0)$	1.826	1.669	1.475	1.291	1.138	1.005	0.883	0.786	0.684
τ/τ_0	52.00	56.00	60.00	70.00	80.00	90.00	100.0	110.0	120.0
$W_{2-}(\tau,x_0)$	0.604	0.532	0.468	0.341	0.248	0.181	0.132	0.096	0.070

Table A2.4 Gaussian Random Process

Rectangular power spectrum excursions below level $(x_0/\sigma) = 0.5$.									
τ/τ_0	0.00	0.20	0.40	0.60	0.80	1.00	1.20	1.40	1.60
$P_0(\tau, (x_0/\sigma)) * 100$	100.0	94.28	88.58	82.91	77.31	71.80	66.44	61.28	56.43
τ/τ_0	1.80	2.00	2.40	2.80	3.20	3.60	4.00	4.40	4.80
$P_0(\tau, (x_0/\sigma)) * 100$	51.98	48.00	41.43	36.11	31.43	27.18	23.36	20.08	17.36
τ/τ_0	5.20	5.60	6.00	6.40	6.80	7.20	7.60	8.00	8.40
$P_0(\tau, (x_0/\sigma)) * 100$	15.10	13.13	11.39	9.820	8.468	7.326	6.361	5.530	4.797
τ/τ_0	8.80	9.20	9.60	10.00	12.00	14.00	16.00	18.00	20.00
$P_0(\tau, (x_0/\sigma)) * 100$	4.149	3.582	3.100	2.694	1.313	0.644	0.317	0.155	0.076
τ/τ_0	0.00	0.20	0.40	0.60	0.80	1.00	1.20	1.40	1.60
$W_{2-}(\tau, x_0)$	28.62	28.53	28.34	28.03	27.55	26.83	25.77	24.26	22.26
τ/τ_0	1.80	2.00	2.40	2.80	3.20	3.60	4.00	4.40	4.80
$W_{2-}(\tau, x_0)$	19.90	17.47	13.83	11.98	10.88	9.875	8.543	7.134	5.889
τ/τ_0	5.20	5.60	6.00	6.40	6.80	7.20	7.60	8.00	8.40
$W_{2-}(\tau, x_0)$	5.076	4.478	4.051	3.478	2.985	2.516	2.151	1.893	1.658
τ/τ_0	8.80	9.20	9.60	10.00	12.00	14.00	16.00	18.00	20.00
$W_{2-}(\tau, x_0)$	1.466	1.261	1.068	0.919	0.454	0.232	0.111	0.054	0.027
Rectangular power spectrum excursions below level $(x_0/\sigma) = 1.0$.									
τ/τ_0	0.00	0.20	0.40	0.60	0.80	1.00	1.20	1.40	1.60
$P_0(\tau, (x_0/\sigma)) * 100$	100.0	96.78	93.56	90.34	87.14	83.95	80.79	77.67	74.62
τ/τ_0	2.00	2.40	2.80	3.20	3.60	4.00	4.40	4.80	5.20
$P_0(\tau, (x_0/\sigma)) * 100$	68.86	63.79	59.36	55.27	51.36	47.63	44.13	40.95	38.08
τ/τ_0	5.60	6.00	6.40	6.80	7.20	7.60	8.00	10.00	12.00
$P_0(\tau, (x_0/\sigma)) * 100$	35.44	32.95	30.61	28.39	26.35	24.50	22.79	15.78	10.96
τ/τ_0	14.00	16.00	18.00	20.00	22.00	24.00	26.00	28.00	30.00
$P_0(\tau, (x_0/\sigma)) * 100$	7.612	5.293	3.666	2.552	1.777	1.237	0.861	0.599	0.417

(Continued)

Table A2.4 (Continued)

τ/τ_0	0.00	0.20	0.40	0.60	0.80	1.00	1.20	1.40	1.60
$W_{2-}(\tau, x_0)$	16.13	16.12	16.09	16.03	15.95	15.81	15.61	15.28	14.79
τ/τ_0	2.00	2.40	2.80	3.20	3.60	4.00	4.40	4.80	5.20
$W_{2-}(\tau, x_0)$	13.15	11.38	10.36	9.896	9.453	8.897	8.183	7.359	6.720
τ/τ_0	5.60	6.00	6.40	6.80	7.20	7.60	8.00	10.00	12.00
$W_{2-}(\tau, x_0)$	6.282	5.914	5.627	5.203	4.743	4.390	4.025	2.808	2.017
τ/τ_0	14.00	16.00	18.00	20.00	22.00	24.00	26.00	28.00	30.00
$W_{2-}(\tau, x_0)$	1.359	0.968	0.676	0.477	0.336	0.237	0.168	0.119	0.084

Rectangular power spectrum excursions below level $(x_0/\sigma) = 1.5$.

τ/τ_0	0.00	0.40	0.80	1.20	1.60	2.00	2.40	2.80	3.20
$P_0(\tau, (x_0/\sigma)) * 100$	100.0	96.90	93.81	90.72	87.65	84.65	81.81	79.16	76.64
τ/τ_0	3.60	4.00	4.40	4.80	5.20	5.60	6.00	6.40	6.80
$P_0(\tau, (x_0/\sigma)) * 100$	74.17	71.74	69.37	67.08	64.91	62.84	60.82	58.85	56.93
τ/τ_0	7.20	7.60	8.00	8.40	8.80	9.20	9.60	10.00	14.00
$P_0(\tau, (x_0/\sigma)) * 100$	55.07	53.28	51.58	49.92	48.32	46.75	45.22	43.76	31.55
τ/τ_0	18.00	22.00	26.00	30.00	34.00	38.00	42.00	46.00	50.00
$P_0(\tau, (x_0/\sigma)) * 100$	22.74	16.39	11.81	8.513	6.137	4.423	3.189	2.298	1.657
τ/τ_0	0.00	0.40	0.80	1.20	1.60	2.00	2.40	2.80	3.20
$W_{2-}(\tau, x_0)$	7.756	7.750	7.734	7.694	7.574	7.246	6.724	6.365	6.204
τ/τ_0	3.60	4.00	4.40	4.80	5.20	5.60	6.00	6.40	6.80
$W_{2-}(\tau, x_0)$	6.113	5.988	5.785	5.524	5.251	5.081	4.954	4.839	4.694
τ/τ_0	7.20	7.60	8.00	8.40	8.80	9.20	9.60	10.00	14.00
$W_{2-}(\tau, x_0)$	4.525	4.305	4.176	4.027	3.953	3.845	3.660	3.556	2.563
τ/τ_0	18.00	22.00	26.00	30.00	34.00	38.00	42.00	46.00	50.00
$W_{2-}(\tau, x_0)$	1.804	1.270	0.894	0.631	0.444	0.315	0.224	0.159	0.113

Rectangular power spectrum excursions below level $(x_0/\sigma) = 2.0$.

τ/τ_0	0.00	1.00	2.00	3.00	4.00	5.00	6.00	7.00	8.00
$P_0(\tau,(x_0/\sigma)) * 100$	100.0	96.94	93.88	90.96	88.17	85.43	82.81	80.26	77.79
τ/τ_0	9.00	10.00	11.00	12.00	13.00	14.00	15.00	16.00	17.00
$P_0(\tau,(x_0/\sigma)) * 100$	75.41	73.08	70.85	68.68	66.57	64.53	62.55	60.63	58.78
τ/τ_0	18.00	20.00	24.00	28.00	32.00	36.00	40.00	44.00	48.00
$P_0(\tau,(x_0/\sigma)) * 100$	56.98	53.54	47.28	41.76	36.87	32.47	27.80	25.28	22.35
τ/τ_0	52.00	56.00	60.00	70.00	80.00	90.00	100.0	110.0	120.0
$P_0(\tau,(x_0/\sigma)) * 100$	19.75	17.39	15.35	11.27	8.248	6.035	4.480	3.294	2.422
τ/τ_0	0.00	1.00	2.00	3.00	4.00	5.00	6.00	7.00	8.00
$W_{2-}(\tau,x_0)$	3.070	3.067	3.018	2.818	2.772	2.672	2.570	2.527	2.423
τ/τ_0	9.00	10.00	11.00	12.00	13.00	14.00	15.00	16.00	17.00
$W_{2-}(\tau,x_0)$	2.356	2.272	2.219	2.157	2.074	2.004	1.940	1.880	1.822
τ/τ_0	18.00	20.00	24.00	28.00	32.00	36.00	40.00	44.00	48.00
$W_{2-}(\tau,x_0)$	1.765	1.660	1.468	1.299	1.148	1.016	0.898	0.795	0.703
τ/τ_0	52.00	56.00	60.00	70.00	80.00	90.00	100.0	110.0	120.0
$W_{2-}(\tau,x_0)$	0.622	0.550	0.486	0.357	0.263	0.193	0.142	0.104	0.077

Table A2.5 Rayleigh Random Process

Gaussian power spectrum excursions above level $(x_0/\sigma) = 0.5$.

τ/τ_0	0.00	0.20	0.40	0.60	0.80	1.00	1.20	1.40	1.60
$P_0(\tau,(x_0/\sigma)) * 100$	100.0	94.42	88.92	83.57	78.43	73.55	68.94	64.62	60.57
τ/τ_0	1.80	2.00	2.20	2.40	2.60	2.80	3.00	3.20	3.40
$P_0(\tau,(x_0/\sigma)) * 100$	56.78	53.22	49.90	46.78	43.86	41.11	38.54	36.14	33.87
τ/τ_0	3.80	4.20	4.60	5.00	5.40	5.80	6.20	6.60	7.00
$P_0(\tau,(x_0/\sigma)) * 100$	29.77	26.25	23.00	20.22	17.77	15.62	13.72	12.06	10.60
τ/τ_0	8.00	9.00	10.00	11.00	12.00	13.00	14.00	15.00	16.00
$P_0(\tau,(x_0/\sigma)) * 100$	7.672	5.554	4.155	2.915	2.109	1.531	1.108	0.806	0.580
τ/τ_0	0.00	0.20	0.40	0.60	0.80	1.00	1.20	1.40	1.60
$W_{2+}(\tau,x_0)$	27.94	27.69	27.12	26.22	25.07	23.74	22.33	20.93	19.59
τ/τ_0	1.80	2.00	2.20	2.40	2.60	2.80	3.00	3.20	3.40
$W_{2+}(\tau,x_0)$	18.35	17.20	16.12	15.11	14.17	13.28	12.43	11.68	10.97
τ/τ_0	3.80	4.20	4.60	5.00	5.40	5.80	6.20	6.60	7.00
$W_{2+}(\tau,x_0)$	9.614	8.441	7.424	6.556	5.717	5.039	4.435	3.874	3.447
τ/τ_0	8.00	9.00	10.00	11.00	12.00	13.00	14.00	15.00	16.00
$W_{2+}(\tau,x_0)$	2.457	1.778	1.286	0.930	0.673	0.487	0.353	0.255	0.185

Gaussian power spectrum excursions above level $(x_0/\sigma) = 1.0$.

τ/τ_0	0.00	0.10	0.20	0.30	0.40	0.50	0.60	0.70	0.80
$P_0(\tau,(x_0/\sigma)) * 100$	100.0	94.37	88.79	83.30	77.93	72.74	67.74	62.97	58.45
τ/τ_0	0.90	1.00	1.10	1.20	1.30	1.40	1.50	1.60	1.80
$P_0(\tau,(x_0/\sigma)) * 100$	54.20	50.22	46.52	43.09	39.90	36.97	34.26	31.75	27.29
τ/τ_0	2.00	2.20	2.40	2.60	2.80	3.00	3.20	3.40	3.60
$P_0(\tau,(x_0/\sigma)) * 100$	23.45	20.17	17.33	14.89	12.80	11.00	9.462	8.126	6.986
τ/τ_0	4.00	4.40	4.80	5.20	5.60	6.00	6.40	6.80	7.20
$P_0(\tau,(x_0/\sigma)) * 100$	5.161	3.812	2.811	2.150	1.538	1.131	0.836	0.617	0.457
τ/τ_0	0.00	0.10	0.20	0.30	0.40	0.50	0.60	0.70	0.80
$W_{2+}(\tau,x_0)$	56.25	56.03	55.37	54.30	52.82	50.99	48.85	46.45	43.83
τ/τ_0	0.90	1.00	1.10	1.20	1.30	1.40	1.50	1.60	1.80
$W_{2+}(\tau,x_0)$	41.14	38.41	35.67	33.07	30.58	28.23	26.08	24.13	20.61

τ/τ_0	2.00	2.20	2.40	2.60	2.80	3.00	3.20	3.40	3.60
$W_{2+}(\tau, x_0)$	17.81	15.29	13.12	11.27	9.679	8.405	7.124	6.193	5.314
τ/τ_0	4.00	4.40	4.80	5.20	5.60	6.00	6.40	6.80	7.20
$W_{2+}(\tau, x_0)$	3.892	2.918	2.111	1.592	1.221	0.879	0.628	0.471	0.348

Gaussian power spectrum excursions above level $(x_0/\sigma) = 1.5$.

τ/τ_0	0.00	0.10	0.20	0.30	0.40	0.50	0.60	0.70	0.80
$P_0(\tau, (x_0/\sigma)) * 100$	100.0	91.55	83.23	75.13	67.37	60.03	53.19	46.91	41.22
τ/τ_0	0.90	1.00	1.10	1.20	1.30	1.40	1.50	1.60	1.70
$P_0(\tau, (x_0/\sigma)) * 100$	36.11	31.59	27.62	24.14	21.12	18.48	16.19	14.20	12.46
τ/τ_0	1.80	1.90	2.00	2.20	2.40	2.60	2.80	3.00	3.20
$P_0(\tau, (x_0/\sigma)) * 100$	10.93	9.598	8.416	6.474	4.989	3.827	2.950	2.282	1.750
τ/τ_0	3.40	3.60	3.80	4.00	4.20	4.40	4.60	4.80	5.00
$P_0(\tau, (x_0/\sigma)) * 100$	1.344	1.036	0.798	0.613	0.471	0.365	0.284	0.219	0.168
τ/τ_0	0.00	0.10	0.20	0.30	0.40	0.50	0.60	0.70	0.80
$W_{2+}(\tau, x_0)$	84.61	83.87	82.13	79.29	75.50	70.87	65.57	59.88	54.00
τ/τ_0	0.90	1.00	1.10	1.20	1.30	1.40	1.50	1.60	1.70
$W_{2+}(\tau, x_0)$	48.13	42.48	37.24	32.48	28.30	24.64	21.42	18.65	16.36
τ/τ_0	1.80	1.90	2.00	2.20	2.40	2.60	2.80	3.00	3.20
$W_{2+}(\tau, x_0)$	14.31	12.55	11.13	8.567	6.597	5.033	3.884	3.014	2.304
τ/τ_0	3.40	3.60	3.80	4.00	4.20	4.40	4.60	4.80	5.00
$W_{2+}(\tau, x_0)$	1.760	1.358	1.044	0.807	0.623	0.480	0.370	0.285	0.220

Gaussian power spectrum excursions above level $(x_0/\sigma) = 2.0$.

τ/τ_0	0.00	0.05	0.10	0.15	0.20	0.25	0.30	0.35	0.40
$P_0(\tau, (x_0/\sigma)) * 100$	100.0	94.36	88.75	83.21	77.76	72.42	67.24	62.29	57.41
τ/τ_0	0.45	0.50	0.55	0.60	0.65	0.70	0.75	0.80	0.90
$P_0(\tau, (x_0/\sigma)) * 100$	52.80	48.43	44.29	40.41	36.77	33.40	30.27	27.39	22.35
τ/τ_0	1.00	1.10	1.20	1.30	1.40	1.50	1.60	1.70	1.80
$P_0(\tau, (x_0/\sigma)) * 100$	18.18	14.80	12.03	9.790	8.003	6.553	5.371	4.384	3.618

(Continued)

Table A2.5 (Continued)

τ/τ_0	1.90	2.00	2.10	2.20	2.30	2.40	2.50	2.60	2.70
$P_0(\tau, (x_0/\sigma)) * 100$	2.963	2.438	1.997	1.635	1.339	1.097	0.901	0.739	0.606
τ/τ_0	0.00	0.05	0.10	0.15	0.20	0.25	0.30	0.35	0.40
$W_{2+}(\tau, x_0)$	112.8	112.5	111.5	110.0	107.8	105.2	101.9	98.25	94.26
τ/τ_0	0.45	0.50	0.55	0.60	0.65	0.70	0.75	0.80	0.90
$W_{2+}(\tau, x_0)$	89.87	85.15	80.19	75.19	70.17	64.98	60.03	55.08	45.82
τ/τ_0	1.00	1.10	1.20	1.30	1.40	1.50	1.60	1.70	1.80
$W_{2+}(\tau, x_0)$	37.72	30.75	24.65	20.06	16.27	13.07	10.51	8.819	7.072
τ/τ_0	1.90	2.00	2.10	2.20	2.30	2.40	2.50	2.60	2.70
$W_{2+}(\tau, x_0)$	5.778	4.813	4.040	3.307	2.614	2.212	1.868	1.550	1.273

Gaussian power spectrum excursions above level $(x_0/\sigma) = 2.5$.

τ/τ_0	0.00	0.04	0.08	0.12	0.16	0.20	0.24	0.28	0.32
$P_0(\tau, (x_0/\sigma)) * 100$	100.0	94.36	88.75	83.21	77.76	72.39	67.19	62.22	57.29
τ/τ_0	0.36	0.40	0.44	0.48	0.52	0.56	0.60	0.64	0.68
$P_0(\tau, (x_0/\sigma)) * 100$	52.64	48.23	43.99	40.06	36.32	32.85	29.61	26.67	23.90
τ/τ_0	0.72	0.76	0.80	0.88	0.96	1.04	1.12	1.20	1.28
$P_0(\tau, (x_0/\sigma)) * 100$	21.41	19.11	17.11	13.54	10.71	8.451	6.706	5.268	4.159
τ/τ_0	1.36	1.44	1.52	1.60	1.68	1.76	1.84	1.92	2.00
$P_0(\tau, (x_0/\sigma)) * 100$	3.329	2.651	2.141	1.729	1.373	1.097	0.887	0.715	0.578
τ/τ_0	0.00	0.04	0.08	0.12	0.16	0.20	0.24	0.28	0.32
$W_{2+}(\tau, x_0)$	141.1	140.6	139.5	137.6	134.8	131.9	127.8	123.7	119.2
τ/τ_0	0.36	0.40	0.44	0.48	0.52	0.56	0.60	0.64	0.68
$W_{2+}(\tau, x_0)$	114.0	108.1	102.1	96.01	89.90	83.58	77.31	71.11	65.02
τ/τ_0	0.72	0.76	0.80	0.88	0.96	1.04	1.12	1.20	1.28
$W_{2+}(\tau, x_0)$	59.25	53.67	49.16	40.10	31.32	24.76	19.58	15.48	12. 24
τ/τ_0	1.36	1.44	1.52	1.60	1.68	1.76	1.84	1.92	2.00
$W_{2+}(\tau, x_0)$	9.678	7.592	5.956	4.672	3.665	2.875	2.256	1.775	1.388

Gaussian power spectrum excursions above level $(x_0/\sigma) = 3.0$.

τ/τ_0	0.00	0.04	0.08	0.12	0.16	0.20	0.24	0.28	0.32
$P_0(\tau,(x_0/\sigma)) * 100$	100.0	93.23	86.51	79.89	73.42	67.14	61.09	55.22	49.79
τ/τ_0	0.36	0.40	0.44	0.48	0.52	0.56	0.60	0.64	0.68
$P_0(\tau,(x_0/\sigma)) * 100$	44.65	39.69	35.29	31.06	27.39	23.99	20.86	18.09	15.61
τ/τ_0	0.72	0.76	0.80	0.84	0.88	0.92	0.96	1.00	1.04
$P_0(\tau,(x_0/\sigma)) * 100$	13.48	11.58	9.942	8.465	7.210	6.122	5.210	4.453	3.781
τ/τ_0	1.08	1.12	1.16	1.20	1.28	1.36	1.44	1.52	1.60
$P_0(\tau,(x_0/\sigma)) * 100$	3.233	2.775	2.327	1.999	1.503	1.082	0.793	0.601	0.456
τ/τ_0	0.00	0.04	0.08	0.12	0.16	0.20	0.24	0.28	0.32
$W_{2+}(\tau,x_0)$	169.4	168.6	166.6	163.7	159.7	154.1	147.7	141.3	134.0
τ/τ_0	0.36	0.40	0.44	0.48	0.52	0.56	0.60	0.64	0.68
$W_{2+}(\tau,x_0)$	125.9	117.2	108.1	98.79	89.68	81.58	73.75	65.59	57.60
τ/τ_0	0.72	0.76	0.80	0.88	0.96	1.04	1.12	1.20	1.28
$W_{2+}(\tau,x_0)$	50.93	44.25	38.21	29.28	21.13	15.79	11.79	8.811	6.582
τ/τ_0	1.36	1.44	1.52	1.60	1.68	1.76	1.84	1.92	2.00
$W_{2+}(\tau,x_0)$	5.035	3.814	2.889	2.189	1.658	1.256	0.952	0.721	0.546

Table A2.6 Rayleigh Random Process

Rectangular power spectrum excursions above level $(x_0/\sigma) = 0.5$.

τ/τ_0	0.00	0.20	0.40	0.60	0.80	1.00	1.20	1.40	1.60
$P_0(\tau,(x_0/\sigma)) * 100$	100.0	93.95	87.94	82.04	76.30	70.78	65.57	60.73	56.32
τ/τ_0	1.80	2.00	2.20	2.40	2.60	2.80	3.00	3.20	3.40
$P_0(\tau,(x_0/\sigma)) * 100$	52.33	48.71	45.37	42.26	39.37	36.65	34.15	31.80	29.65
τ/τ_0	3.80	4.20	4.60	5.00	5.40	5.80	6.20	6.60	7.00
$P_0(\tau,(x_0/\sigma)) * 100$	25.74	22.40	19.43	16.90	14.69	12.78	11.11	9.664	8.400
τ/τ_0	7.40	8.00	9.00	10.00	11.00	12.00	13.00	14.00	15.00
$P_0(\tau,(x_0/\sigma)) * 100$	7.310	5.931	4.187	2.955	2.086	1.473	1.040	0.734	0.518
τ/τ_0	0.00	0.20	0.40	0.60	0.80	1.00	1.20	1.40	1.60
$W_{2+}(\tau,x_0)$	30.31	30.15	29.77	29.12	28.15	26.82	25.12	23.11	20.99
τ/τ_0	1.80	2.00	2.20	2.40	2.60	2.80	3.00	3.20	3.40
$W_{2+}(\tau,x_0)$	19.04	17.41	16.13	15.00	14.02	13.05	12.12	11.25	10.44
τ/τ_0	3.80	4.20	4.60	5.00	5.40	5.80	6.20	6.60	7.00
$W_{2+}(\tau,x_0)$	9.077	7.881	6.814	5.902	5.155	4.426	3.921	3.391	2.913
τ/τ_0	7.40	8.00	9.00	10.00	11.00	12.00	13.00	14.00	15.00
$W_{2+}(\tau,x_0)$	2.542	2.056	1.476	1.024	0.723	0.510	0.360	0.254	0.179

Rectangular power spectrum excursions above level $(x_0/\sigma) = 1.0$.

τ/τ_0	0.00	0.10	0.20	0.30	0.40	0.50	0.60	0.70	0.80
$P_0(\tau,(x_0/\sigma)) * 100$	100.0	94.23	88.49	82.80	77.20	71.71	66.37	61.21	56.25
τ/τ_0	0.90	1.00	1.10	1.20	1.30	1.40	1.50	1.60	1.80
$P_0(\tau,(x_0/\sigma)) * 100$	51.54	47.11	42.98	39.17	35.69	32.56	29.73	27.21	22.90
τ/τ_0	2.00	2.20	2.40	2.60	2.80	3.00	3.20	3.40	3.60
$P_0(\tau,(x_0/\sigma)) * 100$	19.36	16.38	13.84	11.67	9.829	8.278	6.994	5.906	4.984
τ/τ_0	4.00	4.40	4.80	5.20	5.60	6.00	6.40	6.80	7.20
$P_0(\tau,(x_0/\sigma)) * 100$	3.554	2.533	1.806	1.287	0.917	0.659	0.466	0.332	0.236
τ/τ_0	0.00	0.10	0.20	0.30	0.40	0.50	0.60	0.70	0.80
$W_{2+}(\tau,x_0)$	57.74	57.57	57.15	56.44	55.45	54.14	52.53	50.58	48.31

τ/τ_0	0.90	1.00	1.10	1.20	1.30	1.40	1.50	1.60	1.80
$W_{2+}(\tau, x_0)$	45.72	42.83	39.69	36.41	33.07	29.80	26.74	23.97	19.46
τ/τ_0	2.00	2.20	2.40	2.60	2.80	3.00	3.20	3.40	3.60
$W_{2+}(\tau, x_0)$	16.29	13.80	11.75	10.02	8.419	7.064	5.917	4.991	4.236
τ/τ_0	4.00	4.40	4.80	5.20	5.60	6.00	6.40	6.80	7.20
$W_{2+}(\tau, x_0)$	3.036	2.166	1.545	1.089	0.784	0.564	0.398	0.284	0.205
Rectangular power spectrum excursions above level $(x_0/\sigma) = 1.5$.									
τ/τ_0	0.00	0.05	0.10	0.15	0.20	0.25	0.30	0.35	0.40
$P_0(\tau, (x_0/\sigma)) * 100$	100.0	95.72	91.46	87.21	83.00	78.83	74.71	70.66	66.67
τ/τ_0	0.45	0.50	0.55	0.60	0.65	0.70	0.75	0.80	0.85
$P_0(\tau, (x_0/\sigma)) * 100$	62.77	58.96	55.26	51.67	48.20	44.87	41.66	38.61	35.71
τ/τ_0	0.90	0.95	1.00	1.10	1.20	1.30	1.40	1.60	1.80
$P_0(\tau, (x_0/\sigma)) * 100$	32.96	30.38	27.95	23.59	19.88	16.75	14.18	10.31	7.630
τ/τ_0	2.00	2.20	2.40	2.60	2.80	3.00	3.20	3.40	3.60
$P_0(\tau, (x_0/\sigma)) * 100$	5.703	4.218	3.155	2.308	1.707	1.259	0.937	0.697	0.519
τ/τ_0	0.00	0.05	0.10	0.15	0.20	0.25	0.30	0.35	0.40
$W_{2+}(\tau, x_0)$	85.52	85.41	85.10	84.57	83.84	82.90	81.74	80.39	78.84
τ/τ_0	0.45	0.50	0.55	0.60	0.65	0.70	0.75	0.80	0.85
$W_{2+}(\tau, x_0)$	77.09	75.11	72.89	70.60	68.06	65.39	62.53	59.55	56.57
τ/τ_0	0.90	0.95	1.00	1.10	1.20	1.30	1.40	1.60	1.80
$W_{2+}(\tau, x_0)$	53.32	50.07	46.86	40.33	34.11	28.47	23.53	15.94	11.29
τ/τ_0	2.00	2.20	2.40	2.60	2.80	3.00	3.20	3.40	3.60
$W_{2+}(\tau, x_0)$	8.283	6.328	4.830	3.568	2.636	1.947	1.438	1.062	0.785
Rectangular power spectrum excursions above level $(x_0/\sigma) = 2.0$.									
τ/τ_0	0.00	0.04	0.08	0.12	0.16	0.20	0.24	0.28	0.32
$P_0(\tau, (x_0/\sigma)) * 100$	100.0	95.47	90.95	86.45	82.01	77.60	73.26	69.00	64.81
τ/τ_0	0.36	0.40	0.44	0.48	0.52	0.56	0.60	0.64	0.68
$P_0(\tau, (x_0/\sigma)) * 100$	60.72	56.70	52.88	49.20	45.54	42.08	38.76	35.61	32.62
τ/τ_0	0.72	0.80	0.88	0.96	1.04	1.12	1.20	1.28	1.36
$P_0(\tau, (x_0/\sigma)) * 100$	29.79	24.64	20.16	16.37	13.17	10.55	8.452	6.784	5.483

(Continued)

Table A2.6 (Continued)

τ/τ_0	1.44	1.52	1.60	1.68	1.76	1.92	2.08	2.24	2.40
$P_0(\tau,(x_0/\sigma))*100$	4.447	3.663	3.017	2.508	2.087	1.466	1.031	0.713	0.487
τ/τ_0	0.00	0.04	0.08	0.12	0.16	0.20	0.24	0.28	0.32
$W_{2+}(\tau,x_0)$	113.3	113.1	112.5	111.8	110.7	109.3	107.6	105.7	103.7
τ/τ_0	0.36	0.40	0.44	0.48	0.52	0.56	0.60	0.64	0.68
$W_{2+}(\tau,x_0)$	101.2	98.07	94.89	91.70	88.22	84.73	80.84	76.79	72.78
τ/τ_0	0.72	0.80	0.88	0.96	1.04	1.12	1.20	1.28	1.36
$W_{2+}(\tau,x_0)$	68.53	60.16	51.62	43.72	36.19	29.25	23.57	18.57	14.28
τ/τ_0	1.44	1.52	1.60	1.68	1.76	1.92	2.08	2.24	2.40
$W_{2+}(\tau,x_0)$	11.45	8.997	7.044	5.809	4.708	3.158	2.194	1.646	1.237
Rectangular power spectrum excursions above level $(x_0/\sigma)=2.5$.									
τ/τ_0	0.00	0.04	0.08	0.12	0.16	0.20	0.24	0.28	0.32
$P_0(\tau,(x_0/\sigma))*100$	100.0	94.35	88.73	83.16	77.66	72.26	66.98	61.87	56.86
τ/τ_0	0.36	0.40	0.44	0.48	0.52	0.56	0.60	0.64	0.68
$P_0(\tau,(x_0/\sigma))*100$	52.06	47.50	43.08	38.90	35.00	31.31	27.88	24.69	21.77
τ/τ_0	0.72	0.76	0.80	0.84	0.88	0.92	0.96	1.00	1.04
$P_0(\tau,(x_0/\sigma))*100$	19.09	16.71	14.47	12.50	10.77	9.236	7.868	6.697	5.686
τ/τ_0	1.08	1.12	1.16	1.20	1.28	1.36	1.44	1.52	1.60
$P_0(\tau,(x_0/\sigma))*100$	4.836	4.082	3.446	2.915	2.105	1.554	1.147	0.872	0.663
τ/τ_0	0.00	0.04	0.08	0.12	0.16	0.20	0.24	0.28	0.32
$W_{2+}(\tau,x_0)$	141.3	140.9	139.9	138.4	136.3	133.5	130.3	126.5	122.4
τ/τ_0	0.36	0.40	0.44	0.48	0.52	0.56	0.60	0.64	0.68
$W_{2+}(\tau,x_0)$	117.7	112.3	106.7	101.0	95.18	89.06	82.79	76.29	70.00
τ/τ_0	0.72	0.76	0.80	0.84	0.88	0.92	0.96	1.00	1.04
$W_{2+}(\tau,x_0)$	63.80	57.72	51.92	46.20	40.73	36.24	31.74	27.27	23.26
τ/τ_0	1.08	1.12	1.16	1.20	1.28	1.36	1.44	1.52	1.60
$W_{2+}(\tau,x_0)$	20.05	17.38	14.58	12.02	8.480	5.976	4.183	2.953	2.221

Rectangular power spectrum excursions above level $(x_0/\sigma) = 3.0$.

τ/τ_0	0.00	0.04	0.08	0.12	0.16	0.20	0.24	0.28	0.32
$P_0(\tau,(x_0/\sigma)) * 100$	100.0	93.23	86.51	79.88	73.37	67.00	60.91	55.01	49.39
τ/τ_0	0.36	0.40	0.44	0.48	0.52	0.56	0.60	0.64	0.68
$P_0(\tau,(x_0/\sigma)) * 100$	44.08	39.07	34.39	30.09	26.13	22.52	19.24	16.36	13.75
τ/τ_0	0.72	0.76	0.80	0.84	0.88	0.92	0.96	1.00	1.04
$P_0(\tau,(x_0/\sigma)) * 100$	11.52	9.520	7.833	6.379	5.157	4.173	3.345	2.629	2.155
τ/τ_0	1.08	1.12	1.16	1.20	1.24	1.28	1.32	1.36	1.40
$P_0(\tau,(x_0/\sigma)) * 100$	1.658	1.310	1.030	0.827	0.666	0.527	0.423	0.339	0.283
τ/τ_0	0.00	0.04	0.08	0.12	0.16	0.20	0.24	0.28	0.32
$W_{2+}(\tau,x_0)$	169.3	168.7	167.2	164.5	160.6	155.9	150.0	144.3	136.8
τ/τ_0	0.36	0.40	0.44	0.48	0.52	0.56	0.60	0.64	0.68
$W_{2+}(\tau,x_0)$	129.0	121.0	112.8	103.9	94.64	85.19	77.05	69.10	60.12
τ/τ_0	0.72	0.76	0.80	0.84	0.88	0.92	0.96	1.00	1.04
$W_{2+}(\tau,x_0)$	53.02	46.42	39.85	33.44	27.58	22.50	17.99	14.80	12.17
τ/τ_0	1.08	1.12	1.16	1.20	1.24	1.28	1.32	1.36	1.40
$W_{2+}(\tau,x_0)$	9.787	7.847	6.291	5.043	4.043	3.241	2.599	2.083	1.670

Table A2.7 Rayleigh Random Process

Gaussian power spectrum excursions below level $(x_0/\sigma) = 0.5$.

τ/τ_0	0.00	0.04	0.08	0.12	0.16	0.20	0.24	0.28	0.32
$P_0(\tau,(x_0/\sigma)) * 100$	100.0	91.56	83.20	74.96	67.06	59.49	52.46	46.30	40.58
τ/τ_0	0.36	0.40	0.44	0.48	0.52	0.56	0.60	0.64	0.68
$P_0(\tau,(x_0/\sigma)) * 100$	35.80	31.80	28.19	25.20	22.56	20.26	18.27	16.51	14.89
τ/τ_0	0.72	0.76	0.80	0.88	0.96	1.04	1.12	1.20	1.28
$P_0(\tau,(x_0/\sigma)) * 100$	13.45	12.12	10.95	8.916	7.247	5.879	4.752	3.876	3.134
τ/τ_0	1.36	1.44	1.52	1.60	1.68	1.76	1.84	1.92	2.00
$P_0(\tau,(x_0/\sigma)) * 100$	2.551	2.085	1.685	1.382	1.115	0.906	0.742	0.601	0.492
τ/τ_0	0.00	0.04	0.08	0.12	0.16	0.20	0.24	0.28	0.32
$W_{2-}(\tau,x_0)$	211.3	210.0	207.3	201.8	192.5	182.5	166.0	148.0	128.0
τ/τ_0	0.36	0.40	0.44	0.48	0.52	0.56	0.60	0.64	0.68
$W_{2-}(\tau,x_0)$	110.2	95.09	82.73	70.37	62.00	53.62	46.88	42.35	38.25
τ/τ_0	0.72	0.76	0.80	0.88	0.96	1.04	1.12	1.20	1.28
$W_{2-}(\tau,x_0)$	34.77	31.30	28.35	23.26	18.72	15.31	12.78	10.30	8.375
τ/τ_0	1.36	1.44	1.52	1.60	1.68	1.76	1.84	1.92	2.00
$W_{2-}(\tau,x_0)$	6.411	5.061	4.066	3.354	2.730	2.232	1.809	1.472	1.198

Gaussian power spectrum excursions below level $(x_0/\sigma) = 1.0$.

τ/τ_0	0.00	0.10	0.20	0.30	0.40	0.50	0.60	0.70	0.80
$P_0(\tau,(x_0/\sigma)) * 100$	100.0	91.33	82.76	74.43	66.47	59.12	52.57	46.86	41.91
τ/τ_0	0.90	1.00	1.10	1.20	1.30	1.40	1.50	1.60	1.70
$P_0(\tau,(x_0/\sigma)) * 100$	37.55	33.68	30.20	27.09	24.28	21.76	19.51	17.48	15.66
τ/τ_0	1.80	1.90	2.00	2.20	2.40	2.60	2.80	3.00	3.20
$P_0(\tau,(x_0/\sigma)) * 100$	14.04	12.57	11.28	9.101	7.275	5.904	4.698	3.785	3.033
τ/τ_0	3.40	3.60	3.80	4.00	4.20	4.40	4.60	4.80	5.00
$P_0(\tau,(x_0/\sigma)) * 100$	2.436	1.956	1.571	1.266	1.017	0.811	0.651	0.525	0.430
τ/τ_0	0.00	0.10	0.20	0.30	0.40	0.50	0.60	0.70	0.80
$W_{2-}(\tau,x_0)$	86.86	86.32	84.68	81.69	76.87	69.67	61.18	53.15	46.45

τ/τ_0	0.90	1.00	1.10	1.20	1.30	1.40	1.50	1.60	1.70
$W_{2-}(\tau, x_0)$	41.02	36.73	32.95	29.53	26.67	23.87	21.43	19.25	17.08
τ/τ_0	1.80	1.90	2.00	2.20	2.40	2.60	2.80	3.00	3.20
$W_{2-}(\tau, x_0)$	15.34	13.80	12.38	9.991	7.858	6.341	5.117	4.129	3.332
τ/τ_0	3.40	3.60	3.80	4.00	4.20	4.40	4.60	4.80	5.00
$W_{2-}(\tau, x_0)$	2.711	2.177	1.748	1.404	1.127	0.905	0.727	0.584	0.469
Gaussian power spectrum excursions below level $(x_0/\sigma) = 1.5$.									
τ/τ_0	0.00	0.20	0.40	0.60	0.80	1.00	1.20	1.40	1.60
$P_0(\tau, (x_0/\sigma)) * 100$	100.0	91.89	83.91	76.23	69.11	62.72	56.99	51.80	47.08
τ/τ_0	1.80	2.00	2.20	2.40	2.60	2.80	3.00	3.20	3.40
$P_0(\tau, (x_0/\sigma)) * 100$	42.78	38.88	35.32	32.10	29.17	26.51	24.07	21.89	19.89
τ/τ_0	3.60	3.80	4.00	4.40	4.80	5.20	5.60	6.00	6.40
$P_0(\tau, (x_0/\sigma)) * 100$	18.08	16.42	14.91	12.31	10.17	8.405	6.947	5.730	4.733
τ/τ_0	6.80	7.20	7.60	8.00	8.40	8.80	9.20	9.60	10.00
$P_0(\tau, (x_0/\sigma)) * 100$	3.912	3.233	2.672	2.208	1.822	1.503	1.240	1.028	0.852
τ/τ_0	0.00	0.20	0.40	0.60	0.80	1.00	1.20	1.40	1.60
$W_{2-}(\tau, x_0)$	40.65	40.30	39.26	37.18	33.82	30.19	27.23	24.77	22.55
τ/τ_0	1.80	2.00	2.20	2.40	2.60	2.80	3.00	3.20	3.40
$W_{2-}(\tau, x_0)$	20.51	18.64	16.89	15.35	13.92	12.63	11.48	10.44	9.472
τ/τ_0	3.60	3.80	4.00	4.40	4.80	5.20	5.60	6.00	6.40
$W_{2-}(\tau, x_0)$	8.573	7.778	7.089	5.862	4.785	3.906	3.188	2.602	2.124
τ/τ_0	6.80	7.20	7.60	8.00	8.40	8.80	9.20	9.60	10.00
$W_{2-}(\tau, x_0)$	1.734	1.433	1.184	0.979	0.809	0.669	0.553	0.457	0.378
Gaussian power spectrum excursions below level $(x_0/\sigma) = 2.0$.									
τ/τ_0	0.00	0.40	0.80	1.20	1.60	2.00	2.40	2.80	3.20
$P_0(\tau, (x_0/\sigma)) * 100$	100.0	93.35	86.80	80.80	75.51	70.56	65.94	61.71	57.77
τ/τ_0	3.60	4.00	4.80	5.60	6.40	7.20	8.00	9.60	11.20
$P_0(\tau, (x_0/\sigma)) * 100$	54.09	50.67	44.50	39.17	34.49	30.43	26.87	21.06	16.61
τ/τ_0	12.80	14.40	16.00	17.60	19.20	20.80	22.40	24.00	25.60
$P_0(\tau, (x_0/\sigma)) * 100$	13.15	10.50	8.442	6.853	5.564	4.583	3.788	3.145	2.649

(Continued)

Table A2.7 (Continued)

τ/τ_0	27.20	28.80	30.40	32.00	33.60	35.20	36.80	38.40	40.00
$P_0(\tau,(x_0/\sigma)) * 100$	2.238	1.902	1.632	1.389	1.210	1.043	0.915	0.787	0.695
τ/τ_0	0.00	0.40	0.80	1.20	1.60	2.00	2.40	2.80	3.20
$W_{2-}(\tau,x_0)$	16.66	16.50	15.68	14.12	12.80	11.95	11.07	10.22	9.528
τ/τ_0	3.60	4.00	4.80	5.60	6.40	7.20	8.00	9.60	11.20
$W_{2-}(\tau,x_0)$	8.869	8.264	7.162	6.255	5.445	4.753	4.050	3.167	2.455
τ/τ_0	12.80	14.40	16.00	17.60	19.20	20.80	22.40	24.00	25.60
$W_{2-}(\tau,x_0)$	1.892	1.438	1.078	0.857	0.682	0.548	0.441	0.345	0.263
τ/τ_0	27.20	28.80	30.40	32.00	33.60	35.20	36.80	38.40	40.00
$W_{2-}(\tau,x_0)$	0.219	0.189	0.162	0.140	0.121	0.104	0.090	0.077	0.067
Gaussian power spectrum excursions below level $(x_0/\sigma) = 2.5$.									
τ/τ_0	0.00	1.00	2.00	3.00	4.00	5.00	6.00	7.00	8.00
$P_0(\tau,(x_0/\sigma)) * 100$	100.0	93.57	87.46	81.76	76.42	71.44	66.78	62.42	58.34
τ/τ_0	9.00	10.00	11.00	12.00	13.00	14.00	15.00	16.00	17.00
$P_0(\tau,(x_0/\sigma)) * 100$	54.54	50.99	47.66	44.55	41.65	38.92	36.38	34.02	31.80
τ/τ_0	18.00	20.00	22.00	24.00	28.00	32.00	36.00	40.00	44.00
$P_0(\tau,(x_0/\sigma)) * 100$	29.74	25.96	22.69	19.84	15.13	11.57	8.841	6.756	5.172
τ/τ_0	48.00	52.00	56.00	60.00	64.00	68.00	72.00	76.00	80.00
$P_0(\tau,(x_0/\sigma)) * 100$	3.946	3.008	2.297	1.753	1.347	1.029	0.785	0.599	0.457
τ/τ_0	0.00	1.00	2.00	3.00	4.00	5.00	6.00	7.00	8.00
$W_{2-}(\tau,x_0)$	6.487	6.270	5.905	5.520	5.158	4.822	4.512	4.216	3.939
τ/τ_0	9.00	10.00	11.00	12.00	13.00	14.00	15.00	16.00	17.00
$W_{2-}(\tau,x_0)$	3.675	3.441	3.223	3.003	2.811	2.633	2.453	2.291	2.138
τ/τ_0	18.00	20.00	22.00	24.00	28.00	32.00	36.00	40.00	44.00
$W_{2-}(\tau,x_0)$	2.009	1.762	1.532	1.340	1.015	0.756	0.568	0.454	0.351
τ/τ_0	48.00	52.00	56.00	60.00	64.00	68.00	72.00	76.00	80.00
$W_{2-}(\tau,x_0)$	0.262	0.204	0.157	0.118	0.089	0.070	0.055	0.042	0.032

Table A2.8 Rayleigh Random Process

Rectangular power spectrum excursions below level $(x_0/\sigma) = 0.5$.									
τ/τ_0	0.00	0.04	0.08	0.12	0.16	0.20	0.24	0.28	0.32
$P_0(\tau,(x_0/\sigma)) * 100$	100.0	92.51	85.08	77.78	70.65	63.81	57.39	51.61	46.32
τ/τ_0	0.36	0.40	0.44	0.48	0.52	0.56	0.60	0.68	0.76
$P_0(\tau,(x_0/\sigma)) * 100$	41.78	37.80	34.47	31.55	29.02	26.82	24.89	21.62	18.95
τ/τ_0	0.84	0.92	1.00	1.08	1.16	1.24	1.32	1.40	1.56
$P_0(\tau,(x_0/\sigma)) * 100$	16.68	14.75	13.05	11.53	10.21	9.056	8.040	7.164	5.764
τ/τ_0	1.72	1.88	2.04	2.20	2.36	2.52	2.68	2.84	3.00
$P_0(\tau,(x_0/\sigma)) * 100$	4.638	3.843	3.185	2.656	2.222	1.859	1.568	1.318	1.130
τ/τ_0	0.00	0.04	0.08	0.12	0.16	0.20	0.24	0.28	0.32
$W_{2-}(\tau,x_0)$	187.4	186.5	183.5	180.4	173.4	165.8	152.0	138.3	122.0
τ/τ_0	0.36	0.40	0.44	0.48	0.52	0.56	0.60	0.68	0.76
$W_{2-}(\tau,x_0)$	106.0	91.26	79.22	68.17	59.90	51.63	45.79	36.66	30.48
τ/τ_0	0.84	0.92	1.00	1.08	1.16	1.24	1.32	1.40	1.56
$W_{2-}(\tau,x_0)$	26.39	22.58	20.05	17.68	15.38	13.67	11.54	10.18	7.754
τ/τ_0	1.72	1.88	2.04	2.20	2.36	2.52	2.68	2.84	3.00
$W_{2-}(\tau,x_0)$	5.703	4.608	3.723	3.008	2.406	1.982	1.695	1.346	1.069
Rectangular power spectrum excursions below level $(x_0/\sigma) = 1.0$.									
τ/τ_0	0.00	0.10	0.20	0.30	0.40	0.50	0.60	0.70	0.80
$P_0(\tau,(x_0/\sigma)) * 100$	100.0	91.94	83.96	76.14	68.62	61.62	55.42	50.08	45.54
τ/τ_0	0.90	1.00	1.10	1.20	1.30	1.40	1.50	1.60	1.70
$P_0(\tau,(x_0/\sigma)) * 100$	41.65	38.27	35.24	32.48	29.93	27.58	25.40	23.40	21.56
τ/τ_0	1.80	1.90	2.00	2.20	2.40	2.60	2.80	3.00	3.20
$P_0(\tau,(x_0/\sigma)) * 100$	19.89	18.39	17.02	14.58	12.65	10.96	9.503	8.267	7.215
τ/τ_0	3.40	3.60	3.80	4.00	4.20	4.40	4.60	4.80	5.00
$P_0(\tau,(x_0/\sigma)) * 100$	6.358	5.602	4.936	4.350	3.833	3.423	3.059	2.733	2.442

(Continued)

Table A2.8 (Continued)

τ/τ_0	0.00	0.10	0.20	0.30	0.40	0.50	0.60	0.70	0.80
$W_{2-}(\tau, x_0)$	80.69	80.32	79.15	76.93	72.89	66.23	57.75	49.21	41.90
τ/τ_0	0.90	1.00	1.10	1.20	1.30	1.40	1.50	1.60	1.70
$W_{2-}(\tau, x_0)$	36.21	32.08	28.94	26.42	24.48	22.65	20.90	19.15	17.50
τ/τ_0	1.80	1.90	2.00	2.20	2.40	2.60	2.80	3.00	3.20
$W_{2-}(\tau, x_0)$	15.95	14.30	13.08	10.87	9.214	7.853	6.879	5.720	4.756
τ/τ_0	3.40	3.60	3.80	4.00	4.20	4.40	4.60	4.80	5.00
$W_{2-}(\tau, x_0)$	4.154	3.535	3.029	2.665	2.316	2.063	1.743	1.519	1.323
Rectangular power spectrum excursions below level $(x_0/\sigma) = 1.5$.									
τ/τ_0	0.00	0.20	0.40	0.60	0.80	1.00	1.20	1.40	1.60
$P_0(\tau, (x_0/\sigma)) * 100$	100.0	92.38	84.84	77.51	70.75	64.88	59.83	55.30	51.11
τ/τ_0	1.80	2.00	2.20	2.40	2.60	2.80	3.00	3.20	3.40
$P_0(\tau, (x_0/\sigma)) * 100$	47.20	43.60	40.31	37.34	34.63	32.12	29.80	27.66	25.68
τ/τ_0	3.60	3.80	4.00	4.40	4.80	5.20	5.60	6.00	6.40
$P_0(\tau, (x_0/\sigma)) * 100$	23.87	22.21	20.68	17.94	15.60	13.60	11.89	10.41	9.151
τ/τ_0	6.80	7.20	7.60	8.00	8.40	8.80	9.20	9.60	10.00
$P_0(\tau, (x_0/\sigma)) * 100$	8.053	7.115	6.296	5.588	4.960	4.402	3.907	3.562	3.247
τ/τ_0	0.00	0.20	0.40	0.60	0.80	1.00	1.20	1.40	1.60
$W_{2-}(\tau, x_0)$	38.15	37.90	37.16	35.23	31.59	27.30	23.95	21.79	20.25
τ/τ_0	1.80	2.00	2.20	2.40	2.60	2.80	3.00	3.20	3.40
$W_{2-}(\tau, x_0)$	18.78	17.21	15.65	14.21	13.04	12.06	11.15	10.30	9.462
τ/τ_0	3.60	3.80	4.00	4.40	4.80	5.20	5.60	6.00	6.40
$W_{2-}(\tau, x_0)$	8.678	7.989	7.416	6.347	5.403	4.615	3.977	3.413	2.926
τ/τ_0	6.80	7.20	7.60	8.00	8.40	8.80	9.20	9.60	10.00
$W_{2-}(\tau, x_0)$	2.557	2.189	1.899	1.643	1.444	1.258	1.086	0.965	0.844
Rectangular power spectrum excursions below level $(x_0/\sigma) = 2.0$.									
τ/τ_0	0.00	0.30	0.60	0.90	1.20	1.50	1.80	2.10	2.40
$P_0(\tau, (x_0/\sigma)) * 100$	100.0	94.72	89.50	84.50	79.67	75.18	70.95	66.90	63.18
τ/τ_0	2.70	3.00	3.60	4.20	4.80	5.40	6.00	7.20	8.40
$P_0(\tau, (x_0/\sigma)) * 100$	59.62	56.26	50.10	44.61	39.73	35.38	31.51	24.99	19.82

τ/τ_0	9.60	10.80	12.00	13.20	14.40	15.60	16.80	18.00	19.20
$P_0(\tau,(x_0/\sigma)) * 100$	15.72	12.47	9.885	7.838	6.213	4.929	3.908	3.107	2.456
τ/τ_0	20.40	21.60	22.80	24.00	25.20	26.40	27.60	28.80	30.00
$P_0(\tau,(x_0/\sigma)) * 100$	1.975	1.563	1.250	0.980	0.774	0.621	0.490	0.380	0.303
τ/τ_0	0.00	0.30	0.60	0.90	1.20	1.50	1.80	2.10	2.40
$W_{2-}(\tau,x_0)$	17.65	17.50	17.13	16.39	15.46	14.53	13.73	12.94	12.20
τ/τ_0	2.70	3.00	3.60	4.20	4.80	5.40	6.00	7.20	8.40
$W_{2-}(\tau,x_0)$	11.51	10.86	9.679	8.623	7.682	6.844	6.097	4.838	3.840
τ/τ_0	9.60	10.80	12.00	13.20	14.40	15.60	16.80	18.00	19.20
$W_{2-}(\tau,x_0)$	3.047	2.418	1.919	1.523	1.208	0.959	0.761	0.604	0.479
τ/τ_0	20.40	21.60	22.80	24.00	25.20	26.40	27.60	28.80	30.00
$W_{2-}(\tau,x_0)$	0.380	0.302	0.240	0.190	0.151	0.120	0.095	0.075	0.060

Rectangular power spectrum excursions below level $(x_0/\sigma) = 2.5$.

τ/τ_0	0.00	1.00	2.00	3.00	4.00	5.00	6.00	7.00	8.00
$P_0(\tau,(x_0/\sigma)) * 100$	100.0	93.80	88.08	82.77	77.83	73.21	68.85	64.78	60.97
τ/τ_0	9.00	10.00	11.00	12.00	13.00	14.00	15.00	16.00	17.00
$P_0(\tau,(x_0/\sigma)) * 100$	57.36	54.02	50.85	47.84	45.04	42.42	39.95	37.63	35.38
τ/τ_0	18.00	20.00	22.00	24.00	28.00	32.00	36.00	40.00	44.00
$P_0(\tau,(x_0/\sigma)) * 100$	33.33	29.56	26.15	23.22	18.21	14.28	11.12	8.649	6.681
τ/τ_0	48.00	52.00	56.00	60.00	64.00	68.00	72.00	76.00	80.00
$P_0(\tau,(x_0/\sigma)) * 100$	5.124	3.967	3.061	2.334	1.798	1.385	1.099	0.846	0.642
τ/τ_0	0.00	1.00	2.00	3.00	4.00	5.00	6.00	7.00	8.00
$W_{2-}(\tau,x_0)$	6.319	5.959	5.514	5.127	4.781	4.488	4.217	3.938	3.706
τ/τ_0	9.00	10.00	11.00	12.00	13.00	14.00	15.00	16.00	17.00
$W_{2-}(\tau,x_0)$	3.477	3.256	3.090	2.904	2.724	2.548	2.381	2.284	2.149
τ/τ_0	18.00	20.00	22.00	24.00	28.00	32.00	36.00	40.00	44.00
$W_{2-}(\tau,x_0)$	2.005	1.777	1.572	1.432	1.127	0.875	0.693	0.545	0.419
τ/τ_0	48.00	52.00	56.00	60.00	64.00	68.00	72.00	76.00	80.00
$W_{2-}(\tau,x_0)$	0.295	0.234	0.180	0.140	0.105	0.100	0.096	0.062	0.039

Each simulation case (i.e., each set of parameter values) involved no less than 10^9 digital samples of the random process with 10^7 or more excursions above a given threshold. Due to a large number of samples, the statistical error of measurement was reduced to a few tenths of a percentage point in a wide range of distribution density values. Even in the area of exponential tails where excursions are long and the density drops to one hundredth of its maximum value, the relative error never exceeded a few percentage points.

Another thing that reduced the statistical error of measurement in simulated processes was the right approach to choosing the sampling time: it was chosen in proportion to excursion duration.

Appendix 3: Simulation of Distribution Densities for Areas Enveloped by Above-Threshold or Below-Threshold Excursions of Gaussian and Rayleigh Random Process Curves

Table A3.1 shows the distribution density values obtained by simulation for areas under the curves that represent excursions of a Gaussian random process with a Gaussian power spectrum above the set standardized threshold level. The threshold level was gradually increased from 0 to 3 in 0.5 incremental steps.

The data presented in Table A3.2 are only different from those presented in Table A3.1 in that they represent another kind of power spectrum in a Gaussian process, namely the rectangular one.

Tables A3.3 and A3.4 show the distribution density values obtained by simulation for areas above the curves that represent negative excursions of a Gaussian random process below the standardized thresholds 0.1, 0.5, 1.0, 1.5, and 2.0. The data for a random process with a Gaussian power spectrum are shown in Table A3.3, whereas the data for a random process with a rectangular power spectrum are shown in Table A3.4.

Tables A3.5–A3.8 show the distribution density values obtained by simulation for areas enveloped by the curves that represent excursions of a Rayleigh random process with two kinds of a power spectrum, a Gaussian one and a rectangular one.

The positive excursions data shown in Tables A3.5 and A3.6 correspond to a Gaussian power spectrum and a rectangular power spectrum, respectively, the standardized thresholds (upper boundaries) being 0.1, 0.2, 0.5, 1.0, 1.5, 2.0, 2.5, and 3.0.

Tables A3.7 and A3.8 show values obtained by simulation for areas above the curves that represent negative excursions of a Rayleigh process with the two above-mentioned kinds of power spectrum. In this case the following standardized threshold levels were used: 0.5, 1.0, 1.5, 2.0, and 2.5.

It should be noted that the distribution of areas enveloped by excursion curves is quite difficult to simulate compared to the distribution of time intervals between threshold crossings (see Appendices 1 and 2).

Table A3.1 Gaussian Random Process

Gaussian power spectrum distribution of areas under excursions above level $(x_0/\sigma) = 0.0$.

$(s/\tau_0 \cdot \sigma)$	0.00	0.002	0.004	0.006	0.008	0.01	0.02	0.04
$W_{ar}(s/\tau_0 \cdot \sigma) * 100$	00.00	350.17	253.05	209.25	185.85	170.97	130.95	101.45
$(s/\tau_0 \cdot \sigma)$	0.06	0.08	0.10	0.14	0.20	0.30	0.40	0.50
$W_{ar}(s/\tau_0 \cdot \sigma) * 100$	87.179	78.242	72.200	63.954	55.904	47.911	42.280	38.621
$(s/\tau_0 \cdot \sigma)$	0.60	0.80	1.00	1.20	1.40	1.60	1.80	2.00
$W_{ar}(s/\tau_0 \cdot \sigma) * 100$	35.597	31.065	27.142	24.172	21.818	19.690	17.601	16.153
$(s/\tau_0 \cdot \sigma)$	2.20	2.40	2.60	2.80	3.00	3.20	3.40	3.60
$W_{ar}(s/\tau_0 \cdot \sigma) * 100$	14.433	12.959	11.737	10.611	9.6693	8.7422	8.0116	7.1899
$(s/\tau_0 \cdot \sigma)$	3.80	4.00	4.20	4.40	4.60	4.80	5.00	5.20
$W_{ar}(s/\tau_0 \cdot \sigma) * 100$	6.5207	5.9818	5.3433	4.8226	4.3851	3.9444	3.5607	3.2117
$(s/\tau_0 \cdot \sigma)$	5.40	5.60	5.80	6.00	6.20	6.40	6.60	6.80
$W_{ar}(s/\tau_0 \cdot \sigma) * 100$	2.9404	2.6610	2.4153	2.2441	1.9738	1.8159	1.6381	1.4560
$(s/\tau_0 \cdot \sigma)$	7.00	7.20	7.40	7.60	7.80	8.00	8.20	8.40
$W_{ar}(s/\tau_0 \cdot \sigma) * 100$	1.3144	1.2623	1.1214	1.0160	0.8761	0.8032	0.7381	0.6714
$(s/\tau_0 \cdot \sigma)$	8.60	8.80	9.00	9.20	9.40	9.60	9.80	10.00
$W_{ar}(s/\tau_0 \cdot \sigma) * 100$	0.6158	0.5385	0.5118	0.4246	0.4046	0.3842	0.3455	0.3107

$M_1 = 1.7724\ (\sigma\ \tau_0)$; $D = 3.9089\ (\sigma\ \tau_0)^2$; the number of excursions = 112454212.

Gaussian power spectrum distribution of areas under excursions above level $(x_0/\sigma) = 0.5$.

$(s/\tau_0 \cdot \sigma)$	0.00	0.002	0.004	0.006	0.008	0.01	0.012	0.014
$W_{ar}(s/\tau_0 \cdot \sigma) * 100$	00.00	554.03	398.38	330.72	292.57	268.44	248.46	234.73
$(s/\tau_0 \cdot \sigma)$	0.016	0.018	0.020	0.024	0.028	0.032	0.036	0.040
$W_{ar}(s/\tau_0 \cdot \sigma) * 100$	223.60	214.36	204.99	190.24	180.62	171.30	164.28	157.74
$(s/\tau_0 \cdot \sigma)$	0.050	0.060	0.080	0.10	0.12	0.14	0.16	0.18
$W_{ar}(s/\tau_0 \cdot \sigma) * 100$	144.50	134.40	120.04	110.12	101.89	95.244	90.571	85.657
$(s/\tau_0 \cdot \sigma)$	0.20	0.30	0.40	0.50	0.60	0.80	1.00	1.20
$W_{ar}(s/\tau_0 \cdot \sigma) * 100$	81.863	67.396	57.689	50.642	44.926	36.167	29.959	25.182
$(s/\tau_0 \cdot \sigma)$	1.40	1.60	1.80	2.00	2.20	2.40	2.60	2.80
$W_{ar}(s/\tau_0 \cdot \sigma) * 100$	20.962	17.665	14.936	12.580	10.554	8.9845	7.6904	6.3666

$(s/\tau_0 \cdot \sigma)$	3.00	3.20	3.40	3.60	3.80	4.00	4.20	4.40
$W_{ar}(s/\tau_0 \cdot \sigma) * 100$	5.5154	4.6562	3.8621	3.3354	2.8611	2.4964	2.0080	1.7270
$(s/\tau_0 \cdot \sigma)$	4.60	4.80	5.00	5.20	5.40	5.60	5.80	6.00
$W_{ar}(s/\tau_0 \cdot \sigma) * 100$	1.4550	1.1977	1.0146	0.8528	0.7394	0.6392	0.5651	0.4309
$(s/\tau_0 \cdot \sigma)$	6.20	6.40	6.60	6.80	7.00	7.20	7.40	7.60
$W_{ar}(s/\tau_0 \cdot \sigma) * 100$	0.3925	0.3395	0.2940	0.2423	0.1989	0.1756	0.1518	0.1307

$M_1 = 0.997997\ (\sigma\ \tau_0)$; $D = 1.3423745\ (\sigma\ \tau_0)^2$; the number of excursions = 99103715.

Gaussian power spectrum distribution of areas under excursions above level $(x_0/\sigma) = 1.0$.

$(s/\tau_0 \cdot \sigma)$	0.00	0.002	0.004	0.006	0.008	0.010	0.012	0.014
$W_{ar}(s/\tau_0 \cdot \sigma) * 100$	00.00	820.79	586.26	485.45	429.90	393.73	363.05	342.46
$(s/\tau_0 \cdot \sigma)$	0.016	0.018	0.020	0.024	0.028	0.032	0.036	0.040
$W_{ar}(s/\tau_0 \cdot \sigma) * 100$	323.91	311.07	297.08	275.27	259.95	246.84	235.37	224.53
$(s/\tau_0 \cdot \sigma)$	0.050	0.060	0.070	0.080	0.090	0.100	0.120	0.140
$W_{ar}(s/\tau_0 \cdot \sigma) * 100$	205.70	189.70	177.06	167.71	158.15	150.72	138.64	128.06
$(s/\tau_0 \cdot \sigma)$	0.160	0.180	0.200	0.300	0.400	0.500	0.600	0.700
$W_{ar}(s/\tau_0 \cdot \sigma) * 100$	119.18	111.91	105.42	83.013	67.455	56.574	47.892	41.021
$(s/\tau_0 \cdot \sigma)$	0.800	0.900	1.000	1.100	1.200	1.300	1.400	1.500
$W_{ar}(s/\tau_0 \cdot \sigma) * 100$	35.521	30.858	26.423	23.299	20.577	17.724	15.699	13.704
$(s/\tau_0 \cdot \sigma)$	1.600	1.700	1.800	1.900	2.000	2.100	2.200	2.300
$W_{ar}(s/\tau_0 \cdot \sigma) * 100$	12.177	10.562	9.2718	8.2061	7.2985	6.2982	5.5630	4.9662
$(s/\tau_0 \cdot \sigma)$	2.400	2.500	2.600	2.700	2.800	2.900	3.000	3.200
$W_{ar}(s/\tau_0 \cdot \sigma) * 100$	4.3807	3.8121	3.4047	2.9773	2.6630	2.3566	2.0857	1.6358
$(s/\tau_0 \cdot \sigma)$	3.400	3.600	3.800	4.000	4.200	4.400	4.600	4.800
$W_{ar}(s/\tau_0 \cdot \sigma) * 100$	1.2688	0.9591	0.7355	0.5913	0.4494	0.3507	0.2740	0.2140

$M_1 = 0.613042\ (\sigma\ \tau_0)$; $D = 0.546775\ (\sigma\ \tau_0)^2$; the number of excursions = 67980097.

Gaussian power spectrum distribution of areas under excursions above level $(x_0/\sigma) = 1.5$.

$(s/\tau_0 \cdot \sigma)$	0.00	0.002	0.004	0.006	0.008	0.010	0.012	0.014
$W_{ar}(s/\tau_0 \cdot \sigma) * 100$	00.00	1146.9	813.48	671.26	591.81	541.36	498.79	467.93

(Continued)

Table A3.1 (Continued)

Gaussian power spectrum distribution of areas under excursions above level $(x_0/\sigma) = 1.5$.

$(s/\tau_0 \cdot \sigma)$	0.016	0.018	0.020	0.022	0.024	0.026	0.028	0.030
$W_{ar}(s/\tau_0 \cdot \sigma) * 100$	443.87	421.45	404.65	386.57	372.86	360.28	348.79	339.80
$(s/\tau_0 \cdot \sigma)$	0.032	0.034	0.036	0.038	0.040	0.042	0.044	0.046
$W_{ar}(s/\tau_0 \cdot \sigma) * 100$	330.85	322.46	314.03	305.48	298.24	292.04	286.05	279.52
$(s/\tau_0 \cdot \sigma)$	0.048	0.052	0.056	0.060	0.07	0.08	0.10	0.12
$W_{ar}(s/\tau_0 \cdot \sigma) * 100$	274.33	263.79	255.43	246.75	228.67	214.00	189.24	171.54
$(s/\tau_0 \cdot \sigma)$	0.14	0.16	0.18	0.20	0.30	0.40	0.50	0.60
$W_{ar}(s/\tau_0 \cdot \sigma) * 100$	156.22	143.14	132.49	123.47	89.858	69.454	54.778	43.603
$(s/\tau_0 \cdot \sigma)$	0.70	0.80	0.90	1.00	1.10	1.20	1.30	1.40
$W_{ar}(s/\tau_0 \cdot \sigma) * 100$	35.511	29.155	23.478	19.590	16.271	13.620	11.139	9.1326
$(s/\tau_0 \cdot \sigma)$	1.50	1.60	1.70	1.80	1.90	2.00	2.10	2.20
$W_{ar}(s/\tau_0 \cdot \sigma) * 100$	7.8945	6.4456	5.5875	4.6538	3.9128	3.2119	2.6140	2.2383
$(s/\tau_0 \cdot \sigma)$	2.30	2.40	2.50	2.60	2.70	2.80	2.90	3.00
$W_{ar}(s/\tau_0 \cdot \sigma) * 100$	1.8924	1.5728	1.3194	1.1225	0.9531	0.8053	0.6856	0.5765

$M_1 = 0.40369604\ (\sigma \cdot \tau_0)$; $D = 0.255068641\ (\sigma \cdot \tau_0)^2$; the number of excursions = 36303964.

Gaussian power spectrum distribution of areas under excursions above level $(x_0/\sigma) = 2.0$.

$(s/\tau_0 \cdot \sigma)$	0.00	0.002	0.004	0.006	0.008	0.010	0.012	0.014
$W_{ar}(s/\tau_0 \cdot \sigma) * 100$	00.00	1534.0	1079.0	885.84	776.60	701.84	649.84	604.80
$(s/\tau_0 \cdot \sigma)$	0.016	0.018	0.020	0.022	0.024	0.026	0.028	0.030
$W_{ar}(s/\tau_0 \cdot \sigma) * 100$	574.31	544.17	517.60	494.69	475.71	456.35	441.48	428.60
$(s/\tau_0 \cdot \sigma)$	0.032	0.034	0.036	0.038	0.040	0.042	0.044	0.046
$W_{ar}(s/\tau_0 \cdot \sigma) * 100$	413.94	403.01	391.92	381.12	373.96	365.28	357.40	350.09
$(s/\tau_0 \cdot \sigma)$	0.048	0.050	0.054	0.058	0.062	0.066	0.070	0.074
$W_{ar}(s/\tau_0 \cdot \sigma) * 100$	339.98	332.09	317.72	307.88	294.42	285.67	277.34	267.85
$(s/\tau_0 \cdot \sigma)$	0.078	0.082	0.086	0.090	0.094	0.10	0.12	0.14
$W_{ar}(s/\tau_0 \cdot \sigma) * 100$	258.64	251.39	244.13	237.28	231.10	223.49	197.69	176.09
$(s/\tau_0 \cdot \sigma)$	0.16	0.18	0.20	0.30	0.40	0.50	0.60	0.70
$W_{ar}(s/\tau_0 \cdot \sigma) * 100$	159.32	146.30	132.82	89.240	64.500	46.443	35.641	26.467

$(s/\tau_0 \cdot \sigma)$	0.80	0.90	1.00	1.10	1.20	1.30	1.40	1.50
$W_{ar}(s/\tau_0 \cdot \sigma) * 100$	21.182	16.064	12.685	9.3737	7.7627	5.8893	4.6769	3.5675
$(s/\tau_0 \cdot \sigma)$	1.60	1.70	1.80	1.90	2.00	2.10	2.20	2.30
$W_{ar}(s/\tau_0 \cdot \sigma) * 100$	2.8766	2.2687	1.7605	1.4615	1.2091	0.8869	0.7241	0.6178

$M_1 = 0.281321466\ (\sigma\ \tau_0)$; $D = 0.132101142\ (\sigma\ \tau_0)^2$; the number of excursions = 15052704.

Gaussian power spectrum distribution of areas under excursions above level $(x_0/\sigma) = 2.5$.

$(s/\tau_0 \cdot \sigma)$	0.00	0.002	0.004	0.006	0.008	0.010	0.012	0.014
$W_{ar}(s/\tau_0 \cdot \sigma) * 100$	00.00	1964.6	1367.2	1119.5	973.71	873.80	807.11	752.33
$(s/\tau_0 \cdot \sigma)$	0.016	0.018	0.020	0.022	0.024	0.026	0.028	0.030
$W_{ar}(s/\tau_0 \cdot \sigma) * 100$	703.24	666.76	629.65	603.38	577.45	557.48	534.02	513.43
$(s/\tau_0 \cdot \sigma)$	0.032	0.034	0.036	0.038	0.040	0.042	0.044	0.046
$W_{ar}(s/\tau_0 \cdot \sigma) * 100$	499.06	482.38	467.48	451.63	442.89	434.58	426.45	410.45
$(s/\tau_0 \cdot \sigma)$	0.048	0.050	0.052	0.054	0.056	0.058	0.060	0.062
$W_{ar}(s/\tau_0 \cdot \sigma) * 100$	400.90	391.83	382.96	374.30	365.83	357.32	349.23	341.11
$(s/\tau_0 \cdot \sigma)$	0.064	0.066	0.068	0.072	0.076	0.080	0.084	0.088
$W_{ar}(s/\tau_0 \cdot \sigma) * 100$	333.18	325.43	319.79	308.92	298.43	288.28	278.48	269.01
$(s/\tau_0 \cdot \sigma)$	0.092	0.096	0.10	0.12	0.14	0.16	0.18	0.20
$W_{ar}(s/\tau_0 \cdot \sigma) * 100$	260.77	252.18	245.51	213.40	187.54	164.72	146.55	133.66
$(s/\tau_0 \cdot \sigma)$	0.30	0.40	0.50	0.60	0.70	0.80	0.90	1.00
$W_{ar}(s/\tau_0 \cdot \sigma) * 100$	82.599	53.563	38.255	25.368	17.917	13.136	9.9456	6.4641
$(s/\tau_0 \cdot \sigma)$	1.10	1.20	1.30	1.40	1.50	1.60	1.70	1.80
$W_{ar}(s/\tau_0 \cdot \sigma) * 100$	5.2482	3.8972	2.9211	2.1097	1.5818	1.1640	0.8418	0.6132

$M_1 = 0.2052451\ (\sigma\ \tau_0)$; $D = 0.07415787\ (\sigma\ \tau_0)^2$; the number of excursions = 4811193.

Gaussian power spectrum distribution of areas under excursions above level $(x_0/\sigma) = 3.0$.

$(s/\tau_0 \cdot \sigma)$	0.000	0.001	0.002	0.003	0.004	0.005	0.006	0.007
$W_{ar}(s/\tau_0 \cdot \sigma) * 100$	0000.0	2481.0	2266.4	1834.3	1547.8	1354.6	1283.9	1192.6

(Continued)

Table A3.1 (Continued)

$(s/\tau_0 \cdot \sigma)$	0.008	0.009	0.010	0.011	0.012	0.013	0.014	0.015
$W_{ar}(s/\tau_0 \cdot \sigma) * 100$	1148.3	1081.6	1034.8	996.52	971.50	915.97	879.57	852.31
$(s/\tau_0 \cdot \sigma)$	0.016	0.017	0.018	0.019	0.020	0.022	0.024	0.026
$W_{ar}(s/\tau_0 \cdot \sigma) * 100$	821.81	797.24	779.84	753.96	729.67	694.23	665.09	640.70
$(s/\tau_0 \cdot \sigma)$	0.028	0.030	0.032	0.034	0.036	0.038	0.040	0.042
$W_{ar}(s/\tau_0 \cdot \sigma) * 100$	615.79	591.01	569.38	550.93	531.04	515.05	500.30	486.72
$(s/\tau_0 \cdot \sigma)$	0.044	0.046	0.048	0.050	0.060	0.070	0.080	0.090
$W_{ar}(s/\tau_0 \cdot \sigma) * 100$	473.94	461.43	449.79	438.47	389.67	342.61	305.74	280.71
$(s/\tau_0 \cdot \sigma)$	0.10	0.15	0.20	0.25	0.30	0.35	0.40	0.45
$W_{ar}(s/\tau_0 \cdot \sigma) * 100$	259.86	179.81	126.77	95.890	71.629	53.506	42.257	34.822
$(s/\tau_0 \cdot \sigma)$	0.50	0.55	0.60	0.65	0.70	0.75	0.80	0.85
$W_{ar}(s/\tau_0 \cdot \sigma) * 100$	28.691	22.713	17.737	13.903	11.070	8.8054	7.4049	6.2354
$(s/\tau_0 \cdot \sigma)$	0.90	0.95	1.00	1.05	1.10	1.15	1.20	1.25
$W_{ar}(s/\tau_0 \cdot \sigma) * 100$	5.0136	4.1450	3.4267	2.8328	2.3420	1.9361	1.6006	1.3232

$M_1 = 0.15524712\ (\sigma \cdot \tau_0)$; $D = 0.04460401\ (\sigma \cdot \tau_0)^2$; the number of excursions = 2373554.

Table A3.2 Gaussian Random Process

Rectangular power spectrum distribution of areas under excursions above level $(x_0/\sigma) = 0.0$.

$(s/\tau_0 \cdot \sigma)$	0.00	0.004	0.008	0.012	0.016	0.020	0.024	0.028
$W_{ar}(s/\tau_0 \cdot \sigma) * 100$	00.00	155.50	113.07	95.273	85.474	79.011	74.630	71.011
$(s/\tau_0 \cdot \sigma)$	0.032	0.036	0.040	0.060	0.080	0.10	0.20	0.40
$W_{ar}(s/\tau_0 \cdot \sigma) * 100$	68.159	65.536	63.393	56.645	52.532	49.640	42.917	37.917
$(s/\tau_0 \cdot \sigma)$	0.60	0.80	1.00	1.20	1.40	1.60	1.80	2.00
$W_{ar}(s/\tau_0 \cdot \sigma) * 100$	35.463	33.428	31.372	29.379	27.191	25.074	22.968	20.664
$(s/\tau_0 \cdot \sigma)$	2.20	2.40	2.60	2.80	3.00	3.20	3.40	3.60
$W_{ar}(s/\tau_0 \cdot \sigma) * 100$	18.678	16.684	14.842	13.136	11.622	10.175	8.9907	7.8693
$(s/\tau_0 \cdot \sigma)$	3.80	4.00	4.20	4.40	4.60	4.80	5.00	5.20
$W_{ar}(s/\tau_0 \cdot \sigma) * 100$	6.9026	6.1096	5.3357	4.6777	4.0741	3.5761	3.1720	2.8029
$(s/\tau_0 \cdot \sigma)$	5.40	5.60	5.80	6.00	6.20	6.40	6.60	6.80
$W_{ar}(s/\tau_0 \cdot \sigma) * 100$	2.4222	2.1895	1.9195	1.6777	1.4865	1.3067	1.1792	1.0168
$(s/\tau_0 \cdot \sigma)$	7.00	7.20	7.40	7.60	7.80	8.00	8.20	8.40
$W_{ar}(s/\tau_0 \cdot \sigma) * 100$	0.9158	0.7966	0.7247	0.6312	0.5564	0.4742	0.4366	0.3768
$(s/\tau_0 \cdot \sigma)$	8.60	8.80	9.00	9.20	9.40	9.60	9.80	10.00
$W_{ar}(s/\tau_0 \cdot \sigma) * 100$	0.3231	0.2972	0.2567	0.2307	0.2093	0.1747	0.1574	0.1446

$M_1 = 1.77609657\ (\sigma\ \tau_0)$; $D = 2.7571917\ (\sigma\ \tau_0)^2$; the number of excursions = 224796920.

Rectangular power spectrum distribution of areas under excursions above level $(x_0/\sigma) = 0.5$.

$(s/\tau_0 \cdot \sigma)$	0.00	0.004	0.008	0.012	0.016	0.020	0.024	0.028
$W_{ar}(s/\tau_0 \cdot \sigma) * 100$	00.00	308.14	223.26	186.51	167.25	154.18	144.37	136.98
$(s/\tau_0 \cdot \sigma)$	0.032	0.036	0.040	0.044	0.048	0.52	0.056	0.060
$W_{ar}(s/\tau_0 \cdot \sigma) * 100$	130.86	125.77	121.53	117.77	114.44	111.60	109.01	106.85
$(s/\tau_0 \cdot \sigma)$	0.064	0.072	0.080	0.10	0.12	0.14	0.16	0.18
$W_{ar}(s/\tau_0 \cdot \sigma) * 100$	104.80	100.73	97.580	91.083	86.211	82.298	79.185	76.487

(Continued)

Table A3.2 (Continued)

$(s/\tau_0 \cdot \sigma)$	0.20	0.30	0.40	0.50	0.60	0.80	1.00	1.20
$W_{ar}(s/\tau_0 \cdot \sigma) * 100$	73.743	65.396	58.618	54.068	49.517	42.249	36.043	30.588
$(s/\tau_0 \cdot \sigma)$	1.40	1.60	1.80	2.00	2.20	2.40	2.60	2.80
$W_{ar}(s/\tau_0 \cdot \sigma) * 100$	25.854	21.656	17.966	14.906	12.232	9.9368	8.0593	6.5501
$(s/\tau_0 \cdot \sigma)$	3.00	3.20	3.40	3.60	3.80	4.00	4.20	4.40
$W_{ar}(s/\tau_0 \cdot \sigma) * 100$	5.1932	4.2342	3.3623	2.6839	2.1267	1.7108	1.3570	1.0965
$(s/\tau_0 \cdot \sigma)$	4.60	4.80	5.00	5.20	5.40	5.60	5.80	6.00
$W_{ar}(s/\tau_0 \cdot \sigma) * 100$	0.8644	0.7044	0.5563	0.4505	0.3632	0.2944	0.2384	0.1874
$(s/\tau_0 \cdot \sigma)$	6.20	6.40	6.60	6.80	7.00	7.20	7.40	7.60
$W_{ar}(s/\tau_0 \cdot \sigma) * 100$	0.1601	0.1251	0.0977	0.0777	0.0610	0.0505	0.0405	0.0310

$M_1 = 1.000104076\ (\sigma\ \tau_0)$; $D = 0.97260\ (\sigma\ \tau_0)^2$; the number of excursions = 198108912.

Rectangular power spectrum distribution of areas under excursions above level $(x_0/\sigma) = 1.0$.

$(s/\tau_0 \cdot \sigma)$	0.00	0.002	0.004	0.006	0.008	0.010	0.012	0.014
$W_{ar}(s/\tau_0 \cdot \sigma) * 100$	00.00	655.14	476.10	396.38	353.07	324.54	301.18	284.68
$(s/\tau_0 \cdot \sigma)$	0.016	0.018	0.020	0.024	0.028	0.032	0.036	0.040
$W_{ar}(s/\tau_0 \cdot \sigma) * 100$	271.70	261.79	250.05	234.31	221.67	211.15	202.23	195.43
$(s/\tau_0 \cdot \sigma)$	0.050	0.060	0.070	0.080	0.090	0.100	0.120	0.140
$W_{ar}(s/\tau_0 \cdot \sigma) * 100$	180.07	167.53	159.23	151.19	144.51	138.75	128.87	121.05
$(s/\tau_0 \cdot \sigma)$	0.160	0.180	0.200	0.300	0.400	0.500	0.600	0.700
$W_{ar}(s/\tau_0 \cdot \sigma) * 100$	114.75	109.08	104.04	84.991	72.013	61.907	53.355	46.458
$(s/\tau_0 \cdot \sigma)$	0.800	0.900	1.000	1.100	1.200	1.300	1.400	1.500
$W_{ar}(s/\tau_0 \cdot \sigma) * 100$	40.516	35.524	31.045	26.929	23.753	20.317	17.764	15.377
$(s/\tau_0 \cdot \sigma)$	1.600	1.700	1.800	1.900	2.000	2.100	2.200	2.300
$W_{ar}(s/\tau_0 \cdot \sigma) * 100$	13.308	11.552	9.7695	8.4570	7.2658	6.2289	5.3294	4.4622
$(s/\tau_0 \cdot \sigma)$	2.400	2.500	2.600	2.700	2.800	2.900	3.000	3.200
$W_{ar}(s/\tau_0 \cdot \sigma) * 100$	3.8162	3.2379	2.8521	2.3957	2.0437	1.6763	1.4124	1.0619
$(s/\tau_0 \cdot \sigma)$	3.400	3.600	3.800	4.000	4.200	4.400	4.600	4.800
$W_{ar}(s/\tau_0 \cdot \sigma) * 100$	0.7092	0.5218	0.3652	0.2682	0.1999	0.1448	0.0882	0.06761

$M_1 = 0.612585684\ (\sigma\ \tau_0)$; $D = 0.424932882\ (\sigma\ \tau_0)^2$; the number of excursions = 68038031.

Rectangular power spectrum distribution of areas under excursions above level $(x_0/\sigma) = 1.5$.

$(s/\tau_0 \cdot \sigma)$	0.00	0.002	0.004	0.006	0.008	0.010	0.012	0.014
$W_{ar}(s/\tau_0 \cdot \sigma) * 100$	00.00	1007.4	720.75	599.33	530.95	486.29	447.51	420.98
$(s/\tau_0 \cdot \sigma)$	0.016	0.018	0.020	0.022	0.024	0.026	0.028	0.030
$W_{ar}(s/\tau_0 \cdot \sigma) * 100$	400.82	385.63	367.71	355.22	340.95	330.45	321.06	313.04
$(s/\tau_0 \cdot \sigma)$	0.032	0.034	0.036	0.038	0.040	0.042	0.044	0.046
$W_{ar}(s/\tau_0 \cdot \sigma) * 100$	304.50	297.70	290.72	283.60	278.13	271.95	266.73	261.32
$(s/\tau_0 \cdot \sigma)$	0.048	0.052	0.056	0.060	0.07	0.08	0.10	0.12
$W_{ar}(s/\tau_0 \cdot \sigma) * 100$	257.63	248.99	241.02	233.81	218.72	206.21	184.48	168.03
$(s/\tau_0 \cdot \sigma)$	0.14	0.16	0.18	0.20	0.30	0.40	0.50	0.60
$W_{ar}(s/\tau_0 \cdot \sigma) * 100$	154.77	144.23	133.79	126.28	94.736	74.437	59.323	47.908
$(s/\tau_0 \cdot \sigma)$	0.70	0.80	0.90	1.00	1.10	1.20	1.30	1.40
$W_{ar}(s/\tau_0 \cdot \sigma) * 100$	39.074	31.524	26.327	21.449	17.268	14.137	11.610	9.5208
$(s/\tau_0 \cdot \sigma)$	1.50	1.60	1.70	1.80	1.90	2.00	2.10	2.20
$W_{ar}(s/\tau_0 \cdot \sigma) * 100$	7.6161	6.2371	5.0191	4.1205	3.3250	2.7717	2.2254	1.7919
$(s/\tau_0 \cdot \sigma)$	2.30	2.40	2.50	2.60	2.70	2.80	2.90	3.00
$W_{ar}(s/\tau_0 \cdot \sigma) * 100$	1.4423	1.0968	0.8523	0.6861	0.5524	0.4446	0.3579	0.2882

$M_1 = 0.40340036\ (\sigma\ \tau_0)$; $D = 0.2131107\ (\sigma\ \tau_0)^2$; the number of excursions = 36330875.

Rectangular power spectrum distribution of areas under excursions above level $(x_0/\sigma) = 2.0$.

$(s/\tau_0 \cdot \sigma)$	0.00	0.002	0.004	0.006	0.008	0.010	0.012	0.014
$W_{ar}(s/\tau_0 \cdot \sigma) * 100$	00.00	1418.5	1005.4	823.47	730.30	665.09	611.89	571.06
$(s/\tau_0 \cdot \sigma)$	0.016	0.018	0.020	0.022	0.024	0.026	0.028	0.030
$W_{ar}(s/\tau_0 \cdot \sigma) * 100$	539.89	518.35	494.19	474.93	454.01	439.04	423.59	411.23
$(s/\tau_0 \cdot \sigma)$	0.032	0.034	0.036	0.038	0.040	0.042	0.044	0.046
$W_{ar}(s/\tau_0 \cdot \sigma) * 100$	400.61	388.31	378.55	368.14	359.34	350.81	343.62	336.42
$(s/\tau_0 \cdot \sigma)$	0.048	0.050	0.054	0.058	0.062	0.066	0.070	0.074
$W_{ar}(s/\tau_0 \cdot \sigma) * 100$	328.66	321.99	309.83	299.57	288.90	279.57	271.85	263.57
$(s/\tau_0 \cdot \sigma)$	0.078	0.082	0.086	0.090	0.094	0.10	0.12	0.14
$W_{ar}(s/\tau_0 \cdot \sigma) * 100$	255.70	248.71	243.33	236.92	229.31	221.40	198.25	179.24

(Continued)

Table A3.2 (Continued)

$(s/\tau_0 \cdot \sigma)$	0.16	0.18	0.20	0.30	0.40	0.50	0.60	0.70
$W_{ar}(s/\tau_0 \cdot \sigma) * 100$	162.22	148.68	136.42	94.108	68.748	50.223	37.769	28.812
$(s/\tau_0 \cdot \sigma)$	0.80	0.90	1.00	1.10	1.20	1.30	1.40	1.50
$W_{ar}(s/\tau_0 \cdot \sigma) * 100$	21.720	16.679	12.916	9.8960	7.5199	6.0332	4.4701	3.4187
$(s/\tau_0 \cdot \sigma)$	1.60	1.70	1.80	1.90	2.00	2.10	2.20	2.30
$W_{ar}(s/\tau_0 \cdot \sigma) * 100$	2.6150	2.0002	1.5133	1.1514	0.8741	0.6740	0.5115	0.3887

$M_1 = 0.281016\ (\sigma\ \tau_0)$; $D = 0.1166003\ (\sigma\ \tau_0)^2$; the number of excursions = 15066728.

Rectangular power spectrum distribution of areas under excursions above level $(x_0/\sigma) = 2.5$.

$(s/\tau_0 \cdot \sigma)$	0.00	0.002	0.004	0.006	0.008	0.010	0.012	0.014
$W_{ar}(s/\tau_0 \cdot \sigma) * 100$	00.00	1873.0	1311.0	1068.3	933.76	845.68	778.82	726.52
$(s/\tau_0 \cdot \sigma)$	0.016	0.018	0.020	0.022	0.024	0.026	0.028	0.030
$W_{ar}(s/\tau_0 \cdot \sigma) * 100$	680.15	650.18	613.83	591.28	558.50	543.88	524.76	505.15
$(s/\tau_0 \cdot \sigma)$	0.032	0.034	0.036	0.038	0.040	0.042	0.044	0.046
$W_{ar}(s/\tau_0 \cdot \sigma) * 100$	491.01	475.73	462.07	449.35	436.57	423.55	414.18	406.06
$(s/\tau_0 \cdot \sigma)$	0.048	0.050	0.052	0.054	0.056	0.058	0.060	0.062
$W_{ar}(s/\tau_0 \cdot \sigma) * 100$	392.23	382.95	372.74	369.81	360.55	350.37	342.53	340.86
$(s/\tau_0 \cdot \sigma)$	0.064	0.066	0.068	0.072	0.076	0.080	0.084	0.088
$W_{ar}(s/\tau_0 \cdot \sigma) * 100$	331.16	325.54	321.32	307.45	296.60	290.62	277.11	269.71
$(s/\tau_0 \cdot \sigma)$	0.092	0.096	0.10	0.12	0.14	0.16	0.18	0.20
$W_{ar}(s/\tau_0 \cdot \sigma) * 100$	260.12	256.98	244.64	215.08	191.56	169.76	150.23	136.37
$(s/\tau_0 \cdot \sigma)$	0.30	0.40	0.50	0.60	0.70	0.80	0.90	1.00
$W_{ar}(s/\tau_0 \cdot \sigma) * 100$	86.479	58.183	40.736	26.916	19.013	13.188	9.2627	6.2641
$(s/\tau_0 \cdot \sigma)$	1.10	1.20	1.30	1.40	1.50	1.60	1.70	1.80
$W_{ar}(s/\tau_0 \cdot \sigma) * 100$	4.6106	3.3606	2.4507	1.7495	1.2489	0.8619	0.6025	0.4292

$M_1 = 0.2050134\ (\sigma\ \tau_0)$; $D = 0.067881\ (\sigma\ \tau_0)^2$; the number of excursions = 4815015.

Rectangular power spectrum distribution of areas under excursions above level $(x_0/\sigma) = 3.0$.

$(s/\tau_0 \cdot \sigma)$	0.000	0.001	0.002	0.003	0.004	0.005	0.006	0.007
$W_{ar}(s/\tau_0 \cdot \sigma) * 100$	0000.0	2447.1	2224.6	1768.0	1461.7	1327.7	1251.4	1184.9
$(s/\tau_0 \cdot \sigma)$	0.008	0.009	0.010	0.011	0.012	0.013	0.014	0.015
$W_{ar}(s/\tau_0 \cdot \sigma) * 100$	1131.5	1060.2	1019.2	981.36	935.65	888.46	852.19	834.79
$(s/\tau_0 \cdot \sigma)$	0.016	0.017	0.018	0.019	0.020	0.022	0.024	0.026
$W_{ar}(s/\tau_0 \cdot \sigma) * 100$	786.67	768.81	753.75	739.99	729.67	691.93	660.32	628.71
$(s/\tau_0 \cdot \sigma)$	0.028	0.030	0.032	0.034	0.036	0.038	0.040	0.042
$W_{ar}(s/\tau_0 \cdot \sigma) * 100$	608.22	590.00	569.38	556.09	531.04	515.05	500.30	485.72
$(s/\tau_0 \cdot \sigma)$	0.044	0.046	0.048	0.050	0.060	0.070	0.080	0.090
$W_{ar}(s/\tau_0 \cdot \sigma) * 100$	468.84	459.44	449.79	440.47	380.07	339.41	306.84	282.62
$(s/\tau_0 \cdot \sigma)$	0.10	0.15	0.20	0.25	0.30	0.35	0.40	0.45
$W_{ar}(s/\tau_0 \cdot \sigma) * 100$	263.86	179.81	127.83	98.638	74.067	56.919	44.954	37.413
$(s/\tau_0 \cdot \sigma)$	0.50	0.55	0.60	0.65	0.70	0.75	0.80	0.85
$W_{ar}(s/\tau_0 \cdot \sigma) * 100$	29.691	23.460	18.537	14.451	11.315	8.8902	6.9877	5.4923
$(s/\tau_0 \cdot \sigma)$	0.90	0.95	1.00	1.05	1.10	1.15	1.20	1.25
$W_{ar}(s/\tau_0 \cdot \sigma) * 100$	4.3170	3.3931	2.6670	2.0963	1.6477	1.2951	1.0179	0.8001

$M_1 = 0.15509417\ (\sigma\ \tau_0)$; $D = 0.041771936\ (\sigma\ \tau_0)^2$; the number of excursions = 2373538.

Table A3.3 Gaussian Random Process

Gaussian power spectrum distribution of areas above excursions below level $(x_0/\sigma) = 0.1$.

$(s/\tau_0 \cdot \sigma)$	0.00	0.004	0.008	0.012	0.016	0.02	0.04	0.08
$W_{ar}(s/\tau_0 \cdot \sigma) * 100$	00.00	253.57	181.06	148.93	132.53	121.02	92.760	71.444
$(s/\tau_0 \cdot \sigma)$	0.12	0.16	0.20	0.24	0.28	0.32	0.36	0.40
$W_{ar}(s/\tau_0 \cdot \sigma) * 100$	61.520	55.546	51.166	47.707	44.669	42.665	40.833	39.276
$(s/\tau_0 \cdot \sigma)$	0.60	0.80	1.00	1.20	1.40	1.60	1.80	2.00
$W_{ar}(s/\tau_0 \cdot \sigma) * 100$	32.995	29.024	26.024	23.463	21.062	19.203	17.493	16.031
$(s/\tau_0 \cdot \sigma)$	2.20	2.40	2.60	2.80	3.00	3.20	3.40	3.60
$W_{ar}(s/\tau_0 \cdot \sigma) * 100$	14.674	13.442	12.291	11.249	10.303	9.4234	8.6296	7.9092
$(s/\tau_0 \cdot \sigma)$	3.80	4.00	4.20	4.40	4.60	4.80	5.00	5.20
$W_{ar}(s/\tau_0 \cdot \sigma) * 100$	7.2272	6.6141	6.0530	5.5490	5.0876	4.6135	4.2446	3.8849
$(s/\tau_0 \cdot \sigma)$	5.40	5.60	5.80	6.00	6.20	6.40	6.60	6.80
$W_{ar}(s/\tau_0 \cdot \sigma) * 100$	3.5537	3.2629	2.9897	2.7305	2.4909	2.2787	2.0937	1.9237
$(s/\tau_0 \cdot \sigma)$	7.00	7.20	7.40	7.60	7.80	8.00	8.40	8.80
$W_{ar}(s/\tau_0 \cdot \sigma) * 100$	1.7675	1.6224	1.4923	1.3710	1.2597	1.1574	0.9771	0.8249
$(s/\tau_0 \cdot \sigma)$	9.20	9.60	10.00	10.40	10.80	11.20	11.60	12.00
$W_{ar}(s/\tau_0 \cdot \sigma) * 100$	0.6924	0.5877	0.4954	0.4173	0.3516	0.2963	0.2496	0.2103

$M_1 = 2.01813\ (\sigma\ \tau_0)$; $D = 5.0006\ (\sigma\ \tau_0)^2$; the number of excursions = 67079898.

Gaussian power spectrum distribution of areas above excursions below level $(x_0/\sigma) = 0.5$.

$(s/\tau_0 \cdot \sigma)$	0.00	0.004	0.008	0.012	0.016	0.02	0.04	0.08
$W_{ar}(s/\tau_0 \cdot \sigma) * 100$	00.00	165.34	117.31	96.491	85.514	78.075	60.076	46.235
$(s/\tau_0 \cdot \sigma)$	0.12	0.16	0.20	0.24	0.28	0.32	0.36	0.40
$W_{ar}(s/\tau_0 \cdot \sigma) * 100$	40.021	36.067	33.414	31.341	29.771	28.237	27.566	26.565
$(s/\tau_0 \cdot \sigma)$	0.60	0.80	1.00	1.20	1.40	1.60	1.80	2.00
$W_{ar}(s/\tau_0 \cdot \sigma) * 100$	23.076	20.963	19.456	17.977	16.965	16.052	15.015	14.190
$(s/\tau_0 \cdot \sigma)$	2.20	2.40	2.60	2.80	3.00	3.20	3.40	3.60
$W_{ar}(s/\tau_0 \cdot \sigma) * 100$	13.457	12.790	12.077	11.451	10.914	10.350	9.7659	9.3178
$(s/\tau_0 \cdot \sigma)$	3.80	4.00	4.40	4.80	5.20	5.60	6.00	6.40
$W_{ar}(s/\tau_0 \cdot \sigma) * 100$	8.8395	8.3867	7.5467	6.7902	6.1176	5.5046	4.9510	4.4507

$(s/\tau_0 \cdot \sigma)$	6.80	7.20	7.60	8.00	8.40	8.80	9.20	9.60
$W_{ar}(s/\tau_0 \cdot \sigma) * 100$	4.0127	3.6094	3.2484	2.9263	2.6312	2.3681	2.1342	1.9181
$(s/\tau_0 \cdot \sigma)$	10.00	10.40	10.80	11.20	11.60	12.00	12.80	13.60
$W_{ar}(s/\tau_0 \cdot \sigma) * 100$	1.7264	1.5537	1.3983	1.2585	1.1327	1.0194	0.8257	0.6688
$(s/\tau_0 \cdot \sigma)$	14.40	15.20	16.00	16.80	17.60	18.40	19.20	20.00
$W_{ar}(s/\tau_0 \cdot \sigma) * 100$	0.5473	0.4381	0.3579	0.2879	0.2332	0.1889	0.1530	0.1239

$M_1 = 3.5173\ (\sigma\ \tau_0)$; $D = 14.3996\ (\sigma\ \tau_0)^2$; the number of excursions = 59528639.

Gaussian power spectrum distribution of areas above excursions below level $(x_0/\sigma) = 1.0$.

$(s/\tau_0 \cdot \sigma)$	0.00	0.004	0.008	0.012	0.016	0.02	0.04	0.08
$W_{ar}(s/\tau_0 \cdot \sigma) * 100$	00.00	89.690	63.495	51.855	45.647	41.606	31.452	24.299
$(s/\tau_0 \cdot \sigma)$	0.12	0.16	0.20	0.24	0.28	0.32	0.36	0.40
$W_{ar}(s/\tau_0 \cdot \sigma) * 100$	20.864	18.823	17.413	16.274	15.437	14.821	14.357	13.819
$(s/\tau_0 \cdot \sigma)$	0.60	0.80	1.00	1.20	1.40	1.60	1.80	2.00
$W_{ar}(s/\tau_0 \cdot \sigma) * 100$	12.199	11.442	10.806	10.357	9.9117	9.5757	9.2447	8.9252
$(s/\tau_0 \cdot \sigma)$	2.40	2.80	3.20	3.60	4.00	4.40	4.80	5.20
$W_{ar}(s/\tau_0 \cdot \sigma) * 100$	8.5060	8.1065	7.7259	7.3652	7.0246	6.6934	6.3834	6.0855
$(s/\tau_0 \cdot \sigma)$	5.60	6.00	6.80	7.60	8.40	9.20	10.00	10.80
$W_{ar}(s/\tau_0 \cdot \sigma) * 100$	5.8095	5.5317	5.0270	4.5755	4.1560	3.7748	3.4314	3.1225
$(s/\tau_0 \cdot \sigma)$	11.60	12.40	13.20	14.00	14.80	15.60	16.40	17.20
$W_{ar}(s/\tau_0 \cdot \sigma) * 100$	2.8356	2.5834	2.3460	2.1327	1.9388	1.7625	1.6022	1.4554
$(s/\tau_0 \cdot \sigma)$	18.00	18.80	19.60	20.40	21.20	22.00	22.80	23.60
$W_{ar}(s/\tau_0 \cdot \sigma) * 100$	1.3231	1.2028	1.0935	0.9941	0.9037	0.8216	0.7469	0.6790
$(s/\tau_0 \cdot \sigma)$	24.40	25.20	26.00	26.80	27.60	28.40	29.20	30.00
$W_{ar}(s/\tau_0 \cdot \sigma) * 100$	0.6172	0.5611	0.5101	0.4637	0.4216	0.3833	0.3484	0.3165

$M_1 = 7.93954\ (\sigma\ \tau_0)$; $D = 69.3177\ (\sigma\ \tau_0)^2$; the number of excursions = 40918707.

Gaussian power spectrum distribution of areas above excursions below level $(x_0/\sigma) = 1.5$.

$(s/\tau_0 \cdot \sigma)$	0.00	0.004	0.008	0.012	0.016	0.02	0.04	0.08
$W_{ar}(s/\tau_0 \cdot \sigma) * 100$	00.00	44.482	30.690	25.061	22.124	20.207	14.830	11.117
$(s/\tau_0 \cdot \sigma)$	0.12	0.16	0.20	0.30	0.40	0.50	0.60	0.70
$W_{ar}(s/\tau_0 \cdot \sigma) * 100$	9.4882	8.4612	7.8393	6.8055	5.9017	5.5404	5.2433	4.9866

(Continued)

Table A3.3 (Continued)

$(s/\tau_0 \cdot \sigma)$	0.80	0.90	1.00	1.40	1.80	2.20	2.60	3.00
$W_{ar}(s/\tau_0 \cdot \sigma) * 100$	4.7857	4.7166	4.6500	4.5071	4.3686	4.2344	4.1044	3.9783
$(s/\tau_0 \cdot \sigma)$	4.00	5.00	6.00	7.00	8.00	9.00	10.00	11.00
$W_{ar}(s/\tau_0 \cdot \sigma) * 100$	3.7889	3.6085	3.4279	3.2700	3.1172	2.9663	2.8268	2.6936
$(s/\tau_0 \cdot \sigma)$	12.00	13.00	14.00	15.00	16.00	18.00	20.00	22.00
$W_{ar}(s/\tau_0 \cdot \sigma) * 100$	2.5641	2.4354	2.3257	2.2141	2.1080	1.8996	1.7428	1.5907
$(s/\tau_0 \cdot \sigma)$	24.00	26.00	28.00	30.00	32.00	34.00	36.00	38.00
$W_{ar}(s/\tau_0 \cdot \sigma) * 100$	1.4515	1.3274	1.2087	1.0991	1.0144	0.9115	0.8295	0.7555
$(s/\tau_0 \cdot \sigma)$	40.00	42.00	44.00	46.00	48.00	50.00	52.00	54.00
$W_{ar}(s/\tau_0 \cdot \sigma) * 100$	0.6892	0.6274	0.5715	0.5207	0.4747	0.4321	0.3936	0.3586
$(s/\tau_0 \cdot \sigma)$	56.00	58.00	60.00	62.00	64.00	66.00	68.00	70.00
$W_{ar}(s/\tau_0 \cdot \sigma) * 100$	0.3267	0.2976	0.2711	0.2470	0.2250	0.2050	0.1870	0.1702

$M_1 = 20.916\ (\sigma\ \tau_0)$; $D = 461.084\ (\sigma\ \tau_0)^2$; the number of excursions = 21908776.

Gaussian power spectrum distribution of areas above excursions below level $(x_0/\sigma) = 2.0$.

$(s/\tau_0 \cdot \sigma)$	0.00	0.004	0.008	0.012	0.016	0.02	0.04	0.08
$W_{ar}(s/\tau_0 \cdot \sigma) * 100$	00.00	20.429	13.933	11.217	9.5270	8.7603	5.9890	4.5540
$(s/\tau_0 \cdot \sigma)$	0.12	0.16	0.20	0.40	0.60	0.80	1.00	1.20
$W_{ar}(s/\tau_0 \cdot \sigma) * 100$	3.7708	3.2670	2.9164	2.1442	1.9032	1.7480	1.6655	1.5846
$(s/\tau_0 \cdot \sigma)$	1.40	1.60	2.00	2.40	2.80	3.20	3.60	4.00
$W_{ar}(s/\tau_0 \cdot \sigma) * 100$	1.5639	1.5435	1.5034	1.4643	1.4263	1.3892	1.3777	1.3662
$(s/\tau_0 \cdot \sigma)$	5.00	6.00	7.00	8.00	9.00	10.00	11.00	12.00
$W_{ar}(s/\tau_0 \cdot \sigma) * 100$	1.3868	1.3662	1.3458	1.3258	1.3060	1.2866	1.2674	1.2486
$(s/\tau_0 \cdot \sigma)$	13.00	14.00	15.00	16.00	17.00	18.00	19.00	20.00
$W_{ar}(s/\tau_0 \cdot \sigma) * 100$	1.2300	1.2117	1.1936	1.1758	1.1583	1.1411	1.1241	1.1074
$(s/\tau_0 \cdot \sigma)$	22.00	24.00	26.00	28.00	30.00	32.00	34.00	36.00
$W_{ar}(s/\tau_0 \cdot \sigma) * 100$	1.0747	1.0429	1.0121	0.9821	0.9531	0.9249	0.8976	0.8711
$(s/\tau_0 \cdot \sigma)$	40.00	44.00	48.00	52.00	56.00	60.00	64.00	68.00
$W_{ar}(s/\tau_0 \cdot \sigma) * 100$	0.8218	0.7753	0.7314	0.6900	0.6509	0.6141	0.5793	0.5465
$(s/\tau_0 \cdot \sigma)$	72.00	76.00	80.00	84.00	88.00	92.00	96.00	100.00
$W_{ar}(s/\tau_0 \cdot \sigma) * 100$	0.5156	0.4864	0.4589	0.4329	0.4084	0.3853	0.3635	0.3423

$M_1 = 65.7916\ (\sigma\ \tau_0)$; $D = 4441.64\ (\sigma\ \tau_0)^2$; the number of excursions = 9129302.

Table A3.4 Gaussian Random Process

Rectangular power spectrum distribution of areas above excursions below level $(x_0/\sigma) = 0.1$.

$(s/\tau_0 \cdot \sigma)$	0.00	0.004	0.008	0.012	0.016	0.02	0.04	0.08
$W_{ar}(s/\tau_0 \cdot \sigma) * 100$	00.00	132.66	96.610	81.443	73.105	67.677	54.532	45.273
$(s/\tau_0 \cdot \sigma)$	0.12	0.16	0.20	0.24	0.28	0.32	0.36	0.40
$W_{ar}(s/\tau_0 \cdot \sigma) * 100$	41.308	38.936	37.449	36.007	35.232	34.618	34.255	33.553
$(s/\tau_0 \cdot \sigma)$	0.60	0.80	1.00	1.20	1.40	1.60	1.80	2.00
$W_{ar}(s/\tau_0 \cdot \sigma) * 100$	32.087	30.968	29.489	28.068	26.485	24.545	22.821	20.925
$(s/\tau_0 \cdot \sigma)$	2.20	2.40	2.60	2.80	3.00	3.20	3.40	3.60
$W_{ar}(s/\tau_0 \cdot \sigma) * 100$	19.252	17.453	15.649	14.167	12.720	11.268	10.144	9.0169
$(s/\tau_0 \cdot \sigma)$	3.80	4.00	4.20	4.40	4.60	4.80	5.00	5.20
$W_{ar}(s/\tau_0 \cdot \sigma) * 100$	8.0536	7.0602	6.4225	5.6469	5.0513	4.4628	3.9588	3.6254
$(s/\tau_0 \cdot \sigma)$	5.40	5.60	5.80	6.00	6.20	6.40	6.60	6.80
$W_{ar}(s/\tau_0 \cdot \sigma) * 100$	3.2876	2.8775	2.5895	2.3038	2.0450	1.8394	1.6740	1.5071
$(s/\tau_0 \cdot \sigma)$	7.00	7.20	7.40	7.60	7.80	8.00	8.40	8.80
$W_{ar}(s/\tau_0 \cdot \sigma) * 100$	1.3454	1.1815	1.0817	0.9651	0.8515	0.7897	0.6150	0.5289
$(s/\tau_0 \cdot \sigma)$	9.20	9.60	10.00	10.40	10.80	11.20	11.60	12.00
$W_{ar}(s/\tau_0 \cdot \sigma) * 100$	0.4004	0.3144	0.2529	0.2078	0.1650	0.1307	0.1035	0.0830

$M_1 = 2.01709\ (\sigma \cdot \tau_0)$; $D = 3.51256\ (\sigma \cdot \tau_0)^2$; the number of excursions = 67115980.

Rectangular power spectrum distribution of areas above excursions below level $(x_0/\sigma) = 0.5$.

$(s/\tau_0 \cdot \sigma)$	0.00	0.004	0.008	0.012	0.016	0.02	0.04	0.08
$W_{ar}(s/\tau_0 \cdot \sigma) * 100$	00.00	65.713	47.826	40.247	36.017	33.565	27.110	22.776
$(s/\tau_0 \cdot \sigma)$	0.12	0.16	0.20	0.24	0.28	0.32	0.36	0.40
$W_{ar}(s/\tau_0 \cdot \sigma) * 100$	20.814	20.080	19.384	19.263	18.900	18.772	18.626	18.595
$(s/\tau_0 \cdot \sigma)$	0.60	0.80	1.00	1.20	1.40	1.60	1.80	2.00
$W_{ar}(s/\tau_0 \cdot \sigma) * 100$	18.828	19.203	19.604	19.695	19.464	19.394	18.928	18.561
$(s/\tau_0 \cdot \sigma)$	2.20	2.40	2.60	2.80	3.00	3.20	3.40	3.60
$W_{ar}(s/\tau_0 \cdot \sigma) * 100$	17.920	17.007	16.076	15.247	14.522	13.381	12.712	11.760

(Continued)

Table A3.4 (Continued)

$(s/\tau_0 \cdot \sigma)$	3.80	4.00	4.40	4.80	5.20	5.60	6.00	6.40
$W_{ar}(s/\tau_0 \cdot \sigma) * 100$	10.959	10.213	8.9569	7.7989	6.7743	5.9718	5.2415	4.6538
$(s/\tau_0 \cdot \sigma)$	6.80	7.20	7.60	8.00	8.40	8.80	9.20	9.60
$W_{ar}(s/\tau_0 \cdot \sigma) * 100$	4.1414	3.6377	3.2545	2.8382	2.4951	2.1876	1.9471	1.7020
$(s/\tau_0 \cdot \sigma)$	10.00	10.40	10.80	11.20	11.60	12.00	12.80	13.60
$W_{ar}(s/\tau_0 \cdot \sigma) * 100$	1.4976	1.3612	1.1908	1.0548	0.9544	0.8436	0.6481	0.4764
$(s/\tau_0 \cdot \sigma)$	14.40	15.20	16.00	16.80	17.60	18.40	19.20	20.00
$W_{ar}(s/\tau_0 \cdot \sigma) * 100$	0.3838	0.3089	0.2472	0.1822	0.1419	0.1117	0.0871	0.0680

$M_1 = 3.51537\ (\sigma \cdot \tau_0)$; $D = 10.5389\ (\sigma \cdot \tau_0)^2$; the number of excursions = 59563197.

Rectangular power spectrum distribution of areas above excursions below level $(x_0/\sigma) = 1.0$.

$(s/\tau_0 \cdot \sigma)$	0.00	0.004	0.008	0.012	0.016	0.02	0.04	0.08
$W_{ar}(s/\tau_0 \cdot \sigma) * 100$	00.00	22.681	16.344	13.843	12.358	11.355	9.1573	7.7629
$(s/\tau_0 \cdot \sigma)$	0.12	0.16	0.20	0.24	0.28	0.32	0.36	0.40
$W_{ar}(s/\tau_0 \cdot \sigma) * 100$	7.0858	6.8660	6.7870	6.7079	6.6163	6.6744	6.7539	6.8361
$(s/\tau_0 \cdot \sigma)$	0.60	0.80	1.00	1.20	1.40	1.60	1.80	2.00
$W_{ar}(s/\tau_0 \cdot \sigma) * 100$	7.1719	7.7849	8.2715	8.9247	9.3698	9.8326	10.219	10.437
$(s/\tau_0 \cdot \sigma)$	2.40	2.80	3.20	3.60	4.00	4.40	4.80	5.20
$W_{ar}(s/\tau_0 \cdot \sigma) * 100$	10.785	10.650	10.326	9.6860	9.0731	8.3923	7.8233	7.2012
$(s/\tau_0 \cdot \sigma)$	5.60	6.00	6.80	7.60	8.40	9.20	10.00	10.80
$W_{ar}(s/\tau_0 \cdot \sigma) * 100$	6.7908	6.4038	5.7544	5.1815	4.6614	4.1754	3.7468	3.3716
$(s/\tau_0 \cdot \sigma)$	11.60	12.40	13.20	14.00	14.80	15.60	16.40	17.20
$W_{ar}(s/\tau_0 \cdot \sigma) * 100$	3.0294	2.7230	2.4484	2.2011	1.9653	1.7789	1.5926	1.4376
$(s/\tau_0 \cdot \sigma)$	18.00	18.80	19.60	20.40	21.20	22.00	22.80	23.60
$W_{ar}(s/\tau_0 \cdot \sigma) * 100$	1.2936	1.1673	1.0465	0.9402	0.8418	0.7584	0.6823	0.6121
$(s/\tau_0 \cdot \sigma)$	24.40	25.20	26.00	26.80	27.60	28.40	29.20	30.00
$W_{ar}(s/\tau_0 \cdot \sigma) * 100$	0.5538	0.4884	0.4457	0.4007	0.3602	0.3238	0.2911	0.2617

$M_1 = 7.93403\ (\sigma \cdot \tau_0)$; $D = 56.4515\ (\sigma \cdot \tau_0)^2$; the number of excursions = 40948066.

Rectangular power spectrum distribution of areas above excursions below level $(x_0/\sigma) = 1.5$.

$(s/\tau_0 \cdot \sigma)$	0.00	0.004	0.008	0.012	0.016	0.02	0.04	0.08
$W_{ar}(s/\tau_0 \cdot \sigma) * 100$	00.00	6.3711	4.4550	3.7513	3.2620	3.1457	2.3381	1.9412
$(s/\tau_0 \cdot \sigma)$	0.12	0.16	0.20	0.30	0.40	0.50	0.60	0.70
$W_{ar}(s/\tau_0 \cdot \sigma) * 100$	1.8203	1.7770	1.7553	1.7425	1.7359	1.8404	1.9332	2.0866
$(s/\tau_0 \cdot \sigma)$	0.80	0.90	1.00	1.40	1.80	2.20	2.60	3.00
$W_{ar}(s/\tau_0 \cdot \sigma) * 100$	2.2469	2.4166	2.5720	3.0875	3.7832	3.9703	4.5953	5.0994
$(s/\tau_0 \cdot \sigma)$	4.00	5.00	6.00	7.00	8.00	9.00	10.00	11.00
$W_{ar}(s/\tau_0 \cdot \sigma) * 100$	4.8873	4.4254	3.7935	3.5460	3.4354	3.2962	3.1126	2.9643
$(s/\tau_0 \cdot \sigma)$	12.00	13.00	14.00	15.00	16.00	18.00	20.00	22.00
$W_{ar}(s/\tau_0 \cdot \sigma) * 100$	2.8231	2.6885	2.5604	2.4384	2.3223	2.0962	1.8751	1.7040
$(s/\tau_0 \cdot \sigma)$	24.00	26.00	28.00	30.00	32.00	34.00	36.00	38.00
$W_{ar}(s/\tau_0 \cdot \sigma) * 100$	1.5455	1.4017	1.2713	1.1538	1.0465	0.9491	0.8608	0.7818
$(s/\tau_0 \cdot \sigma)$	40.00	42.00	44.00	46.00	48.00	50.00	52.00	54.00
$W_{ar}(s/\tau_0 \cdot \sigma) * 100$	0.7081	0.6422	0.5825	0.5283	0.4791	0.4346	0.3941	0.3574
$(s/\tau_0 \cdot \sigma)$	56.00	58.00	60.00	62.00	64.00	66.00	68.00	70.00
$W_{ar}(s/\tau_0 \cdot \sigma) * 100$	0.3242	0.2940	0.2667	0.2419	0.2194	0.1990	0.1805	0.1637

$M_1 = 20.9068\ (\sigma\ \tau_0)$; $D = 419.518\ (\sigma\ \tau_0)^2$; the number of excursions = 21919125.

Rectangular power spectrum distribution of areas above excursions below level $(x_0/\sigma) = 2.0$.

$(s/\tau_0 \cdot \sigma)$	0.00	0.004	0.008	0.012	0.016	0.02	0.04	0.08
$W_{ar}(s/\tau_0 \cdot \sigma) * 100$	00.00	1.3384	0.9744	0.8263	0.7527	0.7007	0.5009	0.3969
$(s/\tau_0 \cdot \sigma)$	0.12	0.16	0.20	0.40	0.60	0.80	1.00	1.20
$W_{ar}(s/\tau_0 \cdot \sigma) * 100$	0.3686	0.3421	0.3202	0.3503	0.3832	0.4280	0.4845	0.6268
$(s/\tau_0 \cdot \sigma)$	1.40	1.60	2.00	2.40	2.80	3.20	3.60	4.00
$W_{ar}(s/\tau_0 \cdot \sigma) * 100$	0.6843	0.7363	1.0538	1.2946	1.4835	1.8092	1.8612	1.9789
$(s/\tau_0 \cdot \sigma)$	5.00	6.00	7.00	8.00	9.00	10.00	11.00	12.00
$W_{ar}(s/\tau_0 \cdot \sigma) * 100$	2.1760	1.8010	1.4906	1.4390	1.3892	1.3412	1.2948	1.2755

(*Continued*)

Table A3.4 (Continued)

$(s/\tau_0 \cdot \sigma)$	13.00	14.00	15.00	16.00	17.00	18.00	19.00	20.00
$W_{ar}(s/\tau_0 \cdot \sigma) * 100$	1.2564	1.2379	1.2194	1.2012	1.1833	1.1657	1.1483	1.1312
$(s/\tau_0 \cdot \sigma)$	22.00	24.00	26.00	28.00	30.00	32.00	34.00	36.00
$W_{ar}(s/\tau_0 \cdot \sigma) * 100$	1.0974	1.0645	1.0327	1.0018	0.9718	0.9428	0.9146	0.8872
$(s/\tau_0 \cdot \sigma)$	40.00	44.00	48.00	52.00	56.00	60.00	64.00	68.00
$W_{ar}(s/\tau_0 \cdot \sigma) * 100$	0.8356	0.7864	0.7401	0.6965	0.6554	0.6168	0.5805	0.5463
$(s/\tau_0 \cdot \sigma)$	72.00	76.00	80.00	84.00	88.00	92.00	96.00	100.00
$W_{ar}(s/\tau_0 \cdot \sigma) * 100$	0.5141	0.4838	0.4553	0.4284	0.4032	0.3794	0.3571	0.3360

$M_1 = 65.7594\ (\sigma \cdot \tau_0)$; $D = 4337.64\ (\sigma \cdot \tau_0)^2$; the number of excursions = 9133811.

Table A3.5 Rayleigh Random Process

Gaussian power spectrum distribution of areas under excursions above level $(x_0/\sigma) = 0.1$.

$(s/\tau_0 \cdot \sigma)$	0.0000	0.0016	0.0032	0.0048	0.0064	0.008	0.016	0.032
$W_{ar}(s/\tau_0 \cdot \sigma) * 100$	00.000	86.245	56.917	45.003	38.989	34.493	24.144	17.648
$(s/\tau_0 \cdot \sigma)$	0.048	0.08	0.16	0.24	0.32	0.40	0.48	0.56
$W_{ar}(s/\tau_0 \cdot \sigma) * 100$	14.184	11.026	8.0925	7.0069	6.3294	5.7654	5.4830	5.2325
$(s/\tau_0 \cdot \sigma)$	0.64	0.72	0.80	0.88	0.96	1.04	1.20	1.36
$W_{ar}(s/\tau_0 \cdot \sigma) * 100$	4.9802	4.8756	4.7903	4.7183	4.6514	4.6331	4.5719	4.5154
$(s/\tau_0 \cdot \sigma)$	1.52	1.68	1.84	2.00	2.16	2.48	2.80	3.12
$W_{ar}(s/\tau_0 \cdot \sigma) * 100$	4.3821	4.2886	4.2036	4.1510	4.1230	4.0432	3.9768	3.8719
$(s/\tau_0 \cdot \sigma)$	3.44	3.76	4.08	4.40	5.20	6.00	6.80	7.60
$W_{ar}(s/\tau_0 \cdot \sigma) * 100$	3.8213	3.7713	3.7220	3.6732	3.5543	3.4240	3.2983	3.1772
$(s/\tau_0 \cdot \sigma)$	8.40	9.20	10.00	10.80	11.60	12.40	13.20	14.00
$W_{ar}(s/\tau_0 \cdot \sigma) * 100$	3.0606	2.9483	2.8521	2.7470	2.6457	2.5482	2.4542	2.3637
$(s/\tau_0 \cdot \sigma)$	16.00	18.00	20.00	22.00	26.00	30.00	34.00	38.00
$W_{ar}(s/\tau_0 \cdot \sigma) * 100$	2.1518	1.9589	1.7834	1.6236	1.3467	1.1170	0.9265	0.7685
$(s/\tau_0 \cdot \sigma)$	42.00	46.00	50.00	54.00	58.00	62.00	66.00	70.00
$W_{ar}(s/\tau_0 \cdot \sigma) * 100$	0.6374	0.5287	0.4385	0.3637	0.3017	0.2502	0.2075	0.1721

$M_1 = 20.8239\ (\sigma\ \tau_0)$; $D = 457.566\ (\sigma\ \tau_0)^2$; the number of excursions = 22049712.

Gaussian power spectrum distribution of areas under excursions above level $(x_0/\sigma) = 0.2$.

$(s/\tau_0 \cdot \sigma)$	0.0000	0.002	0.004	0.006	0.008	0.010	0.012	0.014
$W_{ar}(s/\tau_0 \cdot \sigma) * 100$	00.000	120.48	81.341	65.788	56.603	50.917	46.721	43.744
$(s/\tau_0 \cdot \sigma)$	0.016	0.018	0.020	0.024	0.028	0.032	0.036	0.040
$W_{ar}(s/\tau_0 \cdot \sigma) * 100$	41.232	38.865	37.101	34.118	31.829	29.727	28.463	27.253
$(s/\tau_0 \cdot \sigma)$	0.10	0.20	0.30	0.40	0.50	0.60	0.70	0.80
$W_{ar}(s/\tau_0 \cdot \sigma) * 100$	18.256	14.814	12.821	11.277	10.703	10.152	9.7678	9.3982
$(s/\tau_0 \cdot \sigma)$	0.90	1.00	1.40	1.80	2.20	2.60	3.00	3.40
$W_{ar}(s/\tau_0 \cdot \sigma) * 100$	9.2538	9.1366	8.5552	8.1437	7.6605	7.3548	7.0876	6.8302

(Continued)

Table A3.5 (Continued)

$(s/\tau_0 \cdot \sigma)$	3.80	4.20	4.60	5.00	5.50	6.00	6.50	7.00
$W_{ar}(s/\tau_0 \cdot \sigma) * 100$	6.5607	6.3019	6.0533	5.8145	5.5286	5.2574	4.9994	4.7541
$(s/\tau_0 \cdot \sigma)$	7.50	8.00	8.50	9.00	9.50	10.00	11.00	12.00
$W_{ar}(s/\tau_0 \cdot \sigma) * 100$	4.5676	4.3884	4.1681	3.9262	3.7671	3.5780	3.2356	2.9259
$(s/\tau_0 \cdot \sigma)$	13.00	14.00	15.00	16.00	17.00	18.00	19.00	20.00
$W_{ar}(s/\tau_0 \cdot \sigma) * 100$	2.6458	2.3925	2.1635	1.9564	1.7692	1.5998	1.4467	1.3082
$(s/\tau_0 \cdot \sigma)$	22.00	24.00	26.00	28.00	30.00	32.00	34.00	36.00
$W_{ar}(s/\tau_0 \cdot \sigma) * 100$	1.0697	0.8747	0.7153	0.5849	0.4783	0.3911	0.3198	0.2515

$M_1 = 9.56235\ (\sigma \cdot \tau_0)$; $D = 98.7808\ (\sigma \cdot \tau_0)^2$; the number of excursions = 55036692.

Gaussian power spectrum distribution of areas under excursions above level $(x_0/\sigma) = 0.5$.

$(s/\tau_0 \cdot \sigma)$	0.00	0.004	0.008	0.012	0.016	0.02	0.04	0.08
$W_{ar}(s/\tau_0 \cdot \sigma) * 100$	00.00	193.26	134.37	109.61	96.216	87.564	65.799	49.974
$(s/\tau_0 \cdot \sigma)$	0.12	0.16	0.20	0.24	0.28	0.32	0.36	0.40
$W_{ar}(s/\tau_0 \cdot \sigma) * 100$	42.851	38.663	35.626	33.485	31.733	30.321	29.223	28.150
$(s/\tau_0 \cdot \sigma)$	0.60	0.80	1.00	1.20	1.40	1.60	1.80	2.00
$W_{ar}(s/\tau_0 \cdot \sigma) * 100$	24.659	22.485	20.723	19.202	18.079	16.961	15.907	15.044
$(s/\tau_0 \cdot \sigma)$	2.20	2.40	2.60	2.80	3.00	3.20	3.40	3.60
$W_{ar}(s/\tau_0 \cdot \sigma) * 100$	14.127	13.275	12.620	11.782	11.123	10.494	9.8890	9.3750
$(s/\tau_0 \cdot \sigma)$	3.80	4.00	4.40	4.80	5.20	5.60	6.00	6.40
$W_{ar}(s/\tau_0 \cdot \sigma) * 100$	8.8222	8.3581	7.3909	6.6097	5.8809	5.1656	4.5943	4.0866
$(s/\tau_0 \cdot \sigma)$	6.80	7.20	7.60	8.00	8.40	8.80	9.20	9.60
$W_{ar}(s/\tau_0 \cdot \sigma) * 100$	3.6528	3.2305	2.8836	2.5545	2.2540	2.0297	1.7811	1.5976
$(s/\tau_0 \cdot \sigma)$	10.00	10.40	10.80	11.20	11.60	12.00	12.80	13.60
$W_{ar}(s/\tau_0 \cdot \sigma) * 100$	1.4215	1.2620	1.1283	1.0053	0.9084	0.7872	0.6292	0.4850
$(s/\tau_0 \cdot \sigma)$	14.40	15.20	16.00	16.80	17.60	18.40	19.20	20.00
$W_{ar}(s/\tau_0 \cdot \sigma) * 100$	0.3935	0.3033	0.2475	0.1911	0.1486	0.1185	0.0945	0.0756

$M_1 = 3.12956\ (\sigma \cdot \tau_0)$; $D = 14.3996\ (\sigma \cdot \tau_0)^2$; the number of excursions = 247151554.

Gaussian power spectrum distribution of areas under excursions above level $(x_0/\sigma) = 1.0$.

$(s/\tau_0 \cdot \sigma)$	0.000	0.004	0.008	0.012	0.016	0.020	0.024	0.028
$W_{ar}(s/\tau_0 \cdot \sigma) * 100$	00.00	393.03	278.69	229.88	202.58	185.03	171.90	161.55
$(s/\tau_0 \cdot \sigma)$	0.032	0.036	0.040	0.044	0.048	0.052	0.056	0.060
$W_{ar}(s/\tau_0 \cdot \sigma) * 100$	153.01	145.98	140.12	135.10	130.35	126.37	122.76	119.41
$(s/\tau_0 \cdot \sigma)$	0.070	0.080	0.090	0.10	0.12	0.14	0.16	0.18
$W_{ar}(s/\tau_0 \cdot \sigma) * 100$	112.42	106.84	101.76	97.743	90.627	85.177	80.552	76.850
$(s/\tau_0 \cdot \sigma)$	0.20	0.30	0.40	0.50	0.60	0.80	1.00	1.20
$W_{ar}(s/\tau_0 \cdot \sigma) * 100$	72.397	61.594	53.636	48.042	43.049	35.848	30.230	25.865
$(s/\tau_0 \cdot \sigma)$	1.40	1.60	1.80	2.00	2.20	2.40	2.60	2.80
$W_{ar}(s/\tau_0 \cdot \sigma) * 100$	22.147	19.004	16.397	14.135	12.271	10.537	9.0532	7.8624
$(s/\tau_0 \cdot \sigma)$	3.00	3.20	3.40	3.60	3.80	4.00	4.20	4.40
$W_{ar}(s/\tau_0 \cdot \sigma) * 100$	6.7818	5.8636	5.0433	4.3553	3.7467	3.2678	2.8092	2.4245
$(s/\tau_0 \cdot \sigma)$	4.60	4.80	5.00	5.20	5.40	5.60	5.80	6.00
$W_{ar}(s/\tau_0 \cdot \sigma) * 100$	2.1168	1.8061	1.5894	1.3626	1.1725	0.9999	0.8648	0.7610
$(s/\tau_0 \cdot \sigma)$	6.20	6.40	6.60	6.80	7.00	7.20	7.40	7.60
$W_{ar}(s/\tau_0 \cdot \sigma) * 100$	0.6538	0.5692	0.4961	0.4266	0.3669	0.3156	0.2714	0.2334

$M_1 = 1.16787\ (\sigma \cdot \tau_0)$; $D = 1.75897\ (\sigma \cdot \tau_0)^2$; the number of excursions = 340988781.

Gaussian power spectrum distribution of areas under excursions above level $(x_0/\sigma) = 1.5$.

$(s/\tau_0 \cdot \sigma)$	0.000	0.004	0.008	0.012	0.016	0.020	0.024	0.028
$W_{ar}(s/\tau_0 \cdot \sigma) * 100$	00.00	644.54	457.29	378.03	331.56	301.87	279.91	262.53
$(s/\tau_0 \cdot \sigma)$	0.032	0.036	0.040	0.044	0.048	0.052	0.056	0.060
$W_{ar}(s/\tau_0 \cdot \sigma) * 100$	247.96	236.05	225.34	216.67	208.83	202.04	195.35	189.99
$(s/\tau_0 \cdot \sigma)$	0.064	0.068	0.072	0.076	0.080	0.084	0.088	0.092
$W_{ar}(s/\tau_0 \cdot \sigma) * 100$	184.39	179.67	175.01	170.78	167.05	163.19	159.97	156.69
$(s/\tau_0 \cdot \sigma)$	0.096	0.100	0.104	0.108	0.112	0.116	0.120	0.128
$W_{ar}(s/\tau_0 \cdot \sigma) * 100$	153.68	150.68	147.81	145.39	142.75	140.39	138.25	134.05

(Continued)

Table A3.5 (Continued)

$(s/\tau_0 \cdot \sigma)$	0.136	0.144	0.152	0.160	0.168	0.176	0.184	0.192
$W_{ar}(s/\tau_0 \cdot \sigma)*100$	129.96	126.18	122.61	119.50	116.50	113.82	110.96	108.44
$(s/\tau_0 \cdot \sigma)$	0.200	0.240	0.280	0.320	0.360	0.400	0.600	0.800
$W_{ar}(s/\tau_0 \cdot \sigma)*100$	106.07	95.464	86.919	79.793	73.566	68.181	48.587	35.865
$(s/\tau_0 \cdot \sigma)$	1.000	1.200	1.400	1.600	1.800	2.000	2.200	2.400
$W_{ar}(s/\tau_0 \cdot \sigma)*100$	27.153	20.665	15.866	12.131	9.4272	7.2710	5.6082	4.3106
$(s/\tau_0 \cdot \sigma)$	2.600	2.800	3.000	3.200	3.400	3.600	3.800	4.000
$W_{ar}(s/\tau_0 \cdot \sigma)*100$	3.3489	2.6119	2.0310	1.5628	1.1888	0.9302	0.7354	0.5622

$M_1 = 0.614449\ (\sigma\ \tau_0)$; $D = 0.535499\ (\sigma\ \tau_0)^2$; the number of excursions = 273337751.

Gaussian power spectrum distribution of areas under excursions above level $(x_0/\sigma) = 2.0$.

$(s/\tau_0 \cdot \sigma)$	0.000	0.002	0.004	0.006	0.008	0.010	0.012	0.014
$W_{ar}(s/\tau_0 \cdot \sigma)*100$	00.00	1153.5	850.60	696.35	623.57	566.96	523.22	488.67
$(s/\tau_0 \cdot \sigma)$	0.016	0.018	0.020	0.024	0.028	0.032	0.036	0.040
$W_{ar}(s/\tau_0 \cdot \sigma)*100$	460.87	439.56	420.85	389.83	364.92	343.75	326.49	310.12
$(s/\tau_0 \cdot \sigma)$	0.044	0.048	0.052	0.056	0.060	0.064	0.068	0.072
$W_{ar}(s/\tau_0 \cdot \sigma)*100$	296.95	285.30	274.99	264.60	256.52	248.48	240.41	233.92
$(s/\tau_0 \cdot \sigma)$	0.076	0.080	0.084	0.088	0.092	0.096	0.100	0.120
$W_{ar}(s/\tau_0 \cdot \sigma)*100$	227.73	221.45	215.76	210.48	205.26	200.30	195.92	176.46
$(s/\tau_0 \cdot \sigma)$	0.140	0.160	0.180	0.200	0.300	0.400	0.500	0.600
$W_{ar}(s/\tau_0 \cdot \sigma)*100$	160.48	147.45	136.46	126.50	91.694	69.854	54.483	43.075
$(s/\tau_0 \cdot \sigma)$	0.700	0.800	0.900	1.000	1.100	1.200	1.300	1.400
$W_{ar}(s/\tau_0 \cdot \sigma)*100$	34.556	27.868	22.531	18.373	14.940	12.315	10.099	8.3161
$(s/\tau_0 \cdot \sigma)$	1.500	1.600	1.700	1.800	1.900	2.000	2.100	2.200
$W_{ar}(s/\tau_0 \cdot \sigma)*100$	6.9292	5.6039	4.6470	3.8724	3.1860	2.6192	2.1566	1.7968
$(s/\tau_0 \cdot \sigma)$	2.300	2.400	2.500	2.600	2.700	2.800	2.900	3.000
$W_{ar}(s/\tau_0 \cdot \sigma)*100$	1.5256	1.2109	1.0175	0.8472	0.6775	0.5747	0.4695	0.4119

$M_1 = 0.376522\ (\sigma\ \tau_0)$; $D = 0.220282\ (\sigma\ \tau_0)^2$; the number of excursions = 151741986.

Gaussian power spectrum distribution of areas under excursions above level $(x_0/\sigma) = 2.5$.

$(s/\tau_0 \cdot \sigma)$	0.000	0.002	0.004	0.006	0.008	0.010	0.012	0.014
$W_{ar}(s/\tau_0 \cdot \sigma) * 100$	00.00	1554.8	1161.2	932.21	833.28	758.74	696.95	648.16
$(s/\tau_0 \cdot \sigma)$	0.016	0.018	0.020	0.022	0.024	0.026	0.028	0.030
$W_{ar}(s/\tau_0 \cdot \sigma) * 100$	608.93	577.07	549.38	527.87	508.24	488.63	472.71	456.09
$(s/\tau_0 \cdot \sigma)$	0.032	0.036	0.040	0.044	0.048	0.052	0.056	0.060
$W_{ar}(s/\tau_0 \cdot \sigma) * 100$	440.92	416.32	394.19	375.60	358.38	342.76	330.03	316.45
$(s/\tau_0 \cdot \sigma)$	0.064	0.068	0.072	0.076	0.080	0.084	0.088	0.092
$W_{ar}(s/\tau_0 \cdot \sigma) * 100$	305.38	294.91	284.93	274.61	266.11	259.16	250.93	244.03
$(s/\tau_0 \cdot \sigma)$	0.096	0.100	0.120	0.140	0.160	0.180	0.200	0.300
$W_{ar}(s/\tau_0 \cdot \sigma) * 100$	236.97	230.49	203.13	181.05	162.90	148.02	134.45	89.185
$(s/\tau_0 \cdot \sigma)$	0.400	0.500	0.600	0.700	0.800	0.900	1.000	1.100
$W_{ar}(s/\tau_0 \cdot \sigma) * 100$	61.979	44.799	32.966	24.296	18.328	14.049	10.502	8.0259
$(s/\tau_0 \cdot \sigma)$	1.200	1.300	1.400	1.500	1.600	1.700	1.800	1.900
$W_{ar}(s/\tau_0 \cdot \sigma) * 100$	6.1447	4.7197	3.5636	2.7318	2.1924	1.7240	1.3174	1.0233
$(s/\tau_0 \cdot \sigma)$	2.000	2.100	2.200	2.300	2.400	2.500	2.600	2.700
$W_{ar}(s/\tau_0 \cdot \sigma) * 100$	0.7748	0.5842	0.4742	0.3707	0.2998	0.2330	0.1678	0.1450

$M_1 = 0.254207\ (\sigma\ \tau_0)$; $D = 0.107832\ (\sigma\ \tau_0)^2$; the number of excursions = 61369931.

Gaussian power spectrum distribution of areas under excursions above level $(x_0/\sigma) = 3.0$.

$(s/\tau_0 \cdot \sigma)$	0.000	0.002	0.004	0.006	0.008	0.010	0.012	0.014
$W_{ar}(s/\tau_0 \cdot \sigma) * 100$	00.00	1983.4	1505.3	1184.3	1054.3	956.55	877.42	811.86
$(s/\tau_0 \cdot \sigma)$	0.016	0.018	0.020	0.022	0.024	0.026	0.028	0.030
$W_{ar}(s/\tau_0 \cdot \sigma) * 100$	759.22	712.69	679.11	646.85	616.84	595.22	574.57	550.41
$(s/\tau_0 \cdot \sigma)$	0.032	0.034	0.036	0.038	0.040	0.042	0.044	0.046
$W_{ar}(s/\tau_0 \cdot \sigma) * 100$	532.76	514.03	499.49	482.45	466.85	454.14	442.33	433.20

(*Continued*)

Table A3.5 (Continued)

$(s/\tau_0 \cdot \sigma)$	0.048	0.050	0.052	0.054	0.056	0.058	0.060	0.062
$W_{ar}(s/\tau_0 \cdot \sigma) * 100$	420.11	411.93	399.70	390.28	381.59	373.81	365.87	358.77
$(s/\tau_0 \cdot \sigma)$	0.064	0.066	0.068	0.070	0.072	0.074	0.076	0.078
$W_{ar}(s/\tau_0 \cdot \sigma) * 100$	349.91	343.08	336.08	329.66	323.28	317.10	311.08	305.57
$(s/\tau_0 \cdot \sigma)$	0.080	0.082	0.084	0.088	0.092	0.096	0.100	0.120
$W_{ar}(s/\tau_0 \cdot \sigma) * 100$	300.07	295.44	290.83	279.82	269.91	260.86	251.49	216.07
$(s/\tau_0 \cdot \sigma)$	0.1400	0.160	0.180	0.200	0.300	0.400	0.500	0.600
$W_{ar}(s/\tau_0 \cdot \sigma) * 100$	187.52	165.11	147.84	131.27	79.625	50.194	33.485	22.601
$(s/\tau_0 \cdot \sigma)$	0.700	0.800	0.900	1.000	1.100	1.200	1.300	1.400
$W_{ar}(s/\tau_0 \cdot \sigma) * 100$	15.633	11.087	7.4355	5.2543	3.7326	2.6596	1.9947	1.3406

$M_1 = 0.182928\ (\sigma \cdot \tau_0)$; $D = 0.0591595\ (\sigma \cdot \tau_0)^2$; the number of excursions = 18499069.

Table A3.6 Rayleigh Random Process

Rectangular power spectrum distribution of areas under excursions above level $(x_0/\sigma) = 0.5$.

$(s/\tau_0 \cdot \sigma)$	0.000	0.011	0.022	0.033	0.044	0.055	0.066	0.077
$W_{ar}(s/\tau_0 \cdot \sigma) * 100$	00.00	89.981	76.875	62.345	54.627	50.460	47.778	45.786
$(s/\tau_0 \cdot \sigma)$	0.088	0.099	0.11	0.121	0.132	0.143	0.154	0.165
$W_{ar}(s/\tau_0 \cdot \sigma) * 100$	43.921	42.421	41.067	39.789	38.759	37.823	36.943	36.296
$(s/\tau_0 \cdot \sigma)$	0.176	0.187	0.198	0.220	0.242	0.264	0.286	0.308
$W_{ar}(s/\tau_0 \cdot \sigma) * 100$	35.625	35.053	34.534	33.615	32.891	32.178	31.593	31.039
$(s/\tau_0 \cdot \sigma)$	0.33	0.44	0.55	0.66	0.77	0.88	0.99	1.10
$W_{ar}(s/\tau_0 \cdot \sigma) * 100$	30.543	28.458	26.939	25.728	24.668	23.727	22.821	21.903
$(s/\tau_0 \cdot \sigma)$	1.65	2.20	2.75	3.30	3.85	4.40	4.95	5.50
$W_{ar}(s/\tau_0 \cdot \sigma) * 100$	18.083	14.843	12.299	10.209	8.5129	7.1606	6.0333	5.1051
$(s/\tau_0 \cdot \sigma)$	6.05	6.60	7.15	7.70	8.25	8.80	9.35	9.90
$W_{ar}(s/\tau_0 \cdot \sigma) * 100$	4.3308	3.6848	3.1290	2.6513	2.2598	1.9146	1.6400	1.3889
$(s/\tau_0 \cdot \sigma)$	10.45	11.00	11.55	12.10	12.65	13.20	13.75	14.30
$W_{ar}(s/\tau_0 \cdot \sigma) * 100$	1.1867	1.0043	0.8528	0.7314	0.6158	0.5242	0.4532	0.3816
$(s/\tau_0 \cdot \sigma)$	14.85	15.40	15.95	16.50	17.05	17.60	18.15	18.70
$W_{ar}(s/\tau_0 \cdot \sigma) * 100$	0.3242	0.2779	0.2324	0.2031	0.1724	0.1451	0.1240	0.1071

$M_1 = 3.07316\ (\sigma \cdot \tau_0)$; $D = 10.921\ (\sigma \cdot \tau_0)^2$; the number of excursions = 665571293.

Rectangular power spectrum distribution of areas under excursions above level $(x_0/\sigma) = 1.0$.

$(s/\tau_0 \cdot \sigma)$	0.000	0.0112	0.0224	0.0336	0.0448	0.0560	0.0672	0.0784
$W_{ar}(s/\tau_0 \cdot \sigma) * 100$	00.00	181.29	156.02	133.74	116.46	106.22	100.07	95.658
$(s/\tau_0 \cdot \sigma)$	0.0896	0.1008	0.1120	0.1232	0.1344	0.1456	0.1568	0.1680
$W_{ar}(s/\tau_0 \cdot \sigma) * 100$	91.907	88.754	85.758	82.996	80.630	78.400	76.476	74.646
$(s/\tau_0 \cdot \sigma)$	0.1792	0.1904	0.2016	0.2128	0.2240	0.2352	0.2464	0.2576
$W_{ar}(s/\tau_0 \cdot \sigma) * 100$	112.42	106.84	101.76	97.743	90.627	85.177	80.552	76.850
$(s/\tau_0 \cdot \sigma)$	0.2688	0.2800	0.2912	0.3024	0.3136	0.3248	0.3360	0.3472
$W_{ar}(s/\tau_0 \cdot \sigma) * 100$	72.397	61.594	53.636	48.042	43.049	35.848	30.230	25.865

(Continued)

Table A3.6 (Continued)

$(s/\tau_0 \cdot \sigma)$	0.3584	0.3696	0.3808	0.3920	0.4032	0.4144	0.4256	0.4368
$W_{ar}(s/\tau_0 \cdot \sigma) * 100$	22.147	19.004	16.397	14.135	12.271	10.537	9.0532	7.8624
$(s/\tau_0 \cdot \sigma)$	0.4480	0.4592	0.4816	0.4928	0.5152	0.5376	0.5600	0.6720
$W_{ar}(s/\tau_0 \cdot \sigma) * 100$	6.7818	5.8636	5.0433	3.3553	3.7467	3.2678	2.8092	2.4245
$(s/\tau_0 \cdot \sigma)$	0.7840	0.8960	1.0080	1.1200	1.6800	2.2400	2.8000	3.3600
$W_{ar}(s/\tau_0 \cdot \sigma) * 100$	2.1168	1.8061	1.5894	1.3626	1.1725	0.9999	0.8648	0.7610
$(s/\tau_0 \cdot \sigma)$	3.920	4.480	5.040	5.600	6.160	6.720	7.280	7.840
$W_{ar}(s/\tau_0 \cdot \sigma) * 100$	0.6538	0.5692	0.4961	0.4266	0.3669	0.3156	0.2714	0.2334

$M_1 = 1.15321\ (\sigma\ \tau_0)$; $D = 1.41603\ (\sigma\ \tau_0)^2$; the number of excursions = 462253721.

Rectangular power spectrum distribution of areas under excursions above level $(x_0/\sigma) = 1.5$.

$(s/\tau_0 \cdot \sigma)$	0.000	0.008	0.016	0.024	0.032	0.040	0.048	0.056
$W_{ar}(s/\tau_0 \cdot \sigma) * 100$	00.00	401.61	318.74	253.98	226.94	210.24	196.05	184.04
$(s/\tau_0 \cdot \sigma)$	0.064	0.072	0.080	0.088	0.096	0.104	0.112	0.120
$W_{ar}(s/\tau_0 \cdot \sigma) * 100$	174.04	165.65	158.55	152.41	147.27	142.48	138.34	134.26
$(s/\tau_0 \cdot \sigma)$	0.128	0.136	0.144	0.152	0.160	0.168	0.176	0.184
$W_{ar}(s/\tau_0 \cdot \sigma) * 100$	130.98	127.73	124.43	121.49	118.85	116.17	113.52	111.46
$(s/\tau_0 \cdot \sigma)$	0.192	0.200	0.208	0.216	0.224	0.232	0.240	0.248
$W_{ar}(s/\tau_0 \cdot \sigma) * 100$	109.12	106.95	105.06	102.87	101.11	99.618	97.839	96.092
$(s/\tau_0 \cdot \sigma)$	0.256	0.264	0.272	0.280	0.288	0.296	0.304	0.312
$W_{ar}(s/\tau_0 \cdot \sigma) * 100$	94.621	93.053	91.681	90.201	88.887	87.480	86.150	84.984
$(s/\tau_0 \cdot \sigma)$	0.320	0.328	0.336	0.344	0.352	0.360	0.368	0.376
$W_{ar}(s/\tau_0 \cdot \sigma) * 100$	83.669	82.384	81.328	80.112	78.983	77.851	76.857	75.859
$(s/\tau_0 \cdot \sigma)$	0.384	0.392	0.400	0.480	0.560	0.640	0.720	0.800
$W_{ar}(s/\tau_0 \cdot \sigma) * 100$	74.739	73.868	72.920	64.064	56.721	50.474	44.962	40.061
$(s/\tau_0 \cdot \sigma)$	1.200	1.600	2.000	2.400	2.800	3.200	3.600	4.000
$W_{ar}(s/\tau_0 \cdot \sigma) * 100$	22.745	12.764	7.0523	3.8497	2.0491	1.0994	0.5890	0.3149

$M_1 = 0.610372\ (\sigma\ \tau_0)$; $D = 0.435261\ (\sigma\ \tau_0)^2$; the number of excursions = 261596893.

Rectangular power spectrum distribution of areas under excursions above level $(x_0/\sigma) = 2.0$.

$(s/\tau_0 \cdot \sigma)$	0.000	0.002	0.004	0.006	0.008	0.010	0.012	0.014
$W_{ar}(s/\tau_0 \cdot \sigma) * 100$	00.00	1063.8	762.56	645.73	573.68	519.16	479.54	452.04
$(s/\tau_0 \cdot \sigma)$	0.016	0.018	0.020	0.022	0.024	0.026	0.028	0.030
$W_{ar}(s/\tau_0 \cdot \sigma) * 100$	429.45	409.86	393.38	377.85	365.22	352.86	342.18	332.79
$(s/\tau_0 \cdot \sigma)$	0.032	0.034	0.036	0.038	0.040	0.042	0.044	0.048
$W_{ar}(s/\tau_0 \cdot \sigma) * 100$	323.92	315.11	307.81	301.41	295.69	288.81	283.46	272.88
$(s/\tau_0 \cdot \sigma)$	0.052	0.056	0.060	0.064	0.068	0.072	0.076	0.080
$W_{ar}(s/\tau_0 \cdot \sigma) * 100$	263.22	254.70	246.81	240.07	233.36	226.52	220.78	215.31
$(s/\tau_0 \cdot \sigma)$	0.084	0.088	0.092	0.096	0.100	0.120	0.140	0.160
$W_{ar}(s/\tau_0 \cdot \sigma) * 100$	210.95	206.05	201.92	197.65	193.76	175.59	161.02	149.04
$(s/\tau_0 \cdot \sigma)$	0.180	0.200	0.300	0.400	0.500	0.600	0.700	0.800
$W_{ar}(s/\tau_0 \cdot \sigma) * 100$	138.45	129.44	95.865	74.105	58.382	46.558	36.821	29.633
$(s/\tau_0 \cdot \sigma)$	0.900	1.000	1.100	1.200	1.300	1.400	1.500	1.600
$W_{ar}(s/\tau_0 \cdot \sigma) * 100$	23.854	19.021	15.514	12.598	9.9798	8.0822	6.5426	5.2787
$(s/\tau_0 \cdot \sigma)$	1.700	1.800	1.900	2.000	2.100	2.200	2.300	2.400
$W_{ar}(s/\tau_0 \cdot \sigma) * 100$	4.1921	3.4303	2.6837	2.2141	1.7667	1.4066	1.1053	0.8740

$M_1 = 0.376522\ (\sigma\ \tau_0)$; $D = 0.220282\ (\sigma\ \tau_0)^2$; the number of excursions = 151741986.

Rectangular power spectrum distribution of areas under excursions above level $(x_0/\sigma) = 2.5$.

$(s/\tau_0 \cdot \sigma)$	0.000	0.002	0.004	0.006	0.008	0.010	0.012	0.014
$W_{ar}(s/\tau_0 \cdot \sigma) * 100$	00.00	1554.8	1161.2	932.21	833.28	758.74	696.95	648.16
$(s/\tau_0 \cdot \sigma)$	0.016	0.018	0.020	0.022	0.024	0.026	0.028	0.030
$W_{ar}(s/\tau_0 \cdot \sigma) * 100$	608.93	577.07	549.38	527.87	508.24	488.63	472.71	456.09
$(s/\tau_0 \cdot \sigma)$	0.032	0.036	0.040	0.044	0.048	0.052	0.056	0.060
$W_{ar}(s/\tau_0 \cdot \sigma) * 100$	440.92	416.32	394.19	375.60	358.38	342.76	330.03	316.45
$(s/\tau_0 \cdot \sigma)$	0.064	0.068	0.072	0.076	0.080	0.084	0.088	0.092
$W_{ar}(s/\tau_0 \cdot \sigma) * 100$	305.38	294.91	284.93	274.61	266.11	259.16	250.93	244.03

(*Continued*)

Table A3.6 (Continued)

$(s/\tau_0 \cdot \sigma)$	0.096	0.100	0.120	0.140	0.160	0.180	0.200	0.300
$W_{ar}(s/\tau_0 \cdot \sigma) * 100$	236.97	230.49	203.13	181.05	162.90	148.02	134.45	89.185
$(s/\tau_0 \cdot \sigma)$	0.400	0.500	0.600	0.700	0.800	0.900	1.000	1.100
$W_{ar}(s/\tau_0 \cdot \sigma) * 100$	61.979	44.799	32.966	24.296	18.328	14.049	10.502	8.0259
$(s/\tau_0 \cdot \sigma)$	1.200	1.300	1.400	1.500	1.600	1.700	1.800	1.900
$W_{ar}(s/\tau_0 \cdot \sigma) * 100$	6.1447	4.7197	3.5636	2.7318	2.1924	1.7240	1.3174	1.0233
$(s/\tau_0 \cdot \sigma)$	2.000	2.100	2.200	2.300	2.400	2.500	2.600	2.700
$W_{ar}(s/\tau_0 \cdot \sigma) * 100$	0.7748	0.5842	0.4742	0.3707	0.2998	0.2330	0.1678	0.1450

$M_1 = 0.252668\ (\sigma\ \tau_0)$; $D = 0.0963715\ (\sigma\ \tau_0)^2$; the number of excursions = 57994615.

Rectangular power spectrum distribution of areas under excursions above level $(x_0/\sigma) = 3.0$.

$(s/\tau_0 \cdot \sigma)$	0.000	0.001	0.002	0.003	0.004	0.005	0.006	0.007
$W_{ar}(s/\tau_0 \cdot \sigma) * 100$	00.00	2172.5	1945.6	1529.1	1291.8	1178.3	1115.0	1051.6
$(s/\tau_0 \cdot \sigma)$	0.008	0.009	0.010	0.011	0.012	0.013	0.014	0.015
$W_{ar}(s/\tau_0 \cdot \sigma) * 100$	1000.1	950.60	906.09	863.51	828.75	795.68	765.29	742.78
$(s/\tau_0 \cdot \sigma)$	0.016	0.017	0.018	0.020	0.022	0.024	0.026	0.028
$W_{ar}(s/\tau_0 \cdot \sigma) * 100$	721.75	703.01	687.61	658.41	627.76	601.18	576.57	555.45
$(s/\tau_0 \cdot \sigma)$	0.030	0.032	0.034	0.036	0.038	0.040	0.042	0.044
$W_{ar}(s/\tau_0 \cdot \sigma) * 100$	535.41	517.94	500.60	484.32	470.94	458.59	447.71	438.27
$(s/\tau_0 \cdot \sigma)$	0.046	0.048	0.050	0.060	0.070	0.080	0.090	0.100
$W_{ar}(s/\tau_0 \cdot \sigma) * 100$	429.08	343.08	336.08	361.01	325.00	298.75	272.09	252.25
$(s/\tau_0 \cdot \sigma)$	0.150	0.200	0.250	0.300	0.350	0.400	0.450	0.500
$W_{ar}(s/\tau_0 \cdot \sigma) * 100$	180.67	133.53	104.06	85.047	67.585	54.167	43.211	34.598
$(s/\tau_0 \cdot \sigma)$	0.550	0.600	0.650	0.700	0.750	0.800	0.850	0.900
$W_{ar}(s/\tau_0 \cdot \sigma) * 100$	28.380	23.357	19.349	15.779	12.898	10.668	8.7602	7.2024
$(s/\tau_0 \cdot \sigma)$	0.950	1.000	1.050	1.100	1.150	1.200	1.250	1.300
$W_{ar}(s/\tau_0 \cdot \sigma) * 100$	6.0227	5.0697	4.2199	3.5424	2.9659	2.4540	2.0589	1.7217

$M_1 = 0.181366\ (\sigma\ \tau_0)$; $D = 0.0545343\ (\sigma\ \tau_0)^2$; the number of excursions = 17417346.

Table A3.7 Rayleigh Random Process

Gaussian power spectrum distribution of areas above excursions below level $(x_0/\sigma) = 0.5$.

$(s/\tau_0 \cdot \sigma)$	0.0000	0.0005	0.0010	0.0015	0.0020	0.0025	0.0030	0.0035
$W_{ar}(s/\tau_0 \cdot \sigma) * 100$	00.00	2590.7	2044.1	1694.6	1508.0	1395.0	1310.5	1245.0
$(s/\tau_0 \cdot \sigma)$	0.0040	0.0045	0.0050	0.0055	0.0060	0.0065	0.0070	0.0075
$W_{ar}(s/\tau_0 \cdot \sigma) * 100$	1191.2	1144.6	1108.3	1074.5	1045.7	1020.1	997.97	976.62
$(s/\tau_0 \cdot \sigma)$	0.0080	0.0085	0.0090	0.0095	0.0100	0.0105	0.0110	0.0115
$W_{ar}(s/\tau_0 \cdot \sigma) * 100$	956.41	937.92	922.11	908.97	896.12	884.77	874.75	865.46
$(s/\tau_0 \cdot \sigma)$	0.0120	0.0125	0.0130	0.0135	0.0140	0.0145	0.0150	0.0155
$W_{ar}(s/\tau_0 \cdot \sigma) * 100$	854.22	845.35	837.38	827.98	820.75	813.51	806.68	799.17
$(s/\tau_0 \cdot \sigma)$	0.0160	0.0165	0.0170	0.0175	0.0180	0.0185	0.0190	0.0200
$W_{ar}(s/\tau_0 \cdot \sigma) * 100$	794.06	789.56	783.00	778.38	773.61	768.99	763.89	755.14
$(s/\tau_0 \cdot \sigma)$	0.0250	0.0300	0.0350	0.0400	0.0450	0.0500	0.0750	0.1000
$W_{ar}(s/\tau_0 \cdot \sigma) * 100$	725.58	704.52	696.34	695.61	702.95	718.21	666.56	432.83
$(s/\tau_0 \cdot \sigma)$	0.1250	0.1500	0.1750	0.2000	0.2250	0.2500	0.2750	0.3000
$W_{ar}(s/\tau_0 \cdot \sigma) * 100$	248.94	152.05	104.88	79.358	62.707	50.369	39.179	30.432
$(s/\tau_0 \cdot \sigma)$	0.3250	0.3500	0.3750	0.4000	0.4250	0.4500	0.4750	0.5000
$W_{ar}(s/\tau_0 \cdot \sigma) * 100$	23.243	17.501	13.219	10.035	7.7454	5.9929	4.6369	3.5777

$M_1 = 0.0813083\ (\sigma\ \tau_0)$; $D = 0.00681689\ (\sigma\ \tau_0)^2$; the number of excursions = 124047573.

Gaussian power spectrum distribution of areas above excursions below level $(x_0/\sigma) = 1.0$.

$(s/\tau_0 \cdot \sigma)$	0.000	0.001	0.002	0.003	0.004	0.005	0.006	0.007
$W_{ar}(s/\tau_0 \cdot \sigma) * 100$	00.00	713.21	514.32	429.43	383.89	352.00	329.13	311.65
$(s/\tau_0 \cdot \sigma)$	0.008	0.009	0.010	0.011	0.012	0.013	0.014	0.015
$W_{ar}(s/\tau_0 \cdot \sigma) * 100$	297.91	286.50	276.23	267.18	259.73	253.30	247.66	242.13
$(s/\tau_0 \cdot \sigma)$	0.016	0.017	0.018	0.020	0.022	0.024	0.026	0.028
$W_{ar}(s/\tau_0 \cdot \sigma) * 100$	237.16	232.23	228.53	220.87	214.32	209.07	204.59	200.79
$(s/\tau_0 \cdot \sigma)$	0.030	0.032	0.034	0.036	0.038	0.040	0.050	0.060
$W_{ar}(s/\tau_0 \cdot \sigma) * 100$	196.40	192.62	189.54	186.50	184.03	181.89	171.66	164.57

(Continued)

Table A3.7 (Continued)

$(s/\tau_0 \cdot \sigma)$	0.070	0.080	0.090	0.100	0.150	0.200	0.250	0.300
$W_{ar}(s/\tau_0 \cdot \sigma) * 100$	160.32	156.72	153.40	151.28	150.99	160.98	166.53	146.21
$(s/\tau_0 \cdot \sigma)$	0.350	0.400	0.450	0.500	0.550	0.600	0.650	0.700
$W_{ar}(s/\tau_0 \cdot \sigma) * 100$	117.35	92.121	75.934	65.329	57.542	51.531	46.826	42.502
$(s/\tau_0 \cdot \sigma)$	0.750	0.800	0.850	0.900	0.950	1.000	1.100	1.200
$W_{ar}(s/\tau_0 \cdot \sigma) * 100$	38.518	34.258	30.389	27.271	24.049	21.095	16.933	13.593
$(s/\tau_0 \cdot \sigma)$	1.300	1.400	1.600	1.800	2.000	2.200	2.400	2.600
$W_{ar}(s/\tau_0 \cdot \sigma) * 100$	10.836	8.7250	5.4627	3.5889	2.2044	1.3559	0.8988	0.5717

$M_1 = 0.423162\ (\sigma\ \tau_0)$; $D = 0.184011\ (\sigma\ \tau_0)^2$; the number of excursions = 170887685.

Gaussian power spectrum distribution of areas above excursions below level $(x_0/\sigma) = 1.5$.

$(s/\tau_0 \cdot \sigma)$	0.000	0.004	0.008	0.012	0.016	0.020	0.024	0.028
$W_{ar}(s/\tau_0 \cdot \sigma) * 100$	00.000	180.59	128.92	107.48	95.461	87.812	82.108	77.636
$(s/\tau_0 \cdot \sigma)$	0.032	0.036	0.040	0.044	0.048	0.052	0.056	0.060
$W_{ar}(s/\tau_0 \cdot \sigma) * 100$	74.202	71.507	68.915	66.933	65.061	63.413	62.135	60.813
$(s/\tau_0 \cdot \sigma)$	0.064	0.068	0.072	0.076	0.080	0.084	0.088	0.092
$W_{ar}(s/\tau_0 \cdot \sigma) * 100$	59.838	58.579	57.541	56.790	56.151	55.497	54.744	54.213
$(s/\tau_0 \cdot \sigma)$	0.096	0.100	0.104	0.108	0.112	0.116	0.120	0.124
$W_{ar}(s/\tau_0 \cdot \sigma) * 100$	53.669	53.346	52.735	52.149	51.711	51.414	51.064	50.772
$(s/\tau_0 \cdot \sigma)$	0.128	0.132	0.136	0.140	0.144	0.152	0.160	0.168
$W_{ar}(s/\tau_0 \cdot \sigma) * 100$	50.505	50.276	49.990	49.674	49.445	49.087	48.743	48.325
$(s/\tau_0 \cdot \sigma)$	0.176	0.184	0.192	0.200	0.240	0.280	0.320	0.360
$W_{ar}(s/\tau_0 \cdot \sigma) * 100$	48.133	47.792	47.676	47.497	47.143	47.523	48.394	50.002
$(s/\tau_0 \cdot \sigma)$	0.400	0.600	0.800	1.000	1.200	1.400	1.600	1.800
$W_{ar}(s/\tau_0 \cdot \sigma) * 100$	52.256	53.071	39.022	32.429	28.795	25.774	22.508	19.645
$(s/\tau_0 \cdot \sigma)$	2.000	3.000	4.000	5.000	6.000	7.000	8.000	9.000
$W_{ar}(s/\tau_0 \cdot \sigma) * 100$	17.234	9.0070	4.7115	2.4433	1.2837	0.6537	0.3536	0.1851

$M_1 = 1.51197\ (\sigma\ \tau_0)$; $D = 2.33156\ (\sigma\ \tau_0)^2$; the number of excursions = 274304650.

Gaussian power spectrum distribution of areas above excursions below level $(x_0/\sigma) = 2.0$.

$(s/\tau_0 \cdot \sigma)$	0.00	0.004	0.008	0.012	0.016	0.020	0.024	0.028
$W_{ar}(s/\tau_0 \cdot \sigma) * 100$	00.000	74.105	51.805	42.579	37.386	34.047	31.657	29.642
$(s/\tau_0 \cdot \sigma)$	0.032	0.036	0.040	0.060	0.080	0.100	0.120	0.140
$W_{ar}(s/\tau_0 \cdot \sigma) * 100$	28.154	27. 008	26.035	22.290	20.256	18.864	17.858	17.154
$(s/\tau_0 \cdot \sigma)$	0.160	0.180	0.200	0.240	0.280	0.320	0.360	0.400
$W_{ar}(s/\tau_0 \cdot \sigma) * 100$	16.547	16.108	15.762	15.273	14.887	14.801	14.649	14.711
$(s/\tau_0 \cdot \sigma)$	0.440	0.480	0.520	0.560	0.600	0.640	0.680	0.720
$W_{ar}(s/\tau_0 \cdot \sigma) * 100$	14.847	15.133	15.382	15.649	16.143	16.678	17.303	17.975
$(s/\tau_0 \cdot \sigma)$	0.760	0.800	1.000	1.200	1.400	1.600	1.800	2.000
$W_{ar}(s/\tau_0 \cdot \sigma) * 100$	18.471	18.673	17.631	15.509	14.300	13.592	13.110	12.869
$(s/\tau_0 \cdot \sigma)$	2.200	2.400	2.600	2.800	3.000	3.200	3.400	3.600
$W_{ar}(s/\tau_0 \cdot \sigma) * 100$	12.369	11.896	11.388	11.061	10.547	10.125	9.8597	9.4294
$(s/\tau_0 \cdot \sigma)$	3.800	4.000	4.200	4.400	4.600	4.800	5.000	6.000
$W_{ar}(s/\tau_0 \cdot \sigma) * 100$	9.1656	8.7748	8.4496	8.1504	7.8555	7.5493	7.2749	6.0071
$(s/\tau_0 \cdot \sigma)$	7.000	8.000	10.000	12.000	14.000	16.000	18.000	20.000
$W_{ar}(s/\tau_0 \cdot \sigma) * 100$	4.9734	4.1281	2.8623	1.9543	1.3352	0.9219	0.6400	0.4357

$M_1 = 5.26983\ (\sigma \cdot \tau_0)$; $D = 28.2131\ (\sigma \cdot \tau_0)^2$; the number of excursions = 152510452.

Gaussian power spectrum distribution of areas above excursions below level $(x_0/\sigma) = 2.5$.

$(s/\tau_0 \cdot \sigma)$	0.000	0.004	0.008	0.012	0.016	0.020	0.024	0.028
$W_{ar}(s/\tau_0 \cdot \sigma) * 100$	00.000	28.983	19.703	15.986	13.715	12.477	11.550	10.553
$(s/\tau_0 \cdot \sigma)$	0.032	0.036	0.040	0.080	0.120	0.160	0.200	0.240
$W_{ar}(s/\tau_0 \cdot \sigma) * 100$	10.039	9.5416	9.0461	6.9039	5.7580	5.1926	4.8328	4.5725
$(s/\tau_0 \cdot \sigma)$	0.280	0.320	0.360	0.400	0.440	0.480	0.520	0.560
$W_{ar}(s/\tau_0 \cdot \sigma) * 100$	4.3260	4.2225	4.1643	4.1591	4.0402	4.0220	3.9885	3.9384
$(s/\tau_0 \cdot \sigma)$	0.600	0.640	0.680	0.720	0.760	0.800	1.000	1.200
$W_{ar}(s/\tau_0 \cdot \sigma) * 100$	4.0281	4.0879	4.1453	4.1910	4.2100	4.3385	4.8474	5.2439

(Continued)

Table A3.7 (Continued)

$(s/\tau_0 \cdot \sigma)$	1.400	1.600	1.800	2.000	2.200	2.400	2.600	2.800
$W_{ar}(s/\tau_0 \cdot \sigma) * 100$	5.1319	4.8336	4.4873	4.3159	4.2411	4.2369	4.2338	4.2311
$(s/\tau_0 \cdot \sigma)$	3.000	4.000	5.000	6.000	7.000	8.000	9.000	10.000
$W_{ar}(s/\tau_0 \cdot \sigma) * 100$	4.2100	3.9554	3.8046	3.5188	3.4388	3.2711	3.1133	2.9655
$(s/\tau_0 \cdot \sigma)$	12.000	16.000	20.000	24.000	28.000	32.000	36.000	40.000
$W_{ar}(s/\tau_0 \cdot \sigma) * 100$	2.6927	2.2584	1.8475	1.5183	1.2476	1.0256	0.8420	0.6906
$(s/\tau_0 \cdot \sigma)$	44.000	48.000	52.000	56.000	60.000	64.000	68.000	72.000
$W_{ar}(s/\tau_0 \cdot \sigma) * 100$	0.5698	0.4676	0.3856	0.3209	0.2579	0.2121	0.1742	0.1431

$M_1 = 20.3801\ (\sigma\ \tau_0)$; $D = 419.276\ (\sigma\ \tau_0)^2$; the number of excursions = 61852879.

Table A3.8 Rayleigh Random Process

Rectangular power spectrum distribution of areas above excursions below level $(x_0/\sigma) = 0.5$.

$(s/\tau_0 \cdot \sigma)$	0.000	0.002	0.004	0.006	0.008	0.010	0.012	0.014
$W_{ar}(s/\tau_0 \cdot \sigma) * 100$	00.00	1222.7	1089.1	937.88	847.54	788.18	747.22	717.07
$(s/\tau_0 \cdot \sigma)$	0.016	0.018	0.020	0.022	0.024	0.026	0.028	0.030
$W_{ar}(s/\tau_0 \cdot \sigma) * 100$	694.36	677.48	663.55	652.73	642.47	636.89	630.61	625.48
$(s/\tau_0 \cdot \sigma)$	0.032	0.034	0.036	0.038	0.040	0.042	0.044	0.046
$W_{ar}(s/\tau_0 \cdot \sigma) * 100$	622.31	620.90	620.00	619.89	622.38	625.52	630.09	636.80
$(s/\tau_0 \cdot \sigma)$	0.048	0.050	0.052	0.054	0.056	0.058	0.060	0.062
$W_{ar}(s/\tau_0 \cdot \sigma) * 100$	644.65	654.75	665.61	676.75	689.03	700.05	706.41	706.76
$(s/\tau_0 \cdot \sigma)$	0.064	0.066	0.068	0.070	0.072	0.074	0.076	0.078
$W_{ar}(s/\tau_0 \cdot \sigma) * 100$	699.73	684.53	672.07	670.89	670.48	668.09	664.59	660.53
$(s/\tau_0 \cdot \sigma)$	0.080	0.082	0.084	0.086	0.088	0.090	0.092	0.094
$W_{ar}(s/\tau_0 \cdot \sigma) * 100$	653.19	644.68	634.87	623.51	609.74	594.42	577.23	558.23
$(s/\tau_0 \cdot \sigma)$	0.096	0.098	0.100	0.120	0.140	0.160	0.180	0.200
$W_{ar}(s/\tau_0 \cdot \sigma) * 100$	536.72	518.33	502.91	343.77	231.79	153.39	104.17	74.872
$(s/\tau_0 \cdot \sigma)$	0.300	0.400	0.500	0.600	0.700	0.800	0.900	1.000
$W_{ar}(s/\tau_0 \cdot \sigma) * 100$	30.887	19.604	9.5802	4.5806	3.0418	2.2992	1.5510	0.9781

$M_1 = 0.103167\ (\sigma\ \tau_0)$; $D = 0.0178826\ (\sigma\ \tau_0)^2$; the number of excursions = 258418546.

Rectangular power spectrum distribution of areas above excursions below level $(x_0/\sigma) = 1.0$.

$(s/\tau_0 \cdot \sigma)$	0.000	0.002	0.004	0.006	0.008	0.010	0.012	0.016
$W_{ar}(s/\tau_0 \cdot \sigma) * 100$	00.00	1222.7	1089.1	937.88	847.54	788.18	747.22	717.07
$(s/\tau_0 \cdot \sigma)$	0.020	0.024	0.028	0.032	0.036	0.040	0.044	0.048
$W_{ar}(s/\tau_0 \cdot \sigma) * 100$	694.36	677.48	663.55	652.73	642.47	636.89	630.61	625.48
$(s/\tau_0 \cdot \sigma)$	0.052	0.056	0.060	0.064	0.068	0.072	0.076	0.080
$W_{ar}(s/\tau_0 \cdot \sigma) * 100$	622.31	620.90	620.00	619.89	622.38	625.52	630.09	636.80
$(s/\tau_0 \cdot \sigma)$	0.084	0.088	0.092	0.096	0.100	0.120	0.140	0.160
$W_{ar}(s/\tau_0 \cdot \sigma) * 100$	644.65	654.75	665.61	676.75	689.03	700.05	706.41	706.76

(Continued)

Table A3.8 (Continued)

$(s/\tau_0 \cdot \sigma)$	0.180	0.200	0.300	0.400	0.500	0.600	0.700	0.800
$W_{ar}(s/\tau_0 \cdot \sigma) * 100$	699.73	684.53	672.07	670.89	670.48	668.09	664.59	660.53
$(s/\tau_0 \cdot \sigma)$	0.900	1.000	1.100	1.200	1.300	1.400	1.500	1.600
$W_{ar}(s/\tau_0 \cdot \sigma) * 100$	653.19	644.68	634.87	623.51	609.74	594.42	577.23	558.23
$(s/\tau_0 \cdot \sigma)$	1.700	1.800	1.900	2.00	2.100	2.200	2.300	2.400
$W_{ar}(s/\tau_0 \cdot \sigma) * 100$	536.72	518.33	502.91	343.77	231.79	153.39	104.17	74.872
$(s/\tau_0 \cdot \sigma)$	2.500	2.600	2.700	2.800	2.900	3.000	3.100	3.200
$W_{ar}(s/\tau_0 \cdot \sigma) * 100$	30.887	19.604	9.5802	4.5806	3.0418	2.2992	1.5510	0.9781

$M_1 = 0.485767\ (\sigma\ \tau_0)$; $D = 0.380827\ (\sigma\ \tau_0)^2$; the number of excursions = 335599309.

Rectangular power spectrum distribution of areas above excursions below level $(x_0/\sigma) = 1.5$.

$(s/\tau_0 \cdot \sigma)$	0.00	0.004	0.008	0.012	0.016	0.020	0.024	0.028
$W_{ar}(s/\tau_0 \cdot \sigma) * 100$	00.000	110.09	81.385	69.112	62.761	58.419	55.568	53.139
$(s/\tau_0 \cdot \sigma)$	0.032	0.036	0.040	0.044	0.048	0.052	0.056	0.060
$W_{ar}(s/\tau_0 \cdot \sigma) * 100$	51.494	50.062	48.908	47.924	47.084	46.388	45.884	45.350
$(s/\tau_0 \cdot \sigma)$	0.080	0.100	0.120	0.140	0.160	0.180	0.200	0.220
$W_{ar}(s/\tau_0 \cdot \sigma) * 100$	43.719	42.957	42.710	42.874	43.202	43.838	44.481	45.406
$(s/\tau_0 \cdot \sigma)$	0.240	0.260	0.280	0.300	0.320	0.340	0.360	0.380
$W_{ar}(s/\tau_0 \cdot \sigma) * 100$	46.306	47.501	48.694	50.227	51.765	53.288	55.685	58.056
$(s/\tau_0 \cdot \sigma)$	0.400	0.420	0.440	0.460	0.480	0.500	0.520	0.540
$W_{ar}(s/\tau_0 \cdot \sigma) * 100$	60.445	63.160	65.841	68.346	70.849	72.512	74.187	74.625
$(s/\tau_0 \cdot \sigma)$	0.560	0.580	0.600	0.620	0.640	0.680	0.720	0.760
$W_{ar}(s/\tau_0 \cdot \sigma) * 100$	75.048	73.057	72.935	70.423	67.966	60.897	53.602	46.266
$(s/\tau_0 \cdot \sigma)$	0.800	1.000	1.200	1.400	1.600	1.800	2.000	3.000
$W_{ar}(s/\tau_0 \cdot \sigma) * 100$	39.914	25.936	23.996	23.164	21.395	18.419	15.368	8.4231
$(s/\tau_0 \cdot \sigma)$	4.000	5.000	6.000	7.000	8.000	9.000	10.000	11.000
$W_{ar}(s/\tau_0 \cdot \sigma) * 100$	4.7144	2.7056	1.6021	0.9623	0.5914	0.3912	0.2610	0.1739

$M_1 = 1.67771\ (\sigma\ \tau_0)$; $D = 4.51681\ (\sigma\ \tau_0)^2$; the number of excursions = 263997289.

Rectangular power spectrum distribution of areas above excursions below level $(x_0/\sigma) = 2.0$.

$(s/\tau_0 \cdot \sigma)$	0.00	0.004	0.008	0.012	0.016	0.020	0.024	0.028
$W_{ar}(s/\tau_0 \cdot \sigma) * 100$	00.000	33.702	24.504	20.688	18.586	17.340	16.329	15.607
$(s/\tau_0 \cdot \sigma)$	0.032	0.036	0.040	0.060	0.080	0.100	0.120	0.140
$W_{ar}(s/\tau_0 \cdot \sigma) * 100$	15.017	14. 620	14.121	12.962	12.369	12.096	11.885	11.934
$(s/\tau_0 \cdot \sigma)$	0.160	0.180	0.200	0.240	0.280	0.320	0.360	0.400
$W_{ar}(s/\tau_0 \cdot \sigma) * 100$	11.980	12.146	12.207	12.667	13.151	13.733	14.438	15.221
$(s/\tau_0 \cdot \sigma)$	0.440	0.480	0.520	0.560	0.600	0.640	0.680	0.720
$W_{ar}(s/\tau_0 \cdot \sigma) * 100$	16.091	17.074	18.045	19.341	20.861	22.542	24.248	26.506
$(s/\tau_0 \cdot \sigma)$	0.760	0.800	1.000	1.200	1.400	1.600	1.800	2.000
$W_{ar}(s/\tau_0 \cdot \sigma) * 100$	28.276	30.171	28.364	16.081	11.699	11.347	11.801	12.530
$(s/\tau_0 \cdot \sigma)$	2.200	2.400	2.600	2.800	3.000	3.200	3.400	3.600
$W_{ar}(s/\tau_0 \cdot \sigma) * 100$	12.713	12.209	11.163	10.095	9.3834	8.8687	8.7249	8.5412
$(s/\tau_0 \cdot \sigma)$	3.800	4.000	4.200	4.400	4.600	4.800	5.000	6.000
$W_{ar}(s/\tau_0 \cdot \sigma) * 100$	8.3049	8.0175	7.6261	7.3206	7.0789	6.6508	6.4586	5.4083
$(s/\tau_0 \cdot \sigma)$	7.000	8.000	10.000	12.000	14.000	16.000	18.000	20.000
$W_{ar}(s/\tau_0 \cdot \sigma) * 100$	4.5225	3.7531	2.6829	1.9411	1.3976	1.0019	0.7213	0.5161

$M_1 = 5.77066\ (\sigma\ \tau_0)$; $D = 53.3123\ (\sigma\ \tau_0)^2$; the number of excursions = 145125424.

Rectangular power spectrum distribution of areas above excursions below level $(x_0/\sigma) = 2.5$.

$(s/\tau_0 \cdot \sigma)$	0.000	0.004	0.008	0.012	0.016	0.020	0.024	0.028
$W_{ar}(s/\tau_0 \cdot \sigma) * 100$	00.000	9.0886	6.4555	5.3910	4.7701	4.4360	4.0884	3.9161
$(s/\tau_0 \cdot \sigma)$	0.032	0.036	0.040	0.080	0.120	0.160	0.200	0.240
$W_{ar}(s/\tau_0 \cdot \sigma) * 100$	3.7510	3. 6453	3.5632	2.9769	2.9530	2.7781	2.8880	2.9654
$(s/\tau_0 \cdot \sigma)$	0.280	0.320	0.360	0.400	0.440	0.480	0.520	0.560
$W_{ar}(s/\tau_0 \cdot \sigma) * 100$	3.0334	3.1745	3.3015	3.5384	3.6971	3.8887	4.2167	4.3347
$(s/\tau_0 \cdot \sigma)$	0.600	0.640	0.680	0.720	0.760	0.800	1.000	1.200
$W_{ar}(s/\tau_0 \cdot \sigma) * 100$	4.6905	4.8851	5.2576	5.5523	5.8499	6.4054	9.1959	12.260

(Continued)

Table A3.8 (Continued)

$(s/\tau_0 \cdot \sigma)$	1.400	1.600	1.800	2.000	2.200	2.400	2.600	2.800
$W_{ar}(s/\tau_0 \cdot \sigma) * 100$	9.8622	5.3573	3.8327	3.6663	3.7933	4.1269	4.5404	4.9065
$(s/\tau_0 \cdot \sigma)$	3.000	4.000	5.000	6.000	7.000	8.000	9.000	10.000
$W_{ar}(s/\tau_0 \cdot \sigma) * 100$	5.0707	3.6000	3.7976	3.3665	3.2108	3.0419	2.9299	2.7178
$(s/\tau_0 \cdot \sigma)$	12.000	16.000	20.000	24.000	28.000	32.000	36.000	40.000
$W_{ar}(s/\tau_0 \cdot \sigma) * 100$	2.4411	2.0063	1.6388	1.3667	1.1399	0.9511	0.7928	0.6612
$(s/\tau_0 \cdot \sigma)$	44.000	48.000	52.000	56.000	60.000	64.000	68.000	72.000
$W_{ar}(s/\tau_0 \cdot \sigma) * 100$	0.5523	0.4614	0.3855	0.3221	0.2699	0.2256	0.1886	0.1576

$M_1 = 22.1798\ (\sigma\ \tau_0)$; $D = 771.733\ (\sigma\ \tau_0)^2$; the number of excursions = 58458078.

To simulate the distribution density of excursion durations, the random process is first transformed into a sequence of bipolar rectangular pulses of the same duration as equivalent excursions beyond the set threshold. In this case, the input information on the signal sample amplitudes is discarded as irrelevant.

When the simulation is done to obtain the distribution of areas enveloped by excursion curves, both amplitude and duration of excursions should be taken into account. The dynamic range of under-curve (above-curve) areas approximately equals the product of dynamic ranges for excursion durations and amplitudes. The maximum-to-minimum ratio for areas enveloped by the excursion curve from above and the set threshold line from below can be as high as several thousands.

In the case of small-area excursions, which are quite common, the accuracy of area measurement can only be increased by reducing the sampling time or amplitude quantization step. Such reduction results in a higher statistical error, which, in turn, can be reduced by increasing the volume of data and the measurement period duration.

Thus, simulation of areas enveloped by excursion curves requires a higher computer speed.

References

1. Bykhovsky MA. *Pioneers of the information age. History of Communication Theory.* Moscow: Tekhnosfera; 2006.
2. Rice SO. Mathematical analysis of random noise. *Bell Syst Tech J.* 1944;23(3): 282–332;1945,24(1):46–156.
3. Bunimovich VI, Leontovich MA. Distribution of the number of large deviations in the electrical fluctuations. *Rep Acad Sci USSR.* 1946;53(1).
4. Stratonovich RL. *Selected Issues of Fluctuation Theory in Radio Communication.* Moscow: Sovetskoye Radio; 1961 [Стратонович Р. Л.: Избранные вопросы теории флуктуаций в радиотехнике. – М.: Советское Радио, 1961].
5. Stratonovich RL. *Topics in the Theory of Random Noise.* New York, NY: Gordon and Breach; 1963–1967.
6. Tikhonov VI. *Excursions in Random Processes.* Moscow: Nauka; 1970:392 [В.И.Тихонов. Выбросы случайных процессов. М.:Наука,1970. 392с].
7. Tikhonov VI, Khimenko VI. *Excursions of Random Process Curves.* Moscow: Nauka; 1987 [В. И. Тихонов, В. И. Хименко. Выбросы траекторий случайных процессов. Наука, Москва, 1987].
8. Levin BR. *Theoretical Foundations of Radio Communication Statistics—Book 1.* Moscow: Sovetskoye Radio; 1974 [Левин Б.Р. Теоретические основы статистической радиотехники. – Книга 1. М.: Советское Радио, 1974].
9. Fomin YaA. *The Theory of Random Process Excursions.* Moscow: Sviaz; 1980 [Фомин Я. А. Теория выбросов случайных процессов. – М.: Связь, 1980].
10. Nakagami M. *The m-Distribution: A General Formula of Intensity Distribution of Rapid Fadings in Statistical Methods in Wave Propagation.* Pergamon Press; 1960.
11. Barnett JT, Kedem B. Zero-crossing rates of functions of Gaussian processes. *IEEE Trans Inform Theor.* July 1991;37(4):1188–1194.
12. Yacoub D, Bautista JEV, Guerra de Rezende Guedes L. On higher order statistics of the Nakagami-m distribution. *IEEE Trans Veh Technol.* May 1999;48(3):790–794.
13. Lai J, Mandayam NB. Minimum duration outages in Rayleigh fading channels. *IEEE Trans Commun.* 2001;49:1755–1761.
14. Abdi A. On the second derivative of a Gaussian process envelope. *IEEE Trans Inform Theor.* 2002;48(5):1226–1231.
15. Patzold M. *Mobile Fading Channels.* New York, NY: John Wiley & Sons; 2003.
16. Ramos F, Kontorovich V, Lara M. Minimum duration outage in Nakagami fading channels: an approximate analytical calculation for wireless systems. In: *Proc. ICEEE '04.* Acapulco, Mexico; 2004:102–107.
17. Kratz MF. Level crossings and other level functionals of stationary Gaussian processes. *Prob Surv.* 2006;3:30–288.
18. Morgan DR. On level-crossing excursions of Gaussian low pass random processes. *IEEE Trans Signal Process.* 2007;55(7):3623–3632.
19. Primak S, Kontorovich V. On the second order statistics of generalized gamma process. *IEEE Trans Commun.* 2009;57(4).

20. Hadzi-Velkov Z, Zlatanov N, Karagiannidis GK. On the second order statistics of the multihop Rayleigh fading channel. *IEEE Trans Commun*. 2009;57(6):1815–1823.
21. Hanif MF, Smith PJ. Level crossing rates of interference in cognitive radio networks. *IEEE Trans Wireless Commun*. 2010;9(4).
22. Goot RE. *Performance of Adaptive Frequency Scheduling in a Group of Radio Communication Links*. Center for Industrial and Applied Mathematics, Holon Academic Institute of Technology, Israel; September 2004.
23. Goot RE. Data transmission with group control of the radiated power. *Telecommun Radio Eng*. 1983;10(10):47–49.
24. Djuknic G, Freidenfelds J, Okunev Y. Establishing wireless communications services via high-altitude aeronautical platform: a concept whose time has come. *IEEE Commun Mag*. 1997;35(9):128–135.
25. Goot R, Trigano T, Tapuchi S, Gavan J. *Adaptive allocation of power transmission for high-altitude platforms*. Institut Mines-Télécom and Springer-Verlag; 2012.
26. Bolotin VV, Trifonov OV. Assessment of safety and failure modes for structures under strong seismic and related actions—Computational stochastic mechanics. In: *Proceedings of the Fourth International Conference on Computational Stochastic Mechanics*. Corfu, Greece, June 9–12, 2002. Rotterdam: Millpress; 2003:67–71.
27. Degarmo EP, Black JT, Kohser RA. :p. 223 *Materials and Processes in Manufacturing*. 9th ed. Wiley; 2003
28. Sergeev VA (Ulyanovsk Branch of the Institute of Radio Engineering and Electronics. Kotelnikov RAS), Frolov IV (Ulyanovsk State Technical University). Measurement of a probability-frequency noise characteristics of semiconductor diode discrete method. In: *Proceedings of the Sixth All-Russian Scientific-Practical Conference (with the CIS countries*); 2009.
29. Brainina IS. *Applying the Theory of Random Process Excursions for Adaptive Digital Processing of Telecommunication Signals*. Moscow: Radio i sviaz; 2002 [И. С. Брайнина. Адаптивная цифровая обработка сигналов связи с использованием прикладной теории выбросов случайных процессов. Радио и связь. Москва. 2002].
30. Barlow RE, Proschan F. *Statistical Theory of Reliability and Life Testing Probability Models*. New York, NY: Holt, Rinehart and Winston; 1975.
31. Nayak PR. Contact vibrations. *J Sound Vibr*. 1972;22(3):297–322.
32. Bakhareva NF, Kartashevsky IV, Tarasov VN. Analysis and calculation of non-Poisson simulations of traffic in computer networks, Samara, Russia. *Info Commun Technol*. 2009;7(4):61–66 [Н. Ф. Бахарева, И.В.Карташевский, В.Н.Тарасов. Анализ и расчет непуассоновских моделей трафика в сетях ЭВМ.– Россия, Самара. Инфокоммуникационные технологии, 2009, т.7, №4, с.61-66].
33. Kuznetsov PI, Stratanovich RL, Tikhonov VI. On the duration of random function excursions. *JTF*. 1954;24(1) [Кузнецов П.И., Стратонович Р.Л., Тихонов В. И. О длительности выбросов случайной функции. – ЖТФ, 1954, т.24, вып.1].
34. Andronov AA, Rabinovich MA. On the distribution of abnormal excursions in the frequency curve. In: *Proceedings of the 7th USSR Conference on Encoding and Data Transmission Theory. Part 6. Statistical Theory of Signal Reception*. Moscow: Vilnius; 1978 [Андронов А. А., Рабинович М. А. О распределении длительностей аномальных выбросов частоты. – Труды 7-ой Всесоюзной конференции по теории кодирования и передачи информации. Часть 6. Статистическая теория приема сигналов. – Москва – Вильнюс, 1978].
35 Mimaki T, Munakata T. Experimental results on the level crossings intervals of Gaussian processes. *IEEE Trans Inform Theor*. 1978;24(4):515–519.

36. Korchagin VYa, Kravtsov LYa, Sodomov YB, Khokhlov LM. *Using Stochastic Computation Devices for the Measurement of Random Process Probability Characteristics*. Leningrad: Energoatomizdat; 1982 [Корчагин В. Я., Кравцов Л. Я., Содомов Ю. Б., Хохлов Л. М. Измерение вероятностных характеристик случайных процессов с применением стохастических вычислительных устройств. – Л.: Энергоатомиздат, 1982].
37. Brainina IS. Variance in the total number of times when a pre-determined level is crossed by a stationary random process and variance in the number of excursions on a time interval. *Radiotek Elektron*. 1977;22(9) [Брайнина И. С. Соотношение между дисперсией общего числа пересечений заданного уровня стационарным случайным процессом и дисперсией числа выбросов на интервале т. - Радиотехника и электроника, 1977, т.22, № 9].
38. Suvorov BI. Variance in the number of excursions in a normal noise. *Radiotek Elektron*. 1971;16(1) [Суворов Б. И. Дисперсия числа выбросов нормального шума. – Радиотехника и электроника, 1971, т.16, № 1].
39. Brainina IS. On distribution of time before the next crossing of level zero by a stationary random process Leningrad. In: *Proceedings of Telecommunication University Colleges. Systems and Means of Transmitting Data over Telecommunication Channels*; 1981 [Брайнина И. С. О распределении времени первого пересечения нулевого уровня стационарным случайным процессом. – Л.: Труды учебных институтов связи. Теория передачи информации по каналам связи, 1981].
40. McFadden JA. The axis-crossing intervals of random functions. *IRER Trans Inform Theory*. 1956 and 1958;IT-2(4):146–150:1958; Vol. IT-4, N.1, pp. 14–24.
41. Brainina IS. On distribution of time before the next moment when a stationary random process crosses level zero in given—upward or downward—direction. In: *Proceedings of telecommunication university colleges. Systems and Means of Transmitting Data over Telecommunication Channels*. Leningrad; 1978 [Брайнина И. С. О распределении времени первого достижения нуля с заданным знаком производной стационарным случайным процессом. – Л.: Труды учебных институтов связи. Системы и средства передачи информации по каналам связи, 1978].
42. Wong E. Some results concerning the zero-crossings of Gaussian noise. *SIAM J Appl Math*. 1966;14(6):1246–1254.
43. Denisenko AI. Estimation of the distribution of durations for random process excursions. *Radiotek Elektron*. 1975;21(7) [Денисенко А. И. Оценка распределения длительности выбросов случайных процессов. – Радиотехника и электроника, 1975, т.21, № 7].
44. Brainina IS. Estimation of distribution density of time intervals between zeros of stationary random process. In: *Proceedings of Telecommunication University Colleges. Components of Telecommunication Systems*. Leningrad; 1991 [Брайнина И. С. Оценка плотности распределения интервалов между соседними нулями стационарных случайных процессов. – Л.: Труды учебных институтов связи. Элементы и устройства систем связи, 1991].
45. Brainina IS. Estimation of distribution density of time intervals between moments when a stationary random process crosses level zero in a given (upward or downward) direction. In: *Proceedings of Telecommunication University Colleges. Data Transmission and Signal Processing Systems*. Leningrad; 1991 [Брайнина И.С. Оценка плотности распределения интервалов между нулями стационарного случайного процесса с заданным знаком производной. – Л.: Труды учебных институтов связи. Системы передачи информации и обработки сигналов, 1991].

46. Brainina IS. A way to obtain a more accurate first approximation functions for the distribution density of time intervals between zeros of a stationary random process. In: *Proceedings of Telecommunication University Colleges. Data Transmission and Signal Processing Systems*. Leningrad; 1991 [Брайнина И. С. Об уточнении функций первого приближения плотностей распределения интервалов между пересечениями уровня стационарным случайным процессом. – Л.: Труды учебных институтов связи. Системы передачи информации и обработки сигналов, 1991].
47. Pukhniuk YV. Statistical independence of points where the random process crosses the given threshold level. *Izvestiya VUZov Radioelektr.* 1977;(9) [Пухнюк Ю. В. Статистическая независимость точек пересечения фиксированного уровня случайным процессом. – Известия ВУЗов. Радиоэлектроника, 1977, № 9]
48. Rainal AJ. Approximate and exact results concerning zeros of Gaussian noise. *Ibid.* 1968;47(3):461–463.
49. RF Patent No. 1317453. A device to determine the distribution function for stationary random process probabilities. Brainina IS. Patent Bulletin 1987/22 [Устройство для определения функции распределения вероятностей стационарных случайных процессов. И. С. Брайнина – Б.И.,1987, № 22].
50. Malevich TL. Asymptotic normality of the number of zero level crossing points in a Gaussian process. *Appl Probab Theor.* 1969;14(2) [Малевич Т. Л. Асимптотическая нормальность числа пересечений нулевого уровня гауссовым процессом. – Теория вероятности и ее применение, 1969, т.14, вып.2]
51. Bolychevtsev AD, Bolychevtsev EM, Dobryden VA. Excursions duration distribution estimates. *Radiotek Elektron.* 1972;17(2) [Болычевцев А. Д., Болычевцев Э. М., Добрыдень В. А. Количественные оценки распределения длительности выбросов. - Радиотехника и электроника, 1972, т.17, № 2].
52. Lvovskaya JI. Distribution of excursion durations in a broadcasting signal waveform envelope. Vorprosy radioelektroniki. Broadcasting Signal Reception and Acoustics Technology series; 1, 1970 [Львовская Ж. И. Распределение длительности выбросов огибающей вещательного сигнала.–Вопросы радиоэлектроники. Серия «техника радиовещательного приема и акустики», выпуск 1, 1970].
53. Mirsky GYa. *Using Measuring Equipment to Determine Random Process Characteristics*. Moscow: Energiya; 1972 [Мирский ГЯ. Аппаратурное определение характеристик случайных процессов. – М.: Энергия, 1972].
54. Mimaki T. Zero-crossing intervals of Gaussian processes. *J Appl Phys.* 1973;44(1):S. 477–485.
55. Mimaki T, Shimizu T. Correlation between zero-crossing intervals of Gaussian processes. *J Phys Soc Jap.* 1974;36(6):1690–1692.
56. Mimaki T, Tanabe M, Wolf D. The multipeak property of the distribution densities of the level-crossing intervals of a Gaussian random process. *IEEE Trans Inform Theor.* 1981;27(4):525–527.
57. Munakata T, Wolf D. The probability distribution density of the sum of two adjacent level-crossing time intervals with Gaussian random processes. In: *EUSIPGO-80. First European Signal Processing Conference, Swiss Feel. Inst. Technol (EPFL)*. Lausanne; 1980:55–56.
58. Wolf D, Brehm H. Die Verteilungsdichte der Zeitintervalle zwischen Nulldurchgangen bei Gaußschen stochastischen Signalen. *AEU.* 1973;Bd. 27(11):477–489.
59. Bendat J, Pirsol A. *Measuring and Analyzing Parameters of Random Processes*. Moscow: Mir; 1971 [Бендат Дж., Пирсол А. Измерение и анализ случайных процессов. – М.: Мир, 1971].

60. Blokhin AV. *Using Measuring Equipment to Analyze Random Process Characteristics.* Moscow: Energiya; 1968 [Блохин А. В. Аппаратурный анализ характеристик случайных процессов. – М.: Энергия, 1968].
61. Kulikov YeI. *Measurement Methods Applied for Random Processes.* Moscow: Radio i sviaz; 1986 [Куликов Е. И. Методы измерения случайных процессов. – М.: Радио и связь, 1986].
62. RF Patent No. 1354219. Random process excursions duration analyzer. Brainina IS. Patent Bulletin 1987/43 [Анализатор длительностей выбросов случайных процессов. И. С. Брайнина. – Б.И.,1987, № 43].
63. RF Patent No. 1374245. Time interval distribution analyzer. Brainina IS. Patent Bulletin 1988/6 [Анализатор распределения интервалов. И. С. Брайнина – Б.И.,1988, № 6].
64. RF Patent No. 1430966. Adaptive analyzer of areas under excursions of random process curves. Brainina IS. Patent Bulletin 1988/38 [Адаптивный анализатор площадей выбросов случайных процессов. И. С. Брайнина. – Б.И.,1988, № 38].
65. RF Patent No. 1173426. A method of measuring variance of a centered normal random process and a device which implements the method. Brainina IS. Patent Bulletin 1985/30. [Способ измерения дисперсии центрированного нормального случайного процесса и устройство для его осуществления. И. С. Брайнина – Б.И., 1985, № 30].
66. Rice SO. Distribution of the duration of fades in radio transmission: Gaussian noise model. *Bell Syst Tech J.* May, 1958;37(1):581–635.
67. Longuet-Higgins MS. On the intervals between successive zeros of a random function. *Proc R Soc Lond Ser A Math Phys Sci.* 1958;246(1244):99–118:07/1958
68. Slepian D. On the zeros of Gaussian noise. In: Rosenblatt M, ed. *Time Series Analysis.* New York: Wiley; 1962:104–115.
69. Rainal. AJ. Axis-crossing intervals of Rayleigh processes. *Bull Syst Tech J.* 1965;44:6.
70. Kramer G, Lidbetter M. *Stationary Random Processes.* Moscow: Mir; 1969 [Крамер Г., Лидбеттер М. Стационарные случайные процессы. – М.: Мир, 1969].
71. Brainina I.S. Calculation of Level-Crossing Parameters of Non-Gaussian Stationary Random Processes in Signal Processing. The 7th World Multiconference on Systemics, Cybernetics and Informatics July 27-30, Orlando, Florida; 2003:275–281.
72. Blokhin AV. Distribution of intervals between successive minima or maxima in a random process. *Radiotek Elektron.* 1970;15(8) [Блохин А. В. Распределение длительностей интервалов между соседними минимумами и максимумами случайного процесса. – Радиотехника и электроника, 1970, т.15, № 8].
73. Blake IF, Lindsey W. Level-crossing problems for random processes. *IEEE Trans Inform Theor.* 1973;IT-9(3):295–315.
74. Arnold HW, Bodtman. WF. Interfade interval statistics of a Rayleigh-distributed wave. *IEEE Trans Commun.* 1983;31(9):1114–1116.
75. Andreev GA. Detection of weak signals from excursions of the envelope. *Izvestiya VUZov Radiophys.* 1970;13(7) [Андреев Г. А. Обнаружение слабых сигналов по выбросам огибающей. – Известия вузов. Радиофизика,1970, т.13, №7.].
76. Rubtsov VD. Excursions occurring in a random process after its functional transformation. *Radiotek Elektron.* 1980;25(1):207–208 [В. Д. Рубцов. Выбросы случайного процесса после функционального преобразования. Радиотехника и электроника, 1980. Т.25, №1. С.207–208].
77. Sviridenko VA. On calculating the mean time before the moment when the set boundaries are reached by a normal non-Markovian process. *Radiotek Elektron.* 1972;17(11):2447–2449 [В. А. Свириденко. К расчету среднего времени достижения

заданных границ нормальным процессом немарковского типа. Радиотехника и электроника, 1972. Т.17, №11. С.2447–2449].

78. Brainina IS. Estimation of variance in duration of stationary random process excursions above a given level. In: *Proceedings of Telecommunication University Colleges. Systems and means of Transmitting Data Over Telecommunication channels.* Leningrad; 1990 [Брайнина И. С. Оценка дисперсии длительности выбросов стационарного случайного процесса над фиксированным уровнем. – Л.: Труды учебных институтов связи. Теория передачи информации по каналам связи, 1990].

79. Brainina IS. On the tails of distributions curves for time intervals between above-threshold excursions of a stationary random processs. In: *Proceedings of Telecommunication University Colleges. Analysis and Simulation of Signals Generated by Telecommunication Systems.* Leningrad; 1993 [Брайнина И. С. О предельных распределениях длительности интервалов между выбросами стационарного случайного процесса над заданным уровнем. – Л.: Труды учебных институтов связи. Анализ и моделирование сигналов систем связи, 1993].

80. Brainina IS. The use of parameters of random signal excursions and information on signal waveforms in adaptive processing of telecommunication signals. The post-doctoral thesis. M.A. Boncz-Brujewicz National University of Telecommunications, St. Petersburg; 1997 [Брайнина И. С. Адаптивная обработка сигналов в системах связи, использующая параметры выбросов случайных процессов и информацию о форме сигнала. Докторская диссертация. ГУТ им. проф. М.А.Бонч-Бруевича, Санкт-Петербург, 1997].

81. Brainina IS, Petrov OA. One method to calculate zero level crossing parameters of centered stationary random processes. Samara, Russia. *Inform Tekhnol.* 2009;7(4):55–61 [И.С. Брайнина, О. А. Петров. Метод расчета параметров пересечения нулевого уровня для центрированных стационарных случайных процессов. – Россия, Самара. Инфокоммуникационные технологии. 2009, т. 7, №4, с.55-61].

82. Brainina IS, Petrov OA. One method to calculate level-crossing parameters of a random process for an arbitrarily chosen standardized level. Samara, Russia. *Inform Tekhnol.* 2011;12(4) [И.С. Брайнина, О. А. Петров. Описание методики расчета параметров пересечения случайным процессом произвольного нормированного уровня анализа. – Россия, Самара. Инфокоммуникационные технологии. 2011, т. 12, №4].

83. RF Patent No. 2012052. Random process analyzer. Brainina IS. Patent Bulletin 1994/8 [Анализатор случайных процессов. И. С. Брайнина – Б.И.,1994, № 8].

84. Geranin VA. Variance in the number of excursions of a stationary Gaussian noise over a given time period. *Radiotek Elektron.* 1967;12(5) [Геранин В. А. Дисперсия числа выбросов отрезка стационарного гауссова шума. – Радиотехника и электроника, 1967, т.12, № 5].

85. Golubeva VA, Magarik VA. On calculation of variance in the number of points where a Gaussian process crosses a given threshold. *Radiotek Elektron.* 1971;16(6) [Голубева В. А., Магарик В. А. О вычислении дисперсии числа пересечений некоторого уровня гауссовым процессом. – Радиотехника и электроника, 1971, т.16, № 6].

86. Tikhonov VI. The time before the first moment when a random process reaches the set boundaries. *Izvestiya VUZov SSSR Radiophys.* 1985;28(4) [В. И. Тихонов. Первое время достижения границ случайным процессом. Известия Вузов СССР. Радиофизика, 1985. Т.28, №4].

87. Yershov MP. The limit theorem for the number of points where a stationary Gaussian process crosses a given threshold. *Rep USSR Acad Sci.* 1967;177(6) (Ершов М. П.

Предельная теорема для числа пересечений уровня стационарным гауссовым процессом.– Доклады АН СССР. 1967, т.177, № 6].

88. Dobryden VA. Approximation of distribution density of time before the next point where the random process reaches a given level. *Kibernetika*. 1972;6 [Добрыдень В. А. Аппроксимация плотности распределения времени первого достижения уровня случайным процессом.–Кибернетика, 1972, № 6].
89. Blachman NM. Narrowband Gaussian noise: variance of the number of zero crossings. *Electron Lett*. 1978;14(3):75–76.
90. Rice J, Beer FP. First occurrence of high-level crossings in a continuous random process. *J Acoust Soc Am*. 1966;39(2):323–335.
91. Siegert AJF. On the first passage time probability problem. *Phys Rev*. February 15, 1951;81:617–623.
92. Slepian D. The one-sided barrier problem for Gaussian noise. *Bell Syst Tech J*. 1961;41:463–501.
93. Middleton D. *An Introduction to Statistical Communication Theory*. New York: McGraw-Hill Book Co; 1960.
94 Helstrom CW. *Statistical Theory of Signal Detection*. 2nd ed. Oxford: Pergamon Press, Ltd.; 1968.

Lightning Source UK Ltd.
Milton Keynes UK
UKOW05n1017171114

241674UK00011B/111/P